Mycobacteria

II CHEMOTHERAPY

Mycobacteria
II CHEMOTHERAPY

CHAPMAN & HALL MEDICAL MICROBIOLOGY SERIES

Edited by

Pattisapu R. J. GANGADHARAM, PH.D.

Professor of Medicine, Microbiology, and Pathology
Director of Mycobacteriology Research,
University of Illinois at Chicago College of Medicine, Chicago, Illinois

P. Anthony JENKINS, PH.D.

formerly Director, Public Health Laboratory Service,
University Hospital of Wales, United Kingdom

CHAPMAN & HALL

 **International Thomson Publishing**
Thomson Science
New York • Albany • Bonn • Boston • Cincinnati • Detroit • London • Madrid • Melbourne
Mexico City • Pacific Grove • Paris • San Francisco • Singapore • Tokyo • Toronto • Washington

Cover design: Trudi Gershenov

Copyright © 1998 by Chapman & Hall

Printed in the United States of America

For more information, contact:

Chapman & Hall
115 Fifth Avenue
New York, NY 10003

Chapman & Hall
2-6 Boundary Row
London SE1 8HN
England

Thomas Nelson Australia
102 Dodds Street
South Melbourne, 3205
Victoria, Australia

Chapman & Hall GmbH
Postfach 100 263
D-69442 Weinheim
Germany

International Thomson Editores
Campos Eliseos 385, Piso 7
Col. Polanco
11560 Mexico D. F.
Mexico

International Thomson Publishing - Japan
Hirakawacho-cho Kyowa Building, 3F
1-2-1 Hirakawacho-cho
Chiyoda-ku, 102 Tokyo
Japan

International Thomson Publishing Asia
221 Henderson Road #05-10
Henderson Building
Singapore 0315

1 2 3 4 5 6 7 8 9 10 XXX 01 00 99 98
Library of Congress Cataloging-in-Publication Data

Mycobacteria / edited by P.R.J. Gangadharam, P.A. Jenkins.
        p.   cm. -- (Chapman & Hall medical microbiology series)
      Includes bibliographical references and index.
      Contents: v. 1. Basic aspects — v. 2. Chemotherapy.
      ISBN 0-412-05451-5 (v. 1: alk. paper). ISBN 0-412-05441-8 (v. 2 : alk. paper)
      1. Mycobacterial diseases. I . Gangadharam, Pattisapu Rama Jogi,
   II. Jenkins, P.A., (P. Anthony) III. Series
      {DNLM: 1. Tuberculosis, Pulmonary. 2. Mycobacterium Infections.
   3. Mycobacterium. WF 300 M995 1997}
   QR201.M96M93   1997
   616'.01474--dc21
   DNLM/DLC
   for Library of Congress   97-3970
                                                              CIP

British Library Cataloguing in Pubication Data available

To order this or any other Chapman & Hall book, please contact **International Thomson Publishing, 7625 Empire Drive, Florence, KY 41042.** Phone: (606) 525-6600 or 1-800-842-3636. Fax: (606) 525-7778. e-mail: order@chaphall.com.

For a complete listing of Chapman & Hall's titles, send your request to **Chapman & Hall, Dept. BC, 115 Fifth Avenue, New York, NY 10003.**

To My Parents,
Mr. Pattisapu Venkata Joga Rao
Mrs. Pattisapu Kameswaramma

Who, despite of the great fear that I may become a serious victim to this dreadful disease, inspired me to continue research in this field.

Pattisapu Rama Jogi Gangadharam

To,

The Staff of the Tuberculosis Reference Laboratory (1959–1977) and the Mycobacterium Reference Unit (1977–1995) in the Public Health Laboratory, Cardiff who contributed so much to the quality of clinical mycobacteriology in the United Kingdom.

P. Anthony Jenkins

Contents

Preface

About 30 years ago, Professor Arthur Meyers stated that if we ask the question "Can tuberculosis be eradicated?" the answer is probably "Yes." If, on the other hand, we ask the question, "Will tuberculosis be eradicated?" the answer will probably be "No." This discussion was based on immunological grounds and was written at a time when rapid progress in chemotherapy of tuberculosis was on the horizon. At about the same time, the famous book *Chemotherapy of Tuberculosis,* edited by Vincent Barry presented a series of excellent progress reports and optimistic notes on the treatment of this disease. Unfortunately, this optimism was misplaced and the questions posed by Meyers have become even more pertinent, not only to the prevention or eradication, but to the overall seriousness of the disease. The powerful tools available for the control of tuberculosis were not properly and consistently utilized, resulting in dismal failure. The story became a perfect example, as Frank Ryan puts it, of a major disease which could have been conquered or, more aptly, nearly so and how soon potential victory has been reversed.

Careful analysis of the facts has given us some insight into the successes and failure in the control of this "curable" disease. Besides the discovery of powerful drugs, considerable knowledge on the best possible drug regimens also became available following tedious but excellently conducted controlled clinical trials. All these contributions made chemotherapy the "magic bullet" in our battle against tuberculosis. Indeed, many developed countries benefited from the advances in chemotherapy leading to a consistent reduction in the incidence of disease in the mid-eighties. Authorities in these countries were excited in the dramatic reduction of the prevalence and had even set the year 2000, which was later revised to 2010 as the date for eradication. Unfortunately, this "over-excitement" resulted in a hasty and premature abandonment of the control efforts and failure of the infrastructure, with the tragic consequence of vigorous resurgence of the disease. As Sir John Crofton stated, "it is a sad reflection on society's incompetence to make proper use of the tools and take control of the situation." Had it not been due to our negligence and many pitfalls, we could have pushed tuberculosis as an old "forgotten" disease of mankind and all the tuberculosis literature to the "archives"! Surprisingly, even the developing and the underdeveloped countries, which never had the fortune of witnessing any significant decline in the incidence of tuberculosis, also became complacent and even grossly negligent in their antituberculosis programs. These countries may have mistakenly believed that tuberculosis is no longer serious, as many sanatoria were closed and the treatment was shifted to domiciliary and outpatient settings. The global indifference to this dangerous disease, which has been prevailing for a long time, prompted the International Union Against Tuberculosis, now called the International Union Against Tuberculosis and Lung Disease (IUATLD) to devote major sessions in its global conferences on "How to inform the general public that tuberculosis is still not under control." This sort of public education is consistently being propagated by the World Health Organization (WHO), which is also stressing the need for greater governmental support.

The gravity of this situation is confounded by the fact that most of the new cases of tuberculosis are with multiple-drug resistant tubercle bacilli, making the available drugs useless. This is mostly due to noncompliance on the part of the patients to take the full course of treatment. These "man-made" failures are aggravated by another "man-made" disease, the acquired immune deficiency syndrome (AIDS). The combination of these two dangerous diseases is potentially disastrous not only for certain areas of Africa as feared by Grange and Stanford but will also affect the whole world if we do not act immediately. It is therefore vital that concentrated efforts are urgently made to discover and develop new drugs and treatment strategies to attack this disease. For this purpose, it is necessary to analyze critically the situation, not necessarily to rejoice at our past glories of successful treatment, but more importantly to understand the "how" and the "why" of the failures and to suggest realizable plans for the future.

In this volume, several aspects of chemotherapy are discussed extensively. Such a comprehensive discussion which was contained in a treatise on the same subject was edited by Vincent Barry more than 30 years ago. At that time, the potential of chemotherapy to succeed was evident. The proper use of the powerful drugs (excluding rifampin, which was not discovered at that time) and the proper application of the optimal regimens which were evolved following meticulously conducted controlled trials were elaborated. Subsequently, much more progress has been achieved, thanks to the discovery of rifampin and its use in short-course chemotherapy, which is the mainstay of our present-day approach to the treatment of tuberculosis.

In this volume, the authors have discussed the development of the basic and experimental aspects of chemotherapy, our present-day philosophy of treatment in the context of developed and developing countries, the lessons that were or could have been drawn from the controlled clinical trials, and the reasons for the failure of treatment. Management of patients with drug-resistant bacilli and with nontuberculous mycobacterial infections and, more importantly, those with HIV infection and with AIDS, are dealt with in other chapters. A chapter on the magnitude of the global impact of the present-day situation is given to introduce the readers to the seriousness of the problem and two chapters on leprosy are included, to impress upon the readers that success can be achieved even under difficult situations and with a stubborn disease. This volume also contains a review on the current approaches to discover newer drugs and strategies. It is sincerely hoped that the diverse topics discussed in this and the companion volume will provoke interest and encourage a spirit of enquiry in the reader.

Finally, the editors are extremely grateful to all the authors, but for them the idea of a comprehensive treatise could not have been realized. The editors are respectfully and apologetically aware of the enormous demands on their time and effort our request should have made to the authors. We do realize how busy they are in their own individual professional commitments. To each one of them we offer our sincere thanks. We realize that timeliness is an important aspect, and however much we and the concerned authors tried, considerable delays have occurred. To those authors who have been very kind to submit their manuscripts early, we offer our apologies, but we have taken care to see that no discussion or topic has become outdated. Overall, we are extremely gratified that these two volumes offer an excellent compendium of the available knowledge on this disease. We sincerely hope that the information these books provide will facilitate prompt application of these thoughts for a successful crusade against this disease. Let us prove that tuberculosis is, indeed, a curable disease.

Pattisapu R.J. Gangadharam
P.A. Jenkins

Acknowledgments

We would like to thank Dr. C.A. Reddy, the consulting editor for Chapman & Hall, who recognized the importance and timeliness of doing these volumes on mycobacteria, recruited us as the editors, has read many of the chapters and offered constructive suggestions, and has constantly given the needed encouragement to bring these books to fruition. We are also indebted to Mr. Gregory Payne, the then-publisher of Life Sciences at Chapman & Hall who was invaluable in launching this effort, Mr. Henry Flesh (Editor), Ms. Lisa LaMagna, (Managing Editor), Ms. Kendall Harris (Editorial Assistant), Barbara Tompkins (Copyeditor), and the able production staff at Chapman & Hall.

Contributors

Nils E. Billo, International Union Against Tuberculosis and Lung Disease, Paris, France.

I.A. Campbell, Chest Department, Penarth, Cardiff, CF64XX, UK.

Pierre Chaulet, Clinique de Pneumo- Phtisilogie Matiben, Hôpital de Beni-Messous, Center Hospitalier et Universitaire d'Alger-Ouest Alger, Algeria.

R.J. Coker, St. Mary's Hospital, London W21NY, UK.

Asim K. Dutt, Medical Service, Department of Veterans Affairs, Alvin C. York Medical Center, and Meharry Medical College, Nashville, TN 37208 USA.

Donald A. Enarson, International Union Against Tuberculosis and Lung Disease, Paris, France.

Pattisapu R.J. Gangadharam, Department of Medicine, University of Illinois at Chicago, College of Medicine, Chicago, IL 60612, USA.

Lawrence C. Geiter, American Lung Association, IUATLD, Washington DC 20036-4502, USA.

Jacques Grosset, Service de Medicine Pitie–Salpetriere, Paris, France.

Philip C. Hopewell, Division of Pulmonary and Critical Care Medicine, University of California, San Francisco, CA, USA.

P.A. Jenkins, Formerly, Mycobacterium Reference Unit, Public Health Laboratory Service, University Hospital of Wales, Health Park, Cardiff, CF44XS, UK.

Baohong Ji, Faculte de Medicine Pitie–Salpetriere, Paris, France.

Jay B. Mehta, Division of Preventive Medicine, James H. Quillen College of Medicine, East Tennessee State University, Johnson City, TN 37614 USA.

D.A. Mitchison, Department of Medical Microbiology, St. Georges Hospital Medical School, London SW17 ORE, UK.

S.K. Noordeen, Action Programme for the Elimination of Leprosy, WHO, Geneva, Switzerland.

L.P. Ormerod, Chest Clinic, Blackburn Royal Infirmary, Blackburn, Lanc BB2 3LR, UK.

S. Radhakrishna, Formerly, Institute for Research in Medical Statistics, Madras 600031, India.

Rory Shaw, Chest and Allergy Clinic, St. Mary's Hospital London W21NY, UK.

Tadao Shimao, The Japan Anti-Tuberculosis Association, Advi IUATLD, Tokyo, Japan.

Richard J. Wallace, Jr., Department of Microbiology, The University of Texas Health Center, Tyler, TX 75710, USA.

Noureddine Zidouni, Clinique de Pneumo-Phtisilogie Matiben, Hôpital de Beni-Messous, Center Hospitalier et Universitaire d' Alger-Ouest Alger, Algeria.

BIOGRAPHICAL SKETCHES
Dr. P.R.J. Gangadharam

Dr. Pattisapu Rama Jogi Gangadharam was born in Vijayawada, Andhra Pradesh, India. He obtained his B.Sc. (Hons.) and M.Sc. degrees from the Andhra University and his Ph.D. degree from the Bombay University. His research work, which deals almost entirely with mycobacteria and the diseases they cause, spans well over 4 decades, starting from his doctoral work on the chemotherapy of tuberculosis at the Indian Institute of Science, Bangalore, India.

He had been in the first team of international scientists who had established the Tuberculosis Chemotherapy Centre, now called the Tuberculosis Research Centre, at Madras, India, where he participated as a senior member with a major scientific responsibility in many of the controlled clinical trials in tuberculosis. Subsequently, he worked at the Baylor College of Medicine, Houston, Texas as the Director of Mycobacteriology Laboratories at the Harris County Hospital District. This was followed by a move to the National Jewish Hospital, now called the National Jewish Center for Immunology and Respiratory Medicine, Denver, Colorado to become the Director of Mycobacteriology Research and a Professor of Medicine and Clinical Pharmacology at the University of Colorado School of Medicine. Denver, Colorado. Currently, he is working as the Director of Mycobacteriology Research and Professor of Medicine, Microbiology and Pathology at the University of Illinois at Chicago College of Medicine, Chicago, Illinois.

He is an author of around 250 publications and several textbook chapters and editorials. His other single-author book "Drug Resistance in Mycobacteria" has been a best seller and is widely used all over the world as a reference book on the subject.

Besides participating in several controlled clinical trials, he has made several significant contributions of international recognition. These include the following: urine tests for antimycobacterial drugs to facilitate monitoring compliance in drug intake; serum levels and metabolism of antimycobacterial drugs to guide and monitor clinical response and development of toxicity; discovery and development of animal models for mycobacterial diseases (the beige mouse model discovered by him has been acclaimed as the most valuable animal model for the *Mycobacterium avium* complex); discovery of new antimycobacterial drugs (a drug discovered by him has been named after him); targeted drug delivery using several types of liposomes and sustained drug delivery using several types of biodegradable polymers. Some of his contributions (e.g., peak serum levels) have given the scientific basis for the intermittent chemotherapy of tuberculosis and have been acclaimed by international journals like the *Lancet*.

He has been a recipient of several national and international awards which include the First Robert Koch Centennial Award and the Burroughs Wellcome Award in Microbiology He is listed in the World's Who is Who from antiquity to the present, and among the Notable Americans of the Bi-centennial Era.

He is a member of several scientific societies, which include The American Thoracic Society, American Society for Microbiology, International Union Against Tuberculosis, Tuberculosis Association of India, and the India Chest Society. He has served on several subcommittees, which include Standards for Tuberculosis Laboratories, of the American Thoracic Society, Safety in Tuberculosis Laboratories of the International Union Against Tuberculosis, and the Scientific Advisory Committee of The American Foundation for AIDS Research.

Dr. P.A. Jenkins

Dr. P.A. (Tony) Jenkins was born in Cardiff in 1936, took an honors degree in microbiology at the University of Wales and then commenced a study of Farmer's lung (extrinsic allergic alveolitis) for a Ph.D. at the Brompton Hospital under Professor Jack Pepys. He was awarded a Ph.D. in 1964 by the University of London.

In 1965, he was appointed senior scientist at the Tuberculosis Reference Laboratory (Public Health Laboratory Service) in Cardiff and eventually became head of the laboratory in 1977 when the previous director, Dr Joe Marks, retired. The laboratory, which was renamed the Mycobacterium Reference Unit, remained in Cardiff until Dr. Jenkins retired after 30 years' involvement in mycobacteriological diagnosis and research. He has authored or coauthored over 100 papers and articles and presented papers at conferences throughout the world. He was secretary and chairman of the Bacteriology and Immunology Committee of the IUATLD and founder secretary of the European Society for Mycobacteriology. In 1990, he was made president of the Thoracic Society of Wales.

On his small holding in West Wales he now raises pigs and sheep but also keeps in touch with developments in the mycobacterial field and maintains his contributions to the literature.

1

Global Aspects of Tuberculosis

Nils E. Billo

1. Introduction

At the beginning of this century, tuberculosis was among the most prominent public health problems in industrialized countries, even though the decline in tuberculosis mortality started in the nineteenth century (1). At that time, no drugs were available to treat patients and everybody was at risk of infection with *Mycobacterium tuberculosis* and of developing the disease. After the discovery of *M. tuberculosis* by Robert Koch in 1882, researchers all over the world were challenged to find a cure for this major killer disease. The search for an effective drug was superbly described by F. Ryan in his book *Tuberculosis: The Greatest Story Never Told* (2). It was not until the late 1940s that the discovery of streptomycin gave new hope to the community that the possibility of treating tuberculosis might contribute to a decrease in mortality and morbidity worldwide (3).

Very soon, it was realized that monotherapy, the use of a single drug to treat the disease, causes the emergence of resistant strains. Other drugs such as isoniazid, p. amino salicylic acid (PAS), and thiacetazone were developed and were added to the treatment regimens in order to avoid drug resistance. The introduction of rifampicin in the early 1970s provided a potent addition to the repertoire of antituberculosis drugs. The British Medical Research Council tested a variety of therapeutic regimens in different regions of the world between the late 1950s and the 1970s under the leadership of Professor Wallace Fox and Professor Denis Mitchison.

The International Union Against Tuberculosis and Lung Disease (IUATLD), a nongovernmental organization based in Paris, devoted most of its activities to improving tuberculosis control in the 1970s and 1980s. Under the leadership of Dr. Karel Styblo and Dr. Annik Rouillon, new approaches to treat tuberculosis under difficult conditions were developed. Within the Mutual Assistance Programmes, Dr. Styblo was able to show that it is possible to treat tuberculosis successfully on a large scale (4).

In most industrialized countries, tuberculosis has been steadily declining by 5% or more each year, not only due to the availability of an effective treatment regimen but most probably also due to improving social and economic conditions. In the light of this decline, many of the efforts undertaken in the past by governments, nongovernmental organizations, patient organizations, and the community have been reduced to a minimum; certain governments have become complacent even to the extent of dismantling well-established tuberculosis control programs. Many antituberculosis associations have shifted their activities toward other lung diseases and have even dropped tuberculosis from their association's name.

Effective tools and drugs are available to combat tuberculosis. The aims of tuberculosis control as mentioned in the *Tuberculosis Guide for Low Income Countries* (5) should be as follows:

- *For individual patients:* to cure their disease, to preserve or quickly restore their work capacity, to allow them to remain within their family, ethnic group, and community, and, in this way, to maintain their socioeconomic position
- *For a community:* to decrease the risk of tuberculous infection in the community and, by this means, to improve the situation of tuberculosis and thus the economic and social conditions of the community.

This chapter describes the magnitude of the tuberculosis problem worldwide and tries to give some explanations of why we are now facing difficulties in reducing morbidity and mortality from tuberculosis.

2. The Toll of Tuberculosis

It is very well known that notification of infectious diseases, including tuberculosis, is very often unreliable. Tuberculosis cases reported at country level and data reported to the World Health Organization (WHO) are very often incomplete and do not reflect the real tuberculosis situation in a particular country. The magnitude of the tuberculosis problem has been estimated by several authors in the last two decades (6–9). All these publications represent estimates, but they all indicate very clearly that tuberculosis is still one of the major killer diseases worldwide. Out of a pool of about 1.7 billion individuals infected with *M. tuberculosis,* it is estimated that every year about 8 million new cases of tuberculosis occur and up to 3 million individuals die from tuberculosis (Table 1.1) (8).

In a recent article, Raviglione and co-authors describe the worldwide epidemic of tuberculosis (9). Based on their assumptions, we will soon be facing enormous problems, especially in Africa, Southeast Asia, and the Western Pacific. Between

Table 1.1. The toll of tuberculosis: estimations, WHO 1991

Region	Population infected (millions)	New cases	Deaths
Africa	171	1,400,000	660,000
Americas[a]	117	560,000	220,000
Eastern Mediterranean	52	594,000	160,000
Southeast Asia[b]	426	2,480,000	940,000
Western Pacific	574	2,560,000	890,000
Europe and other Industrialized Countries[c]	382	410,000	40,000
Total	1,722	8,004,000	2,910,000

[a]Excluding United States and Canada.
[b]Excluding Japan, Australia, and New Zealand.
[c]United States, Canada, Japan, Australia, and New Zealand.
Source: Ref. 8.

1990 and the year 2000 an estimated 90 million new cases of tuberculosis will occur worldwide, about 10% of which will be attributable to the human immunodeficiency virus (HIV) infection. Thirty million people will die from tuberculosis in this decade if tuberculosis programs are not improved, particularly in the most affected regions.

The HIV/AIDS epidemic emerged in the early 1980s and it was recognized at the end of this decade that HIV would have an important impact on the tuberculosis situation throughout the world (10). HIV infection represents one of the main risk factors to develop tuberculosis given infection with *M. tuberculosis* and it has become a reality that those communities infected at the same time with *M. tuberculosis* and with HIV are facing or will face serious problems in tackling both diseases, especially in low-income countries (11).

Even though recent estimations are based on many assumptions, the global toll of tuberculosis is enormous; even if the number of new cases and deaths due to tuberculosis is overestimated, it can still be stated that this disease is to be considered one of the major infectious diseases caused by a single agent and one of the major killers, especially in Southeast Asia, Western Pacific, and in sub-Saharan Africa. (See Table 1.2). It is important to improve methods to better estimate the magnitude of the tuberculosis problem and to overcome some methodological issues related to estimating the trends of tuberculous infection in a community (12).

It is known but not sufficiently emphasized that there are also high incidence rates of tuberculosis in deprived areas of large cities in some industrialized countries. It is, therefore, not always appropriate to distinguish between industrialized

Table 1.2. Distribution of individuals who have been infected with tuberculosis and HIV 15–49-year age group, early 1992

Region	HIV infected (thousands)	TB infected (%)	HIV/TB infected (thousands)	Percent of total
Africa[a]	6,500	48	3,120	77.8
Americas[b]	1,000	30	300	7.5
Eastern Mediterranean[a]	50	23	11	0.3
Southeast Asia and Western Pacific[c]	1,020	40	408	10.2
Europe[a] and others[d]	1,550	11	170	4.2
All regions	10,120	34	4,009	100

[a]Includes all countries of WHO region.
[b]Includes all countries of the American Region of WHO, except the United States and Canada.
[c]Includes all countries of the Western Pacific Region, except Japan, Australia, and New Zealand.
[d]United States, Canada, Japan, Australia, and New Zealand.
Source: Ref. 11.

and low-income countries. The recent reemergence of tuberculosis in industrialized countries reminds us of the fact that tuberculosis is a disease which will accompany our society for future decades, especially if we dismantle tuberculosis control programs based on the assumption that the disease will disappear by itself (13). It is important to monitor the tuberculosis situation very carefully in order to adapt tuberculosis control strategies. Tuberculosis has been underestimated and neglected as a health care problem and it is time to improve our surveillance and our commitment in industrialized and low-income countries, and this on a long-term basis (14). Not only the medical profession but, very importantly, politicians and policy-makers and those responsible for distribution of financial resources in the health sector have to be convinced that it is important to consider investment in tuberculosis control a priority.

3. Mortality

Tuberculosis is the fifth most common cause of death worldwide, after cardiovascular diseases (about 12 million deaths annually), acute respiratory infections, cancer, and diarrheal diseases (5 million annual deaths for each). Tuberculosis kills about 700,000 women every year and this represents more than all perinatal causes of death combined (500,000 deaths per year). In developing countries, tuberculosis is responsible for an estimated 7% of all deaths, and among 15–59-year-olds, tuberculosis contributes 25% of all causes of death.

Reliable mortality data are based on death certificates. Such data are available in most industrialized countries. However, death certificates are often incomplete and mortality rates may be underestimated. Several studies have shown that between 3.5% and 6.5% of all tuberculosis cases were only detected postmortem. Mortality has been declining more rapidly than incidence. Several factors such as better housing conditions with less crowding and the availability of drugs and successful treatment have resulted in a marked decrease of tuberculosis mortality with a less impressive decline in tuberculosis morbidity. In Switzerland, as an example of an industrialized country with appropriate social and economic conditions, tuberculosis mortality has sharply declined since World War II (15). (See Fig. 1.1.)

Mortality data on tuberculosis in developing countries are not available on a routine basis. It is estimated that mortality rates range from 2 per 100,000 in industrialized countries, to 25 per 100,000 in the Americas (excluding the United States and Canada), 76 per 100,000 in Africa, and 84 per 100,000 in South East Asia (16).

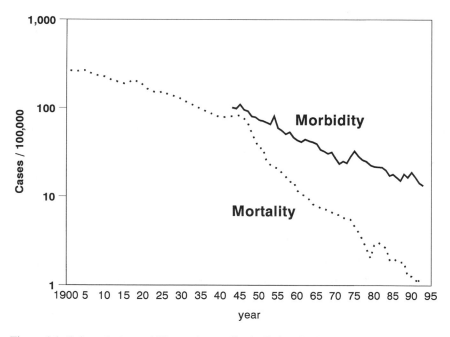

Figure 1.1 Tuberculosis morbidity and mortality in Switzerland.

4. Incidence of Tuberculosis, Impact of HIV

Incidence data are obtained through notification of tuberculosis cases by physicians and laboratories to the health authorities. As mentioned above, these data are very often incomplete and may not give an adequate reflection of the real situation. Data from WHO show that estimated incidence rates per 100,000 population vary between 23 per 100,000 for industrialized countries and 237 per 100,000 in Southeast Asia (191 per 100,000 for Africa) (16). It may be noteworthy that these rates are much lower than those observed at the beginning of the century in Europe.

Of the 1.7 billion of the world's population estimated to be infected with *M. tuberculosis*, two-thirds of these are found in Asia, where HIV infection is spreading at an alarming pace in some of the countries. HIV-related tuberculosis is increasingly becoming a problem in countries such as India, Thailand, Brazil, Haiti, and others and will probably cause similar devastating effects on families and society as it is having now in sub-Saharan Africa. It is important that governments, nongovernmental agencies, and health care professionals realize that tuberculosis and HIV infection must be dealt with in a coordinated way (11). Comprehensive prevention programs have to address both diseases at the same time and train health care professionals accordingly.

Incidence rates of tuberculosis are increasing in sub-Saharan Africa, especially in large cities, even where there is a good tuberculosis control program in place. In Tanzania, a marked increase in cases in the capital, Dar es Salaam, as compared with the rest of the country has been observed since the HIV/AIDS epidemic started in the 1980s. Incidence rates have reached 213.5 per 100,000 in 1993 compared to 77.6 per 100,000 in 1985, before the HIV epidemic had begun to have an important impact on tuberculosis morbidity and mortality (17). (See Fig. 1.2.)

Health care centers are overwhelmed with patients presenting with suspicion of tuberculosis and it is very likely that the health care system may collapse, unable to deal with so many patients who very often may present with other HIV-related conditions. HIV-positive patients very often present to health care services with clinical symptoms suggestive of tuberculosis or other lung diseases. The results of the sputum smear examination are often negative or are not done at all. It is one of the very important challenges of tuberculosis control now and in the future to diagnose each case of tuberculosis according to clear guidelines in order to avoid treatment of smear-negative individuals who actually do not have tuberculosis but another lung disease. Refined, simple, and inexpensive diagnostic tools should be developed in the near future to address the diagnostic problems for smear-negative patients presenting to health care facilities in all affected areas.

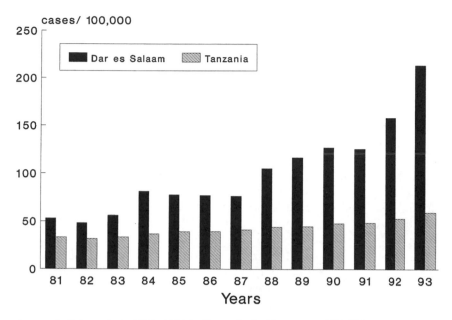

Figure 1.2 Tuberculosis 1981–1994 in Tanzania, incidence rates/100,000.

5. Tuberculosis—A Social Disease

Tuberculosis has been known to be one of the major public health problems in low-income countries, especially among the underprivileged population who of ten live in very unfavorable conditions with little or difficult access to health care facilities. The tuberculosis problem has been neglected by governmental and non-governmental agencies until recently. It is sad to remark that only a few organizations have constantly reminded the medical community that tuberculosis is, indeed, an important public health problem. Only a few donor countries such as Norway and Switzerland had a favorable attitude toward fighting tuberculosis in the 1980s when others had other priorities. Several tuberculosis outbreaks have been described in the United States and elsewhere, and tuberculosis has become an issue at the global level again. WHO declared tuberculosis a global emergency in 1993 and has been very successful in advocating tuberculosis control measures worldwide.

Why has the scientific community neglected tuberculosis? Responsible government officials in ministries of health are still neglecting a disease that is killing more individuals than AIDS, malaria, and other tropical diseases combined. Other public health problems, such as the occurrence of tuberculosis in population groups with little political influence and insufficient financial resources within

Ministries of Health are the main reasons for the neglect of the tuberculosis problem. Table 1.3 shows some reasons why tuberculosis is still not a priority in most of the countries concerned.

It is well accepted that we have the tools to cure tuberculosis with existing drugs in most cases. In industrialized countries, tuberculosis is found among the elderly and in high-risk groups such as the homeless, immigrants, and refugees from high-risk countries and among HIV-infected individuals. The epidemiological situation of tuberculosis in Switzerland shows that most cases occur among Swiss people aged 50 and over. Tuberculosis cases among the foreign born, including migrant workers immigrants from countries with a high risk for tuberculous infection and refugees, are found predominantly in the age group 20–40 years old who very often live in less favorable conditions. This situation is typical for most of the industrialized countries (Fig. 1.3).

It is the responsibility of governments to ensure that everybody has access to health care and adequate treatment (i.e., directly observed therapy). It has been claimed that it is difficult to reach underprivileged patients and to obtain acceptable compliance with treatment. However, it is the responsibility of private and public health care structures to provide appropriate services for tuberculosis patients in order to achieve good treatment results and avoid the emergence of multidrug resistance.

Table 1.3. Tuberculosis—reasons for low priority in low-income and industrialized countries

Low-income countries
  Other public health problems prevalent
  Patients have no lobby
  Lack of political commitment
  Lack of leadership among health policy-makers
  Lack of financial resources for antituberculosis drugs
  Inadequate procurement and distribution of antituberculosis drugs
  Health care structures not adequate
  Health care personnel often not well trained

Industrialized countries
  Patients mostly among older age groups and immigrants with no lobby
  Awareness of tuberculosis among physicians minimal
  Tuberculosis is not a money maker, other lung diseases are more interesting for the medical community
  Health care structures deficient in treating "difficult" patients
  Lack of leadership among health policy-makers and medical associations

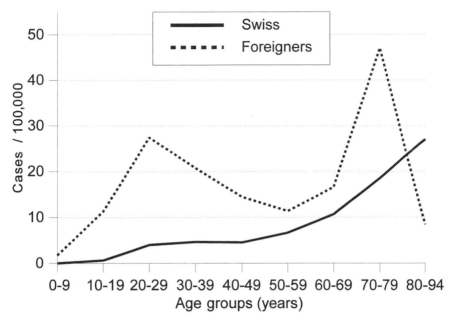

Figure 1.3 Tuberculosis in Switzerland—Swiss and foreigners 1994.

6. Drug-Resistant Tuberculosis

The availability of drugs alone does not guarantee a cure for tuberculosis (18). It is important to mention that a regimen must first be chosen in a correct combination and taken for a sufficient period of time to avoid the emergence of drug resistance. In countries where tuberculosis control is well organized, the drug regimen to be used is recommended by those responsible for the National Tuberculosis Control Programme. It usually comprises four drugs in the first 2 months of the treatment and usually two drugs taken for 4 to 6 or more months in the second phase. It is crucial to supervise the intake of each tablet in the first phase directly (it may be daily or intermittent) and if possible in the second phase, especially if rifampicin-containing drugs are used. In addition to this golden rule of directly supervised chemotherapy, it is crucial that drugs are free of charge for the patient in order to avoid noncompliance due to financial constraints. In most developing countries where the patient is asked to pay for the drugs, there is a grave danger that drug intake may be interrupted due to lack of funds. It is also important for health authorities to make sure that they are purchasing drugs which are of good quality and where bioavailability testing has been shown to be satisfactory. Respecting these guidelines will help to keep drug resistance to a minimum. Efforts to minimize drug resistance are particularly important in low-income countries where it may be impossible to get funds to treat multidrug-resistant tuberculosis cases.

Most countries do not know their situation regarding drug resistance and they may be unaware that they are producing drug resistance. This may happen in countries which start a national tuberculosis program without fulfilling certain conditions required to avoid the emergence of treatment failure and multidrug resistance (see Chapter 6). Currently, there are initiatives underway to improve drug-resistance surveillance throughout the world. A working group of WHO/IUATLD has recently proposed a protocol which is now being used in several countries to obtain better results. It is hoped that in the near future, reliable data on drug resistance will be available in each country in order to monitor the quality of tuberculosis programs.

7. Factors Contributing to the Quality of Tuberculosis Control Programs

Why do we have unfavorable conditions in many developing countries and a worsening situation in some of the industrialized countries? Several factors such as poverty, homelessness, HIV infection as mentioned above, drug and alcohol abuse, and other social factors have been made responsible for this reappearance of tuberculosis. However, the most important factors are inappropriate or nonexisting tuberculosis programs in developing and industrialized countries and the lack of interest by the medical profession. With proper use of the available anti-TB drugs, it is possible to cure most patients and to prevent the emergence of chronic cases and multidrug-resistant tuberculosis. Unfortunately, the improper application of drug regimens and the lack of drugs due to scarce financial resources or due to inadequate drug distribution systems in developing countries are a great threat, creating chronic cases and, thus, new infections and disease.

The IUATLD, a nongovernmental organization with over 100 member associations, tested new treatment models in the 1980s in several developing countries. It was possible to show that tuberculosis control can be organized efficiently and effectively if certain conditions are fulfilled. These conditions also apply to industrialized countries: (1) It is crucial to obtain a strong commitment from the government to fight against tuberculosis and to allocate adequate funding for personnel, infrastructure, and anti-TB drugs; (2) an adequate number of microscopy centers and other diagnostic facilities must be available; (3) drugs should always be available in sufficient quantities to treat all patients; (4) the personnel involved in the tuberculosis program should be trained adequately; (5) directly observed chemotherapy, especially in the first phase of the treatment, is crucial. The IUATLD model was successfully implemented in several low-income countries such as Tanzania, Mozambique, Malawi, Benin, and Nicaragua in the 1980s under very unfavorable conditions (war, famine, high turnover of personnel) respecting the principles mentioned above. In this way, it was possible to achieve high cure ratios of over 80%. The World Health Organization has recognized this model as one of the ways to combat tuberculosis worldwide and has integrated it in its Global Programme Against Tuberculosis which was established in the early 1990s.

The IUATLD model was recently presented in the World Bank's *World Development Report 1993* and proper tuberculosis control has been identified as one of the most cost-effective health interventions (19). This report has been widely distributed and is known in most parts of the world. Policy-makers and politicians should be briefed about the possibility of successfully tackling the tuberculosis problem. In addition, it has been recognized that proper treatment of tuberculosis, especially of sputum smear-positive tuberculosis cases, is the best prevention of new cases. However, political commitment and funding for personnel, education, and drugs is still lacking in many affected countries. It is apparently difficult for health policy-makers and politicians to redirect scarce resources in the health sector to tuberculosis control programs, especially when other diseases such as AIDS, acute respiratory infections, malaria, and so forth are present and causing widespread disease and death. Unfortunately, it is a sad fact that several countries who urgently need technical assistance through WHO and nongovernmental agencies like IUATLD are unable to obtain funding for these activities.

What can be done now and in the future to improve the situation and to avoid the emergence of multidrug-resistant tuberculosis cases worldwide, especially in low-income countries? As mentioned above, the most important point is a strong political will, even in those countries where financial resources are scarce. Governments and those responsible for tuberculosis programs may change quite frequently and it is, therefore, very important to develop a strategy to regularly inform health authorities about the importance and the priority of good tuberculosis control programs. Development agencies from affluent countries should now invest heavily in tuberculosis programs for several years to come. Only this assistance will help prevent new infections and the disease in the younger population. In this way, this productive age group will be able to remain so and contribute to the improvement of social conditions. If we fail to implement adequate tuberculosis programs, transmission of infection will continue and the tuberculosis problem will not diminish in the near future. If bad programs are implemented, it may be even worse; through inadequate treatment procedures, the tuberculosis program could be responsible for an increase of chronic and multidrug-resistant cases. It will by no means be possible to invest the millions of dollars that are currently being allocated in the United States to improve a situation that got out of control in some parts of the United States because tuberculosis control had been neglected. A fraction of this money would be sufficient to initiate good tuberculosis programs worldwide, with a good chance to treat and cure most of the patients. It should be remembered that a course of treatment in a developing country is about US$20–30.

In order to build or maintain a viable tuberculosis control program and to avoid the emergence of drug resistance, it is important to consider the following points.

Before starting a control program, it should be ensured that personnel and financial resources are adequate and are guaranteed by the appropriate health authority, if possible with financial commitment from a donor development agency for a time period of at least 5 years. It is important that the tuberculosis program is managed by an appropriate central unit at the ministry of health level. It is the task of this unit to plan, coordinate, train health care workers, and supervise staff involved in tuberculosis control on a regular basis. Such a unit should be operational not only in low-income countries but definitively also in industrialized countries.

A tuberculosis control program will only get off the ground or be sustained if there are well-trained collaborators who know how to integrate the tuberculosis program within the primary health care system and to convince the diagnostic laboratories responsible for sputum smear examination and other laboratory techniques to closely collaborate with the clinicians or the health care personnel responsible for treating the patient.

A system of quality control for sputum smear microscopy should be set up to ensure proper diagnostic procedures. A secure supply of all recommended drugs used in the tuberculosis control program should be guaranteed to avoid any shortage of drugs.

Records of diagnostic examinations and of each case of tuberculosis should be kept and recorded in the laboratory and tuberculosis register for evaluation, analysis, and reporting. Results of each treatment course should be recorded in such a way that they can be used to judge the quality of the program. The following categories based on the IUATLD tuberculosis control model should be recorded at the end of the treatment course in developing and industrialized countries and be evaluated on a regular basis (5):

- Smear negative: indicates an individual who was smear negative at 1 month prior to the completion of chemotherapy and on at least one previous occasion
- Smear not done: indicates those patients who completed treatment but in whom smear examination results are not available on at least two occasions prior to the completion of treatment
- Smear positive: designates any patient who remains or becomes again smear positive at 5 months or more during chemotherapy
- Died: recorded for patients who die for any reason during the course of their chemotherapy
- Defaulted: recorded for any patient who has failed to collect medication for more than 2 consecutive months after the date of the last attendance during the

course of treatment and who, at the time of evaluation of treatment results (15 months after close of the quarter in which the patient was entered into the tuberculosis register), is still on treatment
- Transferred out: indicates any patient for whom treatment results are unknown and who completed chemotherapy at another center to which the patient had been referred to for continued treatment

Training and supervision should be an integral component of a tuberculosis control strategy, regardless of whether the program is integrated in a health care system controlled mainly by the ministry of health and the regional authorities, or the tuberculosis control measures are applied within the private health care sector. In both systems, personnel in the laboratories and the medical personnel responsible for treating the patients should be trained and supervised. Training and supervision will be particularly difficult in a system where most of the health care delivery and tuberculosis treatment is provided by private physicians who often have other priorities. It is, therefore, the responsibility of the medical associations to inform their members about the importance of treatment guidelines and to ensure that tuberculosis gets enough attention during medical training and at lung health conferences.

All these seemingly obvious points are implemented in well-functioning national tuberculosis control programs following the IUATLD/WHO model. It is very important that other countries who wish to improve tuberculosis control respect these points.

8. Conclusion

Tuberculosis is a disease of the poor and the underprivileged, most of the burden of diseased living in low-income countries. In industrialized countries, tuberculosis has become a disease of the elderly and of immigrants, the homeless, and drug- and alcohol-addicted individuals. It is therefore important to encourage young colleagues to become interested in this disease that is affecting public health. It is also to be hoped that federal, state, and local governments realize that an investment today will be a savings in the future. Nonprofit organizations which were very active in the past and especially effective through a very active group of volunteers still have a key role to play; they need to discuss ways of increasing awareness in the general population, increase efforts to raise funds for tuberculosis control, and design innovative programmes to ensure that patients are adequately treated.

References

1. Grigg ERN (1958) The arcana of tuberculosis. Am Rev Tuberc 78:151.
2. Ryan F. (1992) *Tuberculosis: The Greatest Story Never Told.* Swift, Bromsgrove.

3. Schatz A, Waksman SA (1944) Effect of streptomycin and other antibiotic substances upon *Mycobacterium tuberculosis* and related organisms. Proc Soc Exp Biol Med 57:244.

4. Rouillon A (1991) The Mutual Assistance Programme of the IUATLD. Development, contribution and significance. Bull Int Union Tuberc Lung Dis 66:159.

5. Enarson D, Rieder H, Arnadottir T (1994) *Tuberculosis Guide for Low Income Countries,* 3rd Ed. Paris: IUATLD.

6. Bulla A (1977) Tuberculosis patients—how many now? Bull Int Union Tuberc Lung Dis 52:35.

7. Styblo K, Rouillon A (1981) Estimated global incidence of smear-positive pulmonary tuberculosis. Unreliability of officially reported figures on tuberculosis. Bull Int Union Tuberc Lung Dis 56:118.

8. Kochi A (1991) The global tuberculosis situation and the new control strategy of the World Health Organization. Tubercle 72:1.

9. Raviglione MC, Snider DE, Kochi A (1995) Global epidemiology of tuberculosis. Morbidity and mortality of a worldwide epidemic. J Am Med Assoc 273(3):220.

10. Rieder HL, Snider DE (1986) Tuberculosis and the acquired immunodeficiency syndrome. Chest 90:469.

11. Narain JP, Raviglione MC, Kochi A (1992) HIV-associated tuberculosis in developing countries: epidemiology and strategies for prevention. Tubercle Lung Dis 73:311.

12. Rieder HL (1995) Methodological issues in the estimation of the tuberculosis problem from tuberculin surveys. Tubercle Lung Dis 76(2):114.

13. Brudney K, Dobkin J (1991) Resurgent tuberculosis in New York City. Am Rev Respir Dis 144:745.

14. Rieder HL (1992) Misbehaviour of a dying epidemic: a call for less speculation and better surveillance. Tubercle Lung Dis 73:181.

15. Office fédéral de la santé publique (1995) Tuberculose en Suisse en 1993. Bull OFSP 17.

16. Dolin PJ, Raviglione MC, Kochi A (1994) Global tuberculosis incidence and mortality during 1990–2000. Bull WHO 72:213.

17. International Union Against Tuberculosis and Lung Disease (1994) The National Tuberculosis/Leprosy Programme—United Republic of Tanzania. Progress Reports 1981–1993.

18. Gangadharam PRJ (1993) Drug resistance in tuberculosis: it's not always the patient's fault! Tubercle Lung Dis 74:65.

19. World Bank (1993) *World Development Report 1993: Investing in Health.* Oxford: World Bank/Oxford University Press.

2

Basic Concepts in the Chemotherapy of Tuberculosis

D.A. Mitchison

This chapter deals with the broad issue of the efficacy of the drugs used in chemotherapy. It starts with a brief history of how the randomized controlled clinical trial was used in the development of effective drug treatment, as controlled trials provided the most direct evidence of the efficacy of individual drugs and also indicated the manner in which they should be used. There follow sections dealing with intermittent treatment, the bactericidal activity of drugs, their grading according to different criteria of activity, and their action during different phases of chemotherapy. The molecular mode of action of drugs is only mentioned when knowledge of the mechanism has a direct bearing on the use of the drug in treatment. Even though dosage of the drug is determined by the balance between antibacterial efficacy and toxicity, little will be said about mechanisms of toxicity. Abbreviations for the names of the major antituberculosis drugs will be used as follows: isoniazid (H), rifampin (R), streptomycin (S), pyrazinamide (Z), ethambutol (M), *p*-aminosalicylic acid (P), and thiacetazone (T).

1. History of the Chemotherapy of Pulmonary Tuberculosis

This brief history is based mainly on the clinical studies carried out in association with the British Medical Research Council. The first modern drugs used in the chemotherapy of tuberculosis were thiacetazone (T) and streptomycin (S). T was introduced as Conteben by Domagk in 1946 (1,2) but was found to be too toxic when initially used in high dosage (3). S, a more potent drug, was isolated from *Streptomyces griseus* (4) and was the subject of the first controlled clinical trial in which patients with severe, acute pulmonary tuberculosis were allocated at random to treatment with either bed rest plus S or bed rest alone (5). The initial improvement attributed to S was followed by the emergence of S-resistant *Mycobacterium tuberculosis*. In consequence, the patients treated with or without S eventually had similar high mortalities (6).

1.1. Preventing the Emergence of Drug-Resistant Tubercle Bacilli

The main aim in developing effective regimens subsequently was to prevent the emergence of drug resistance. The first step was the demonstration that adding a second drug, *p*-aminosalicylic acid (P), to S prevented the emergence of S resistance (7,8). This was followed by the exploration of regimens of isoniazid (H) in combination with P or S (9,10). Then came the first surveys of primary drug resistance (11,12), which led to the development of highly effective regimens starting with three-drug combinations (for cases in which patients have been infected with drug-resistant *M. tuberculosis*) and continuing with two-drug combinations (13,14).

1.2. Domiciliary Treatment

The next important stage was the demonstration that the efficacy of chemotherapy was not increased by giving treatment in a sanatorium (15) nor was the risk of infecting household contacts decreased (16). However, although this made it possible to develop national programs for the control of tuberculosis based on domiciliary treatment, there remained the cost of the drugs, a matter of importance for low-income developing countries, and the need for improving compliance. An important conclusion from this work was that putting patients in hospital for an initial period did not improve the regularity of drug-taking, but created problems for families and, therefore, encouraged discharge against medical advice, and made it more difficult to establish domiciliary supervision on discharge from the hospital.

1.3. Low-Cost Thiacetazone–Isoniazid Regimens

The cost issue was addressed by the introduction of the low-cost combination of thiacetazone and isoniazid following a series of clinical trials in East Africa and other countries (Table 2.1). It is evident that the standard combination of 150 mg T with 300 mg H (TH) daily was similar in efficacy to 10 g P with 200 mg H (PH) daily (studies 2, 4, and 8 in Table 2.1). Reducing the dose size of T to 100 mg or of H to 200 mg decreased efficacy substantially (study 2), but an increase in the H dosage to 450 mg daily did not improve results (study 5). The addition of S initially improved results substantially (studies 3, 6, and 9). Curiously, although T was as effective as P in preventing the emergence of H resistance, T was much less effective than P in preventing S resistance (study 7).

Table 2.1. Studies of daily regimens containing thiacetazone in the treatment of pulmonary tuberculosis due to drug-sensitive organisms

Study No.	Place	Date of start	Regimen	No. of patients	Length of treatment (months) Total	Hospital	Favorable status at 12 months (%)	Ref.
1	Africa	1958	PH[a]	38	12	6	84	17
			T150.H200	43	12	6	79	
2	Africa	1960	PH	72	12	6	85	18
			TH	66	12	6	89	
			T150.H200	75	12	6	66	
			T100.H300	65	12	6	54	
3	Africa	1962	TH	190	18	2	79	19
			2STH/TH	170	18	2	90	
4	Madras	1962	PH	72	12	0	82	20
			6P6.H/H	71	12	0	67	
			TH	77	12	0	82	
5	Africa	1963	TH	127	18	0 or 2	77	21
			T150.H450	124	18	0 or 2	68	
6	Africa	1967	TH	147	18	2	88	22
			0.5STH/TH	161	18	2	90	
			1STH/TH	159	18	2	94	
			2STH/TH	162	18	2	96	
7	Rhodesia (Zimbabwe)	1964	TH	58	6	6	74[b]	23
			SP15	55	6	6	85[b]	
			ST	58	6	6	40[b]	
8	Hong Kong	1965	PH	77	12	8	82	24
			TH	87	12	8	75	
9	Singapore	1967	TH	78	12	3	65	25
			6STH/TH	94	12	3	100	
			6SH/H	114	12	3	99	

[a]PH = a regimen of 10g sodium P and 200 mg H in 2 equal daily doses. Any other number at the start of the regimen abbreviation indicates the duration of the initial phase, e.g., 2STH/TH = 2 months of STH followed by TH in the continuation phase. TH = a regimen of 150 mg T and 300 mg H daily. If other dose sizes were given, they are indicated by numbers after the drug initial, e.g., T150.H200 = 150 mg T and 200 mg H, daily and SP15 = S 1 g and P 15 g, daily.
[b]Assessment at 6 months.

International studies showed great regional variation in the incidence of side effects to T (26,27). Low-cost regimens containing TH are now widely used in developing countries in which patients did not encounter many side effects in these studies. However, regimens based on TH are steadily being replaced by short-course chemotherapy.

1.4. Improved Compliance

Compliance was improved by the use of fully supervised drug-taking (28), now renamed directly observed treatment (DOT), and by the use of intermittent regimens with drug doses given less frequently than daily, thus making supervision easier for the patient and the medical services. Clinical studies of intermittent chemotherapy are reviewed in a later section of this chapter (Section 2.1). The greatest improvement was produced by the introduction of short-course chemotherapy.

1.5. Short-Course Regimens

The last stage in the development of modern chemotherapy was the introduction of short-course regimens. The stimulus to try these regimens was provided by earlier work, mainly in experimental murine tuberculosis, which demonstrated the unique sterilizing activity of Z and R. High doses of Z in combination with H was the only regimen capable of sterilizing the organs of mice after a 12-week treatment period (29,30). Similarly, combinations of H with R (31,32) and with Z, given in a dose comparable to the human dose (32), were highly effective in sterilizing mouse organs.

The first large-scale multicenter study of short-course regimens was carried out in East Africa (study 1, Table 2.2). The patients received 6 months of SH in hospital (to assure regular drug-taking) followed by monthly sputum bacteriology during the succeeding 2 years to measure the relapse rate. To the basic SH regimen were added rifampin (R), Z or T, as a control, in separate regimens, while a fifth control 18-month regimen of STH/TH was included. This study showed that the addition of R or Z to the basic SH regimen increased the proportion of sputum cultures that were negative at 2 months and substantially reduced the relapse rate. R was slightly more effective than Z in both assessments. R and Z were therefore established as drugs that sterilized the lesions with particular rapidity and thus reduced the duration of treatment necessary to assure a low relapse rate after discontinuation of chemotherapy. These results showed that the exceptional sterilizing ability of R and Z in animal experiments was also found in human disease. The study led to an accelerated program of clinical studies for the development of short-course chemotherapy for pulmonary tuberculosis in several centers worldwide.

Tables 2.2, 2.3, and 2.4 summarize these studies in East and Central Africa, Hong Kong, Singapore, and Madras, India. The effectiveness of the regimens in all of these studies was assessed by the relapse rate usually determined during a 2-year follow-up after the end of chemotherapy, and also by the proportion of negative sputum cultures at 2 months, which is well correlated with the relapse rate (see Section 3.2).

Table 2.2. Short-course chemotherapy studies in East and Central Africa; Results in patients with pulmonary tuberculosis due to drug-sensitive organisms

Study No.	Date of start	Regimen	Duration (months)	Patients assessed for relapse	Sputum culture negative at 2 m (%)	Relapse rate in 2-yr follow-up (%)	Ref.
1	1970	SHR	6	152	69	3	33–36
		SHZ	6	153	66	8	
		SHT	6	104	42	22	
		SH	6	112	49	29	
		2STH/TH	18	133	56	4	
2	1972	SHR	6	171	70	2	37, 38
		HR	6	164	64	7	
		2SHRZ/TH	6	179	83	7	
		2SHRZ/$S_2H_2Z_2$[a]	6	159	80	4	
3	1974	2SHRZ/TH	6	75 } 87		{ 13	39, 40
		2SHRZ/TH	8	81		0	
		1SHRZ/TH	6	79 } 67		{ 18	
		1SHRZ/TH	8	58		7	
		1SHRZ/$S_2H_2Z_2$	6	75 } 68		{ 9	
		1SHRZ/$S_2H_2Z_2$	8	88		2	
		2SHR/TH		82 } 75		{ 18	
		2SHR/TH	6	77		6	
4	1976	2SHRZ/HRZ	4	104		16	41, 42
		2SHRZ/HR	4	104 } 85		11	
		2SHRZ/HZ	4	98		32	
		2SHRZ/H	4	105		30	
		2HRZ/H	4	100	79	40	
5	1978	2SHRZ/HR	6	166		3	43, 44
		2SHRZ/HZ	6	164 } 84		8	
		2SHRZ/H	6	156		10	
		2SHRZ/H	8	123		3	
6	1978	2SHRZ/TH	6	105 } 94		{ 3	45
		2SHRZ/H	6	100		11	

[a]$S_2H_2Z_2$ indicate streptomycin, isoniazid and pyrazinamide given twice a week (see legend for Table 2.5)

Table 2.3. Short-course chemotherapy studies in Hong Kong (HK) and Singapore (Sing); results in patients with pulmonary tuberculosis due to drug-sensitive organisms

Study No.	Place (date of start)	Regimen	Duration (months)	Patients assessed for relapse	Sputum culture negative at 2 m (%)	Relapse rate (%) follow-up for		Ref.
						2 yr	5 yr	
1	HK 1972	SHZ	6	60 ⎫	77	18 ⎱	—	46, 47
		SHZ	9	65 ⎭		5 ⎰	—	
		$S_3H_3Z_3$	6	68 ⎫	70	24 ⎱	—	
		$S_3H_3Z_3$	9	65 ⎭		6 ⎰	—	
		$S_2H_2Z_2$	6	39 ⎫	72	21 ⎱	—	
		$S_2H_2Z_2$	9	49 ⎭		6 ⎰	—	
2	HK 1974	SHR	6	143	88	6	—	48, 49
		$2SHRZ/S_2H_2Z_2$	6	87 ⎫	95	7 ⎱	—	
		$2SHRZ/S_2H_2Z_2$	8	87 ⎭		3 ⎰	—	
		$2SHRM/S_2H_2M_2$	6	84 ⎫	81	23 ⎱	—	
		$2SHRM/S_2H_2M_2$	8	84 ⎭		10 ⎰	—	
		$4S_3H_3R_3Z_3/S_2H_2Z_2$	6	71 ⎫	94	6 ⎱	—	
		$4S_3H_3R_3Z_3/S_2H_2Z_2$	8	83 ⎭		1[b] ⎰	—	
3	HK 1977	HRZM	6	163	94	1[b]	4	50–52
		$H_3R_3S_3Z_3M_3$	6	152	88	1[b]	4	
		$H_3R_3S_3Z_3$	6	151	90	1[b]	1	
		$H_3R_3S_3M_3$	6	166	76	8[b]	10	
		$H_3R_3Z_3M_3$	6	160	90	2[b]	4	
4	HK 1979	$S_3H_3R_3Z_3-Z_2$[a]	6	220 ⎫		3 ⎱	—	53
		$S_3H_3R_3Z_3-Z_4$	6	205 ⎬	91	5 ⎰	—	
		$S_3H_3R_3Z_3-Z_6$	6	208 ⎭		3	—	
		$H_3R_3Z_3-Z_6$	6	199	88	6	—	
5	Sing 1973	2SHRZ/HRZ	4	⎫		11 ⎱	13	54–56
		2SHRZ/HRZ	6	⎬ 165		0 ⎰	1	
		2SHRZ/HR	4	⎭	98	8 ⎱	14	
		2SHRZ/HR	6	⎭ 165		2 ⎰	3	
6	Sing 1978	$2SHRZ/H_3R_3$	6	97	99	1	2	57–59
		$1SHRZ/H_3R_3$	6	94	85	1	2	
		$2HRZ/H_3R_3$	6	109	90	1	3	
7	Sing 1983	$2SHRZ(C)/H_3R_3$[c]	6	46 ⎫	98	7[b] ⎱	—	
		$2SHRZ(S)/H_3R_3$	6	47 ⎭		0[b] ⎰	—	
		$1SHRZ(C)/H_3R_3$	6	42 ⎫	92	5[b] ⎱	—	
		$1SHRZ(S)/H_3R_3$	6	46 ⎭		2[b] ⎰	—	
		$2HRZ(C)/H_3R_3$	6	40 ⎫	97	8[b] ⎱	—	
		$2HRZ(S)/H_3R_3$	6	44 ⎭		2[b] ⎰	—	

[a] Z_2 = Z for first 2 months; Z_4 = Z for first 4 months; Z_6 = Z for 6 months.

[b] = follow-up for 18 months.

[c] C = combined preparation of HRZ; S = H, R and Z as separate preparations.

1.6. Low-Cost Regimens with an Initial Intensive Phase and a TH Continuation

A major issue in the multicenter African studies was to economize on the use of R, then a very expensive drug, by using it only in an initial intensive phase. The first point was the demonstration that the sterilizing activities of R and Z were additive. Adding Z to SHR increased the sterilizing activity of the initial phase as shown by (i) the increase in negative sputum cultures at 2 months from 70% in the SHR regimen of study 2 of Table 2.2 to 80% and 82% in the two regimens starting with SHRZ; (ii) similar differences in the proportions of 2-month negative cultures in the SHR and SHRZ regimens in study 3 of Table 2.2 and in study 2 of table 2.3; and (iii) the lower relapse rates in the 2SHRZ/TH

Table 2.4. Short-course chemotherapy studies at the Tuberculosis Research Centre, Madras; results in patients with pulmonary tuberculosis due to drug-sensitive organisms

Study No.	Regimen	Duration (months)	Patients assessed for relapse	Sputum culture negative at 2 m (%)	Relapse rate (%) follow-up for		Ref.
					2 yr	5 yr	
1	$2SHRZ/S_2H_2Z_2$	5	129 }	92	{ 5	7	61, 62
	$2SHRZ/S_2H_2Z_2$	7	132 }		{ 0	4	
	$2SHZ/S_2H_2Z_2$	7	129	72	4	7	
	SHRZ, prednisolone		132	92	3 }	7	
	SHZ, prednisolone		129	75	2 }		
	SHRZ, no prednisolone		129	93	2 }	6	
	SHZ, no prednisolone		140	74	4 }		
2	SHRZ	3	200 }	91	{ 20	—	63
	$3SHRZ/S_2H_2Z_2$	5	187 }		{ 4	—	
	$3SHZ/S_2H_2Z_2$	5	199	74	13	—	
3	$2S_3H_3R_3Z_3/S_2H_2R_2$	6	111 }	89	{ 2 }	6 }	64
	$2S_3H_3R_3Z_3/S_1H_1R_1$	6	111		5	7	
	$2S_3II_3R_3Z_3/R_2H_2$	6	101 }		3 } 2.9^a	6 } 6.2^a	
	$2S_3H_3R_3Z_3/R_1H_1$	6	116		2	5	
	$2S_3H_3R_3Z_3/S_2H_2$	6	151 }		3 }	7 }	
	$2S_2H_2R_2Z_2/S_2H_2R_2$	6	108 }	86	{ 3 }	7 }	
	$2S_2H_2R_2Z_2/S_1H_1R_1$	6	117		4	8	
	$2S_2H_2R_2Z_2/R_2H_2$	6	102 }		6 } 6.3^a	7 } 7.9^b	
	$2S_2H_2R_2Z_2/R_1H_1$	6	109		7	7	
	$2S_2H_2R_2Z_2/S_2H_2$	6	155 }		10 }	12 }	

[a]Follow-up for 18 months.
[b]Follow-up for 4.5 years.

regimens than in the corresponding 2SHR/TH regimens in study 3 of Table 2.2. Correspondingly, the addition of R to an SHZ initial phase in study 1 of Table 2.4 increased the proportion of negative 2-month cultures from 72% to 92% and decreased the relapse rate from 7% to 4%. Similar results were obtained in study 2, Table 2.4. The second point was whether the addition of S or ethambutol (M) to HRZ increased the sterilizing activity. Although nonsignificant differences in 2-month culture results and in relapse rates (studies 2 and 4, Table 2.2) suggested that S might have low sterilizing action, larger, later studies (studies 3, 4, and 7 of Table 2.3) failed to confirm any significant differences in relapse rates by the addition of S. The addition of M in the initial phase to an SRH regimen did not increase the proportion of 2-month negative cultures (study 2, Table 2.3), whereas replacement of Z by M decreased the 2-month negative cultures and increased the relapse rate (studies 2 and 3, Table 2.3). These findings showed that neither S nor M have appreciable sterilizing activity. They have been included in the initial phase and sometimes in the continuation phase as well because of their value in the treatment of infections due to initially resistant tubercle bacilli. Thus, standard initial treatment starts with the most actively sterilizing drug combination known—HRZ—to which may be added S or M when epidemiological evidence suggests that initially resistant tubercle bacilli are prevalent.

The third issue in the development of low-cost "African" regimens was the nature of the continuation phase of chemotherapy. TH was the first choice for self-administered regimens because of its low cost and because it had already been used widely in the STH/TH regimen. An alternative of twice-weekly SHZ was also explored as potentially useful as a fully supervised regimen. Studies 2, 3, and 6 of Table 2.2 showed that the TH continuation given for 6 months after an initial phase lasting 2 months gave a very low relapse rate (0% in study 3); but higher rates were found after an initial phase lasting only 1 month. In studies 2 and 3, the twice-weekly SHZ continuation always gave slightly lower relapse rates than TH. The reason for this is unclear because twice-weekly SHZ is slightly less effective than daily SHZ (study 1, Table 2.3), Z should not be effective as a sterilizing drug in the continuation phase (discussed below), and, as we have seen, S has little, if any, sterilizing activity. In countries with a high prevalence of HIV infections, there is a need to replace TH in the continuation phase (when the country can afford the change) because of the high level of serious Stevens–Johnson reactions to T encountered in HIV-positive patients (60).

1.7. High-Cost Short-Course Regimens

Although isoniazid, rifampin, pyrazinamide, and usually a fourth drug were always included in the crucial initial phase, the choice of drugs for the continuation phase was greatly influenced by their cost. After an initial exploration of regimens with Z but without R (study 1, Table 2.3), the role of R and Z in the continuation phase was explored in several studies (studies 4 and 5, Table 2.2; study 5, Table 2.3). The addition of Z in the continuation phase to either H or to HR did not decrease the relapse rate. On the other hand, when R was added to H or to HZ, the relapse rate was reduced substantially. The failure of Z to act as a sterilizing drug was confirmed in the later study in Hong Kong (study 4, Table 2.3) when no difference was found in relapse rates whether Z was given for 2 months initially, for 4 months, or for the whole 6 months. Whereas Z had no sterilizing activity in the continuation phase, it seemed effective in preventing failure in patients with initial H resistance; thus, failure occurred in 25% of 57 patients with initial S or H resistance given H alone in the continuation phase but in only 1 of 36 similar patients given HZ (65).

Regimens were just as effective when drugs were given three times per week as when they were given daily, whether intermittent dosage was given throughout treatment (studies 2 and 3, Table 2.3) or in the continuation phase only (compare studies 5–7, Table 2.3). The results of study 3 of Table 2.4 suggest a slightly less good result when the initial SHRZ combination was given twice weekly rather than thrice weekly. These studies established modern 6-month regimens which have very low relapse rates, whether drugs are given daily or three times weekly. Further studies in Hong Kong showed that similar low 5-year relapse rates of 1–4% were obtained when patients with smear-negative pulmonary disease were treated for a total of 4 months (66). The bacterial content of the sputum had unexpectedly little influence on the duration of therapy, because treatment periods of 3 and 2 months were followed by appreciably higher relapse rates (67).

A point of interest for immunotherapy was the failure of prednisolone, given at the start of treatment, to influence the 2-month conversion rate or the relapse rate (study 1, Table 2.4).

2. Intermittent Chemotherapy

2.1. Clinical Studies

Studies in which systematic appraisals were made of the efficacy and toxicity of drugs given at increasingly wide intervals are listed in Table 2.5.

Studies 1–5 (Table 2.5) show that although twice-weekly SH was more effective than 10PH, once-weekly SH was less effective, particularly in rapid inactivators of H, due to a deficiency in the activity of H. Improved results were not obtained by adding once-weekly Z but did occur when the once-weekly regimen was started with an initial 1 month of daily SH, although not to the level obtained with twice-weekly treatment in rapid inactivators. Study 6 showed that twice-weekly PH was as good as daily PH. Study 7 suggested, on small numbers of patients, that twice-weekly TH was inferior to daily TH, as the proportions with a favorable status (80% and 92%) were lower than after similar regimens with

Table 2.5. Studies of intermittent regimens lasting at least 12 months; results in patients with pulmonary tuberculosis due to drug-sensitive organisms

Study No.	Place (date of start)	Regimen	No. of patients	Favorable status at 12 months (%)			Patients with H-resistant strains (%)	Ref.
				Slow	Rapid	All		
1	Madras	$10PH^a$	71	—	—	85	6	68
	1961	S_2H_2	79	—	—	94	1	
2	Madras	S_2H_2	123	92	91	90	8	69
	1963	S_1H_1	117	82	60	72	19	
3	Madras	S_2H_2	104	97	91	94	5	69
	1964	S_1H_1	79	76	56	68	22	
		$1SH/S_1H_1$	106	95	76	88	9	
		$S_1H_1Z_1$	105	87	53	74	16	
4	Madras	$1SH/S_1H_1$	176	93	72	85	—	70
	1966	$1SPH/S_1P_1H_1$	170	95	76	87	—	
		H13mg/kg	178	—	—	83	—	
		H17mg/kg	168	—	—	89	—	
5	Prague	3SPH/PH	165	—	—	98	—	71
	1967	$3SPH/S_2H_2$	233	—	—	97	—	
6	Madras	0.5SPH/PH	83	—	—	87	10	72
	1968	$0.5SPH/P_2H_2$	90	—	—	88	10	
7	Africa	$1STH/T_2H_2$	25	80	80	80	22	73
	1972	$2STH/T_2H_2$	24	94	88	92	9	
8	Singapore	$0.5SRH/R900_2H_2$	107	—	—	100	—	74
	1973	$0.5SRH/R600_2H_2$	110	—	—	100	—	
		$0.5SRH/R900_1H_1$	112	100	98	97	3	
		$0.5SRH/R600_1H_1$	102	100	95	93	7	
9	Prague	$3SPH/S_2H_2$	119	99^b	97^b	98^b	—	75
	1970	$1.5SPH/S_2H_2$	109	100^b	97^b	99^b	—	
		$3SPH/S_1H_1$	130	100^b	81^b	94^b	—	
10	Madras	$0.5MHS/M_7H_7$	107	95	98	96	5	76
	1971	$0.5MHS/M_2H_2$	101	92	83	88	8	
		$0.5MHS/M_1H_2$	107	95	91	93	6	
		$0.5MHS/M_1H_1$	109	91	57	75	11	

^{a}PH = a regimen of 10 g sodium P and 200 mg H in two equal daily doses. Any number at the start of the regimen abbreviation indicates the duration of the initial phase, e.g., 3SPH/PH = 3 months of SPH followed by PH in the continuation phase. Subscript numbers indicate the number of doses per week, i.e., 1 = once weekly; 3 = three times weekly; no number or 7 = daily. Unusual or important dose sizes are indicated by numbers after the drug letter, e.g., H13 mg/kg = dose of 13 mg/kg H and R900 = dose of 900 mg R.

daily TH in the continuation phase (94% and 96%) in study 6 of Table 2.1. Study 8 (Table 2.5) compared with studies 2–4 showed that once-weekly RH performed better than SH, especially when the higher 900 mg dose of R was given. Study 9 repeated the demonstration that twice-weekly SH was effective in both slow and rapid inactivators, whereas once-weekly SH was less effective in rapid inactivators, even when preceded by 3 months of daily SPH. This study led to the determination of the inactivator status at the start of treatment; the rapid inactivators were then treated with twice-weekly SH, whereas the slow inactivators were given once-weekly SH. Study 10 explored intermittent dosage with ethambutol (M), the best results being obtained with daily administration of M and H or with once-weekly M (at a high dose) plus twice-weekly H.

2.2. Effects of Pulses of Drugs

Doses of antituberculosis drugs are usually given to patients at daily intervals or even less frequently in intermittent regimens. Because most of them have half-lives in humans of 5 h or less (Table 2.6), it is perhaps surprising that, where studied, the same total size of dose was at least as effective and usually more effective when given in one daily dose than when divided into two or more doses daily. How is it that drugs are active even when their blood concentrations may

Table 2.6. Serum concentrations and minimal inhibitory concentrations of antituberculosis drugs

Drug	Dose (mg)	Peak (mg/L)	Half-life (h)	Protein binding (%)	MIC[a] (mg/L)
Isoniazid (slow[b])	300	5	3	0	0.1
(rapid[b])		4	1.3		
Rifampin	600	12	3	85	0.3
Pyrazinamide	2000	40	8	0	20
Streptomycin	750	40	3–5	35	2
Ethambutol	1200	3	3	0	1.5
PAS (Na)	12000	250	1	70	0.5
Thiacetazone	150	2	12	—	0.4
Ethionamide	500	3	2	—	0.6
Ofloxacin	800	11	6	23	1.0

Note: The dose size and peak serum concentration are appropriate for a 60-kg subject.
— = uncertain
[a]minimal inhibitory concentration (MIC) in liquid medium with serum but without Tween 80.
[b]Acetylator status.

be below the minimal inhibitory concentration (MIC) for much of the day or even for several days? These issues have been examined in studies in which serial counts of colony-forming units (CFU) were done on cultures of *M. tuberculosis* that were exposed to drugs for various periods after which the drugs were removed, usually by filtration through a membrane filter, and the bacteria were then resuspended in fresh culture medium (77). After a pulsed exposure to all of the bactericidal antituberculosis drugs, there was a lag period before growth recovered to its prepulse level (Table 2.7). These bacteriopausal effects differed according to the drug studied.

ISONIAZID

Single pulses of 1 μg/ml H lasting over 6 h were followed by lag periods that reached a maximum of about 7 days (Table 2.7). The lag following five successive small pulses at daily intervals was also longer than a single long pulse with the same concentration and total duration, suggesting that the immediate biochemical effect was slowly cumulative (81). Furthermore, bactericidal action was evident in several experiments in vitro only after exposure had continued for 24 h. These findings indicate an immediate cumulative but reversible biochemical change that inhibits growth but does not kill, perhaps due to depletion of nicotinamide adenine dinucleotide (NAD) (83) or formation of isonicotinic acid (84). This is followed by a lethal step such as the inhibition of mycolic acid synthesis (85).

As we have seen, intermittent chemotherapy with H is highly effective in slow and rapid inactivators when doses are given twice weekly, but when given once weekly, the H component fails in rapid inactivators in whose sputum drug-sensitive tubercle bacilli persist unduly long. A comparable situation was found in experimental tuberculosis of the guinea pig, whose metabolism of H resembles

Table 2.7. Lag periods after pulsed exposures to antituberculosis drugs

Drug	Concentration of drug (mg/L)	Lag period (days) after exposure for		Ref.
		6 h	24 h or more (maximal)	
Isoniazid	1	0	6–9	77
Rifampicin	0.2	2–3	2–3	78
Pyrazinamide	50	5–9	(pH dependent)	79
Streptomycin	5	8–10	8–10	77
Ethambutol	10	0	4–5	80
Ethionamide	5	0	10	77
Thiacetazone	10	0	0	77

that of human rapid inactivators; the effect of doses of H given at 8-day intervals was substantially less than with doses given at intervals of 1, 2, or 4 days (77,87). These findings suggest that when H is given at intervals of up to twice weekly, the bacilli are constantly under its influence. However, when the interval is 7 days and the weekly pulse of drug is small, as in rapid inactivators, the bacilli can escape from its influence and start to multiply again. Nevertheless, they remain under its influence if there is a larger weekly pulse, as in slow inactivators.

RIFAMPIN

Rifampin differs from H in being bactericidal during even short exposure periods (Table 2.7). Furthermore, it also differs in having a shorter lag period after a pulse of drug, whether the exposure period is only 6 h or is 24 h or more. Thus, the onset of the effects of a pulse is rapid, as is the recovery from it. If a culture is exposed to successive pulses, the bactericidal effect is small if doses are given at daily intervals, but if they are more widely spaced and the bacilli have time to recover, the effect is much greater (86). Any escape from the influence of H leads to diminished efficacy, but, with R, just such an escape promotes greater bactericidal activity. Although the mechanism is different, R is as good a drug as H for intermittent use both in experimental tuberculosis of the guinea pig (78) and in pulmonary tuberculosis (study 3, Table 2.4; study 8, Table 2.5).

THIACETAZONE

Thiacetazone differs from other antituberculosis drugs in being entirely bacteriostatic. Bacilli recover immediately with no lag period after a pulsed exposure (Table 2.7). These characteristics indicate low efficacy during intermittent dosage, as has been found in experimental guinea pig tuberculosis (77) and is a suggestive finding in pulmonary tuberculosis (study 7, Table 2.5). Furthermore, the absence of a lag period after an exposure suggests that irregularity in drug-taking would lead to a failure of treatment more often with regimens containing T than with other regimens.

OTHER DRUGS

Less is known about the effects of pulsed exposures to other drugs. In the guinea pig, the effectiveness of M increased as the interval between doses was spaced out, provided that the total dose given was the same (by increasing the size of the individual dose proportionately to the interval between doses). This conclusion agrees with the results of study 10 of Table 2.5. On the other hand, similar guinea pig experiments suggested that streptomycin would not be particularly effective and ethionamide would be less effective when given intermittently (77). Fairly short-pulsed exposures to pyrazinamide can inhibit bacterial growth (79).

2.3. Interpretation of Drug Blood Levels

We may wish to relate measures of drug concentrations in serial specimens of blood (usually plasma) to the efficacy or toxicity of the drug. Studies of this sort on H have been particularly informative because genetic polymorphism divides the patients into distinct groups of slow and rapid inactivators of the drug (88). In most parts of the world, about 60% are slow inactivators and 40% rapid inactivators, although in populations of Mongolian descent (Chinese, Japanese, and Eskimos), there is a much higher proportion of rapid inactivators (89). Peak plasma concentrations are only slightly lower in rapid inactivators than in slow inactivators, but rapid inactivators have much shorter half-lives (Table 2.6) and, therefore, much smaller areas under the curve (AUC, "exposure" to the drug pulse). The pharmacokinetic differences make it possible to separate relationships to peak concentrations and exposures (90). The principal toxic effect of high dosage with H is peripheral neuropathy due to excretion of vitamin B_6 in combination with H in the urine and possibly by inhibition of pyridoxal kinase by high H concentrations (91,92).

In a series of studies at the Tuberculosis Chemotherapy Centre, Madras (93–96) in which patients getting a B_6-deficient diet were treated with H alone, often at high dosage, an excellent relationship was found between the incidence of peripheral neuropathy and the exposure, as seen in Table 2.8, part A. Such a relationship would be expected if H depletes body B_6. On the other hand, in daily treatment, the proportion of patients with a favorable bacteriologic response at the end of 1 year of treatment was closely related to the peak concentration (Table 2.8, part B), probably because high peaks coming at regular, frequent intervals prevented the emergence of mutant bacilli with low degrees of resistance (90,97). Finally, in once-weekly intermittent therapy, the proportion of patients with a favorable response was related to the exposure since keeping the organisms under the influence of H (its first cumulative effect) would be more successful following a large rather than a small pulse (31,32,90).

No similar analysis has been possible with other drugs mainly because the peak concentration and the exposure cannot be dissociated by the equivalent of rapid and slow inactivators, so that both are likely to be related to toxicity and efficacy. Furthermore, in the case of H, the associations that have been found seem due to the ability of the bacilli to store the effects of a pulse of exposure. The effects of other drugs, for instance, R, are probably not stored in the same way.

3. Bactericidal Activity of Drugs

3.1. Early Bactericidal Activity

Information on the action of antituberculosis drugs during the early stages of chemotherapy of pulmonary tuberculosis has been obtained from studies first

Table 2.8. A. Peripheral neuropathy (PN) related to exposure to isoniazid per day; B. response to treatment with isoniazid alone related to peak concentration in daily regimens

A. Treatments in order of exposure

Treatment group	Dose of isoniazid (mg/kg)	No. of doses/week	Inactivator status	Exposure per day[a]	Patients Total	PN No.	PN %
1	13.9	2	Rapid	4.0	36	0	0
2	2.2	14	Rapid	4.5	36	0	0
3	4.4	14	Rapid	8.7	28	0	0
4	8.7	7	Rapid	8.7	32	2	6
5	13.9	2	Slow	9.5	36	1	3
6	2.2	14	Slow	10.8	50	0	0
7	13.9	7	Rapid	13.9	42	3	7
8	4.4	14	Slow	20.9	44	6	14
9	8.7	7	Slow	20.9	39	11	28
10	13.9	7	Slow	33.4	37	17	46

B. Treatments in order of peak concentration

Treatment group	Peak concentration (mg/L)	Patients Total	Favorable response (%)
2	0.7	36	44
6	1.2	46	48
3	1.9	27	56
8	2.6	39	59
4	4.2	32	66
9	6.6	36	72
7	8.4	62	66
10	9.2	81	69

[a]The dose of isoniazid in mg/kg body weight multiplied by 1.0 for rapid inactivators and by 2.4 for slow inactivators, i.e., a value proportional to the AUC.

reported in 1980 (98). In these studies, newly diagnosed, previously untreated patients with smear-positive disease were treated with single drugs or with drug combinations. Sputum was collected pretreatment and at intervals of 1 or 2 days thereafter, and their content of viable tubercle bacilli was estimated from counts of colony-forming units (CFU) on plates of selective oleic acid–albumin–agar medium seeded with dithiothreitol digests of the sputum.

In study 1 of (Table 2.9), which was continued for 14 days of treatment, statistically significant differences were found in the rate of fall of the CFU counts between different drugs and drug doses during the first 2 days but not in the following 12 days. In consequence, treatment was only given for the first 2 days in subsequent studies. A summary of values of the early bactericidal activity (EBA) obtained in the first four studies are set out in Table 2.9. Pyrazinamide, despite its sterilizing activity in animal and human disease, had very low EBAs, although viable counts during the entire 14 days of treatment in study 1 showed slow bactericidal activity which, unlike the findings with other drugs, did not slow

Table 2.9. Early bactericidal activity of drugs in pulmonary tuberculosis

Drug	Dose	Early bactericidal activity (fall in CFU/ml sputum/day, no. of patients in parentheses) in study			
		1 Nairobi (Ref. 98)	2 Hong Kong (Ref. 99)	3 S. Africa (Ref. 82)	4 S. Africa (Ref. 100)
Nil		0.02 (4)	0.06 (8)	−0.02 (13)	—
Z	2 g	0.02 (9)	—	—	0.004 (10)
T	150 mg	0.04 (9)	—	—	—
S	1 g	0.07 (8)	—	—	—
R	6 mg/kg (300 mg)	—	0.16 (10)	0.15 (10)	—
	10 mg/kg	0.19 (8)	—	—	—
	12 mg/kg (600 mg)	—	0.29 (11)	0.20 (10)	—
RBU[a]	600 mg	—	0.05 (10)	0.08 (10)	—
M	25 mg/kg	0.31 (8)	—	—	0.25 (9)
P	15 g	0.26 (4)	—	—	—
H	300 mg	0.72 (4)	0.43 (14)	0.50 (11)	—
	2-drug combinations				
H	(with Z, R, S, M)	0.61 (16)	—	—	—
	3-drug combinations				
H	(with SZ, RM, SR, RZ)	0.54 (12)	—	—	0.56 (9)
Variation in EBA between patients:					
Standard deviation		0.19	0.38	0.079	0.13
Degrees of freedom		60	84	67	28

[a]RBU = rifabutin.

down during the period (98). Thiacetazone also had a very low EBA, as would be expected from a purely bacteriostatic drug, although even if bacilli in cavity walls were not killed, they would not be replaced so that an EBA slightly above the nil value would be expected. Then, there are three drugs, R, M, and P, all with moderate EBAs of 0.2–0.3. Finally, H had the highest EBA, usually of about 0.5, which was not appreciably altered when it was given in two-drug or three-drug combinations.

What bacteriological process does the EBA measure? As it is measured only during the first 2 days, or thereabouts, of chemotherapy, one must assume that the EBA measures bactericidal activity against the most rapidly growing bacilli in cavity walls. The bacilli in sputum come from cavities. Furthermore, the slow-down in bactericidal activity during chemotherapy is usually thought to be due to initial killing of actively growing bacilli followed by slower killing of bacilli that metabolize less rapidly. A further important point is that bacilli in the lining of cavities are almost always outside macrophages (102), so that the EBA measures activity against extracellular bacilli. Because drug resistance is likely to originate early in chemotherapy when bacterial populations of actively growing bacilli are large, the ability of a drug to prevent the emergence of resistance to a companion drug (for instance, H) should be related to the EBA.

Unfortunately, the single value of the EBA given by the usual dose of a drug gives little information about the activity of that drug during chemotherapy, as there is little direct correlation between the activity of a drug and the degree of its bactericidal activity. However, much more information is available if several dose sizes of a drug graded from the usual size downward are studied. In Figure 2.1A, the activities of R and rifabutin (RBU) are compared with each rifamycin given in doses graded at twofold intervals (82). The availability of the results with graded doses makes it possible to compare drug potencies with reasonable accuracy using straight lines relating EBA to log dose, which are parallel for R and RBU. The potency (wt/wt) of R was 2.73 times the potency of RBU with confidence limits of 1.96 and 3.78. Another way of looking at the results is to measure the "therapeutic margin," that is to say, the reduction in the size of the dose from its usual size to the size that results in an EBA of nearly 0. The therapeutic margin for R is 4 because a fourfold reduction in dose from the usual 600 mg to 150 mg decreased the EBA to nearly 0. The corresponding therapeutic margin for RBU is 2 if one assumes that the usual dose is 600 mg. Another example of the therapeutic margin is seen in Figure 2.1B, where curves are plotted relating dose size to EBA for H as well as for R (101). H in a usual dose of 300 mg has a therapeutic margin of about 16–32 compared to the margin of 4 for R, probably because of the considerable binding of R but not of H in plasma (Table 2.6). A high therapeutic margin suggests that a drug would continue to be active if the dose were reduced. This is likely to be true for H, where high doses are given because they are active against mutant bacilli with low degrees of resistance (97). A drug with

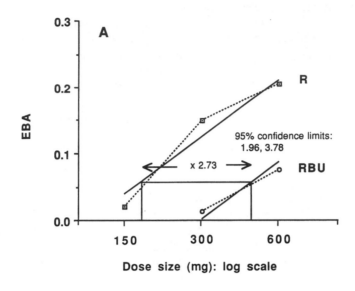

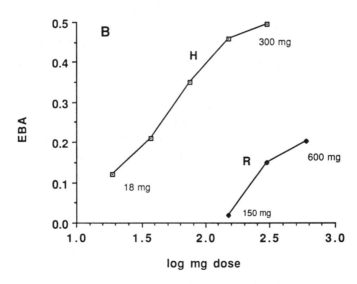

Figure 2.1. Relationship between dose size of drug and early bactericidal activity (EBA). (A) Comparison of the potencies of rifampin (R) and rifabutin (RBU). Observed means joined by dotted lines. Solid fitted regression lines. (Reproduced with permission from the *Journal of Antimicrobial Chemotherapy.*) (B) Comparison of EBAs of isoniazid (H), which has a high therapeutic margin, and rifampin (R), which has a lower therapeutic margin.

a high margin would also continue to be active even in lesions where it penetrates poorly and would retain effectiveness during periods when lesional levels were not bacteriostatic. A low margin suggests that even a small reduction in dose size would lead to lower efficacy, as has been shown for a small reduction in R dosage from 600 mg to only 450 mg in a clinical trial (103). Note that no assumptions about the therapeutic margin can be made without testing a range of dose sizes, as one would then not know the slope of the curve relating EBA to dose size. A drug with low bactericidal activity but a high therapeutic margin might well have a rather flat curve relating EBA to dose size and, therefore, a low EBA at its usual dose size, whereas a highly bactericidal drug with a small therapeutic margin could have a much steeper EBA–dose size curve with a larger EBA at its usual dose size. Determinations of the EBA for graded drug doses have only been made for a few drugs thus far, so that we know little about therapeutic margins or the slope of the EBA–dose curve.

A further point about the comparison of rifabutin and R illustrated in Fig. 2.1A is the apparent low potency of rifabutin. A dose of 600 mg rifabutin has an EBA corresponding to a little below that of 300 mg R, whereas the EBA of 300 mg rifabutin is so low as to suggest inactivity at this dosage. However, several clinical studies have been carried out comparing 2-month bacteriology and relapse rates in short-course regimens in which rifabutin is substituted for R (100,104,105). These studies indicate that rifabutin in a dose of 300 mg is as good a sterilizing drug as 600 mg R. The most likely explanation for this apparent discrepancy is that the EBA measures bactericidal activity against extracellular bacilli, whereas persisting bacilli that are killed by sterilizing drugs lie inside cells (82,99). Rifabutin produces very low plasma concentrations and may, therefore, be relatively ineffective against extracellular organisms in cavity walls, whereas it penetrates into macrophages better than R and would therefore be effective if persisters are intracellular (99).

3.2. Sterilizing Activity

The sterilizing activity of a drug is measured by its ability to kill the bacilli that persist for long periods during the last months of chemotherapy. A good sterilizing drug reduces the relapse rate after chemotherapy has stopped and allows its duration to be shortened. The introduction of the two good sterilizing drugs R and Z made it possible to shorten regimens from at least 12 months to 6 months. Although the main observable effect of a sterilizing drug is to kill persisting bacilli, studies in the mouse indicate that R appears to have a delayed action, in that adding it to an SH regimen from the start of treatment has little effect on the early organ CFU counts (at, say, 1 month) but reduces the counts substantially in later months (106,107). Furthermore, as we have seen, Z is only active as a sterilizing drug during the first 2 months of treatment of pulmonary disease. These findings led to the hypothesis, illustrated in Figure 2.2, that the bulk of the initial bacterial population was growing rapidly and was therefore killed rapidly, mainly by H because it has the largest EBA. However, smaller subpopulations were growing so slowly that they were not killed rapidly by H. The growth of one of these subpopulations was slowed by a very acid microenvironment, which, however, was sufficiently acid for Z to be lethal (79). The acid microenvironment might result from acute inflammation; once this subsides, Z would be inactive, perhaps accounting for its failure to sterilize after the first 2 months (108). Another subpopulation was only metabolizing in spurts which lasted insufficiently long for H to be lethal but long enough for R to kill (109). It is the killing of these Z-specific and R-specific subpopulations that resulted in effective sterilizing activity.

The manner in which Z acts in chemotherapy is still uncertain. In vitro, it is only active when the bacillary environment is markedly acidic. Furthermore, as there is an excellent correlation between sensitivity to Z and the presence of bacterial pyrazinamidase (which converts Z to pyrazinoic acid), the activity of Z is thought to be due to pyrazinoic acid rather than Z itself (110). That the con-

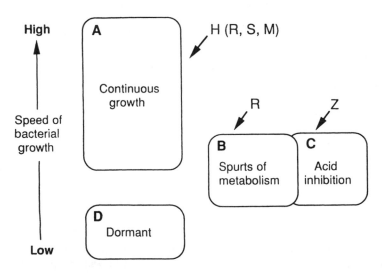

Figure 2.2. Action of drugs on special parts of the bacterial population: (A) rapidly growing bacteria killed mainly by isoniazid (H); (B) bacilli only metabolizing in spurts killed mainly by rifampin (R); (C) bacilli inhibited by an acid environment killed mainly by pyrazinamide (Z); (D) dormant bacilli.

version to pyrazinoic acid occurs at the surface of the bacterial cell or within it is indicated by the much higher MICs of pyrazinoic acid than of Z in vitro (111,112). Z is bacteriostatic, but not bactericidal, against tubercle bacilli growing within macrophages (113,114). However, recent evidence suggests that the pH within the macrophage phagosome containing tubercle bacilli is not acid, so that inhibition of growth might be due to accumulation of pyrazinoic acid within the phagosome trapped by the action of proton pumps in the phagosome membrane (115). It has been easier to demonstrate bactericidal activity of Z in vitro than in the macrophage, but the rate of kill is usually slow (79,111). A more rapid rate of kill has been found at pH 4.8, after exposure to Z for several weeks (116), but at this pH, the growth of the bacilli is almost completely inhibited in Z-free medium. Thus, in explaining the slow but steady bactericidal activity of Z during at least the first 14 days of chemotherapy, one has to assume that it acts not on intracellular bacilli (because these are uncommon in cavity walls and because no one has shown bactericidal action against intracellular organisms) but on extracellular bacilli in an environment that is so acidic as to virtually stop growth; because caseous lesions have a neutral pH, the origin of the acidity can only be acute inflammation (108). Activity of Z only during the first 2 months of therapy might be due to resolution of acute inflammation and an increase in local pH.

The sterilizing activity of a drug during the chemotherapy of pulmonary tuberculosis can be measured in two ways. The ultimate measure is the ability of the drug, when added to a regimen, to decrease the relapse rate after the end of chemotherapy. A secondary measure is the proportion of sputum cultures that are negative at 2 months. As set out in Table 2.10, there is an excellent correlation between 2-month sputum culture and the eventual relapse rate in eight studies in which the sterilizing activity of R and Z have been assessed. Table 2.10 also shows that differences among the regimens were smaller in 1-month sputum cultures and, of course, also in 3-month cultures, almost all of which were negative. The greater discriminative ability of the 2-month cultures reflects the delayed effect of R in the mouse, referred to above (106), in which differences due to the addition of R to the SH regimen were small at 1 month and increased thereafter. Presumably at 1 month, the results of culture are dominated by killing of the more actively growing initial subpopulation (A in Fig. 2.2), whereas at 2 months, the residual bacteria are mainly persisters in the R-specific and Z-specific subpopulations (B and C, Fig. 2.2) and are, therefore, particularly sensitive to the addition of R or Z to the regimen. This analysis suggests that measurement of culture positivity at about 2 months, when persisters dominate, is likely to be a more sensitive indicator of sterilizing activity than the broader measure of time to sputum conversion which does not concentrate on the late killing of persisters. However, no data are available to make comparisons between the two methods. The 2-month culture positivity rate provides an estimate of the sterilizing activity of a drug long before the eventual relapse rate is known. Furthermore, it avoids

Table 2.10. Culture results at 1, 2, and 3 months after the start of chemotherapy related to the subsequent relapse rate

Study No.	Regimen	No. of patients at 2 months	Percent patients			Relapse after chemotherapy	Ref.
			Culture negative at month				
			1	2	3		
1	6SH	154	19	49	81	29	34, 35
Africa	6SHZ	150	27	66	91	11	
	6SHR	148	27	69	94	2	
		X^2[2]	3.1	15.3	14.7		
2	6SHR	169	34	70	95	—	37, 38
Africa	6HR	173	23	64	96	—	
	2SHRZ/4TH or SHZ$_2$	347	35	82	—	—	
		X^2[2]	7.9	20.3	—		
3	2SHR/TH	194	34	75	91	13	39, 40
Africa	2SHRZ/TH	179	40	87	93	6	
		X^2[1]	1.6	7.0	0.2		
4	2SHRZ/4HR	146	38	77	97		117, 118
UK	2EHRZ/4HR	141	35	77	99	1.6	
	2EHRZ/7HR	157	29	64	88		
		p	NS	<0.01	<0.0001		
5	6SHRM$_3$	153	51	76	—	8	51, 52
Hong Kong	6SHRZ$_3$	145	50	90	—	1	
	6, all with RZ	607	53	90	—	1.4	
		X^2[1]^a	0.1	21.8	—		
6	2SHZ/SHZ$_2$	129	38	72	86	6.2^b	61, 62
Madras	2SHRZ/SHZ$_2$	261	50	92	96	2.3	
		p	0.05	<0.0001	0.001		
7	2SHRZ/SHZ$_2$	167	67	95	99	7	48, 49
Hong Kong	2SHRM/SHM$_2$	171	50	81	97	23	
		X^2[1]	9.5	13.8	1.7		
8	3SHRZ					20	63
Madras	3SHRZ/2SHZ$_2$	457	37	91	96	4	
	3SHZ/2SHZ$_2$	236	19	74	93	13	
		X^2[1]	22.6	34.0	NS		

^{a}6SHRM$_3$ compared to all four regimens with RZ.
^{b}Data for the 2SHZ/5SHZ$_2$ and 2SHRZ/5SHZ$_2$ regimens only.

problems caused by the loss of patients to follow-up that often greatly reduces the patient populations available for analysis of a clinical trial.

4. Grading the Efficacy of Drugs

Concepts of what constitutes a good antituberculosis drug were first expressed during the early development period when the main aim was preventing the emergence of drug resistance. A good drug prevented the growth of tubercle bacilli throughout the lesions of a patient at all times during therapy, whether or not bacteriostatic concentrations were present at the time. A less than good performance was expected if some growth occurred because (i) the therapeutic margin was so low (usually because drug dosage was limited by toxicity) that drug penetration might not be adequate in some lesions, or (ii) the potency of the drug was greatly influenced by pH, such that it was less active under acid conditions (S and other aminoglycosides) or only active in very acid conditions (Z), or (iii) there was little or no bacteriopausal effect after a pulse so that growth might occur between doses, particularly if there was some irregularity in drug-taking (T). The performance of a drug could be gauged by giving it in a two-drug combination with another drug, usually H, and measuring the proportion of patients who developed strains resistant to the second drug. Table 2.11 lists gradings of the main antituberculosis drugs according to their types of activity. The performance of drugs according to these mechanisms is far more important as a cause of failure of chemotherapy than the proportion of mutants resistant to the drug concerned in "wild," sensitive bacillary populations. If bacterial multiplication occurs during chemotherapy, mutants will eventually grow out (later on in treatment if there

Table 2.11. Grading of antituberculosis drugs according to type of activity

Drug	Prevention of drug-resistance			Early bactericidal activity			Sterilizing activity,		
	No. of patients	Failure (%)	Grade	No. of patients	Mean EBA	Grade	negative 2-month culture (%)	Relapse (%)	Grade
Isoniazid	—[a]	0	7	29	0.50	7	49	29	2
Rifampin	183[b1]	0.5	6	8	0.19	4	73	3	5
Streptomycin	180[b2]	1	5	8	0.09	3	—	—	1?
Ethambutol	105[b3]	4	4	17	0.28	6	—	—	0
PAS	461[b4]	14	3	4	0.26	5	—	—	—
Thiacetazone	665[b5]	18	2	9	0.04	2	—	—	0
Pyrazinamide	219	—	1	19	0.01	1	64	8	4
Nil	108[b6]	50	0	25	0.01	0	—	—	0

[a]Not known.
[b]References: (1) 37; (2) 34; (3) 76; (4) 15, 17, 20, 24, 119, and 120; (5) 18–22, and 24; (6) 120.

were fewer of them initially), provided, of course, that the original bacterial population was large enough to contain resistant mutants at all.

The three columns under the heading "Prevention of drug resistance" give results obtained in clinical studies in which the drug was given with H. The proportion of patients whose treatment failed with the emergence of H resistance is listed. H itself must be considered the best drug with a grading of 7 because when it is given in these two-drug combinations, persisting sputum positivity is always accompanied by H resistance but not invariably by resistance to any partner drug. The grading of Z as the least effective drug is based on the report of early Veterans Administration studies comparing treatment with ZH and PH (121), which concludes that Z at a 3-g daily dose is no better than P. An assessment of the emergence of H resistance using modern criteria of H resistance indicates that there were significantly more patients with H resistance during the first 4 months of treatment with ZH than with PH. The proportion of failures when the second drug is "nil" is based on the results of treatment with H alone in a 200-mg daily dose shown in Table 2.4.

The next three columns in Table 2.11 show values of the EBA obtained by taking weighted means of the estimates in Table 2.9, the weights being the numbers of patients. The values fall roughly into three groups. H has the highest EBA. It is followed by M, P, and R, but it should be noted that the estimate for P is based on only four patients and is therefore very uncertain. S and T have lower EBAs still; Z has the lowest of all, although it should be noted that Z has a steady slow bactericidal action throughout the first 14 days, unlike other drugs (98,108). These gradings show some but by no means a close relationship to the gradings for the prevention of drug resistance. As noted earlier, single values of the EBA are of relatively little value in grading the activity of a drug. The therapeutic margin obtained from estimates of the EBA with graded doses of the drug may prove a better guide.

The last three columns of Table 2.11 grade the same drugs according to their sterilizing activities, as shown by their effect on the proportion of negative cultures at 2 months and the eventual relapse rate after the end of chemotherapy. R and Z are the most active sterilizing drugs, R being slightly superior to Z. H is also capable of sterilizing lesions, but much more slowly. Adding on the other drugs listed has not altered either the 2-month bacteriology or the relapse rates in clinical studies, although there is no evidence on the sterilizing activity of P. There are striking differences between the gradings of the drugs for their sterilizing activity and their gradings for prevention of resistance or for their EBA. R moves from the best grade for sterilizing to an only moderate EBA. Z also moves from very effective sterilization to a very poor performance in preventing resistance or in EBA. M has no sterilizing activity but is moderately good at preventing resistance and has quite a high EBA.

5. Drug Action During Chemotherapy and the Construction of Regimens

5.1. Penetration of Drugs into Lesions

One of the oldest questions to be asked about chemotherapy concerns the penetration of drugs into tuberculous lesions, particularly those with extensive fibrous walls and those containing large volumes of fluid, pus, or caseous material. The first issue is that where there are large volumes into which the drug must diffuse, the peak (C_{max}) concentration will tend to be a little lower than in serum and will also tend to occur later. This smoothing of the concentration–time curve is often less evident than might be imagined and is only likely to influence the activity of the drug when C_{max} values are close to the MIC, as for instance with ethambutol (M). The second issue concerns the extent to which the drug is bound by plasma protein. From Table 2.6, it is evident that among the major five drugs (H, R, Z, S, and M), 85% of R and 35% of S are bound, whereas there is no binding of H, Z, or M. If we examine the influence of binding on the activity and lesional concentrations of R, the drug with the highest proportion bound, we find that (i) concentrations after a 600-mg dose are much lower in pulmonary lesions, where the mean concentration in 21 specimens of caseum was 2.3 mg/L (range 0.3–6.2 mg/L) (122) and in the cerebrospinal fluid (CSF), where the mean peak concentration was 0.78 mg/L (123), than in serum and (ii) there is evidence that the dosage of rifampin is marginal, as shown by the decrease in efficacy found when the dose size was reduced from 600 to 450 mg daily (103) and also by the small therapeutic margin shown in EBA studies (Fig 2.1). The probable explanation for these findings is that the bound fraction (85%) of R does not diffuse into the CSF and only penetrates partially into other lesions. In any case, any claim that R does not penetrate adequately into particular types of lesions can only be justified if the concentrations found are lower than those already measured in lesions (122), particularly in pulmonary disease, in which we know R to be effective. The influence of plasma protein binding should be smaller for S. Evidence that lesional concentrations are adequate has been obtained in pulmonary lesions (124) and also in spinal tuberculosis where lesions may contain large collections of pus (125). In short, there is little evidence that drugs fail to produce adequate concentrations in a wide variety of lesions, except perhaps when the dosage of the drug is known to be marginal or the sequestered lesion very large.

5.2. Phases of Chemotherapy

The rate at which tubercle bacilli are killed in the lesions, reflected by the content of viable bacilli in the sputum, is very rapid during the first few days of conventional H-containing chemotherapy, when there is more than a 10-fold fall in CFU sputum counts during the first 2 days. The rate of kill then slows progressively until it takes months to kill the last surviving bacilli. Because a log phase culture is killed exponentially by exposure to a drug in vitro, without any slowing of the rate of kill, we have to assume that there is great initial heterogeneity in the growth rate of bacilli in lesions, as shown in Fig. 2.2. We can now visualize three phases in chemotherapy:

1. The first 2 days during which actively growing bacilli are killed, mainly by H. During this phase, the rate of kill (the EBA) is closely related to the bactericidal activity against log phase bacilli of the drugs being given.
2. The next period of 4–6 weeks during which bacilli that initially grew more slowly are being killed. The rate of killing is determined mainly by the physiologic state of the bacilli and less by the bactericidal activity of the drug. Hence, the fall in sputum counts is not influenced by the drug (except perhaps that it should be bactericidal and not just bacteriostatic) or its dose size.
3. The period from about 6 weeks to the end of chemotherapy, during which bacilli that grew well or fairly well initially have been killed and the survivors are populations of persisters that were growing very slowly or intermittently at the start of treatment.

The rate of kill is now determined by the sterilizing activity of the drugs. We have seen that the two drugs with high sterilizing activity have special properties that allow them to kill slowly metabolizing bacteria effectively: R because it starts to kill very rapidly and is, therefore, effective against bacilli that metabolize in spurts, and Z because it is only bactericidal when bacilli are highly inhibited in growth by acidity. If we are examining new drugs for their sterilizing activity, it would seem sensible to see whether they could kill semidormant or, indeed, completely dormant organisms with similar or greater activity than R or Z. A further point of interest is that the great majority of the bacilli that must be killed by drugs are extracellular in cavity walls during the first two phases of chemotherapy but may well be intracellular during the third phase of sterilization. If this is so, and we need evidence additional to the findings with rifabutin, then the capacity to penetrate macrophages could be important in defining the role for a particular drug. Thus, rifabutin, which produces low plasma concentrations but is good at cellular penetration, might be an inadequate drug in the early stages of chemotherapy but a good drug later on, so that chemotherapy could be optimal if it started with R in the early phases but finished with rifabutin. Again, a penicillin given with a penicillinase inhibitor (amoxicillin + clavulanic acid or ampicillin + sulbactam) is unlikely to penetrate cells well but might still be of value in preventing the emergence of resistance to a second drug during the early phases of treatment.

5.3. Rationale of Chemotherapy Regimens

In constructing regimens of chemotherapy, it is desirable to (i) make the duration of treatment as short as possible, (ii) make the regimen effective in the presence of current levels of initial drug resistance, and (iii) keep drug costs sufficiently low to be affordable. Furthermore, the regimen must meet the requirement of the domiciliary treatment service for full supervision of drug-taking and might, therefore, use an intermittent frequency of drug dosage. A short duration of treatment, currently 6 months, is obtained by including R throughout and Z during the initial 2 months. Omission of Z from the initial phase increased the total duration necessary to obtain a low relapse rate from 6 months to 9 months (117,126). Replacement of RH in the continuation phase by TH increased the total duration from 6 months to 8 months (39,40). It seems that giving R thrice weekly in the initial phase or twice weekly in the continuation phase does not reduce the sterilizing activity of the regimen (50,51). That R and Z are responsible for most of the sterilizing activity in a multidrug regimen is shown by the similarity in 2-month culture results and only slightly higher relapse rates in patients with initial resistance to H or to S and H as compared to those with initially sensitive organisms (65).

The need for treatment to be effective when strains are initially drug resistant is met by the inclusion of drugs additional to R and Z. H is always added because of its high therapeutic margin, low toxicity, and low cost, even though it does not have high sterilizing ability. The HRZ combination in the initial phase may be adequate if drug resistant strains are infrequently encountered, for instance, if less than 3% of strains from previously untreated patients are resistant to H. If resistant strains are found more often, for instance, in 3–10% of untreated patients, a fourth drug, usually M but sometimes S, is added. Patients with initial resistance to H or S or SH respond well to such a four-drug regimen, although much less good results are obtained when there is initial resistance to R, which is often accompanied by resistance to other drugs (65).

The cost of short-course regimens is considerably greater than the older STH/ TH regimen. Regimens of the "African" type with R and Z in the initial phase but TH in the continuation phase are effective and only require a 2-month prolongation of treatment. However, when HIV infection is common, an alternative to T is necessary. If drug-taking can be supervised, intermittent RH is effective and fairly cheap, but it might be necessary to give M in place of T, despite its high cost.

References

1. Domagk G, Behnisch R, Mietzsch F, Schmidt H (1946) Concerning a new class of compounds active against tubercle bacilli in vitro. Naturwissenschaften 33:315 (in German).

2. Domagk G (1950) Investigations on antituberculosis activity of thiosemicarbazones. Am Rev Tuberc 61:8–19.

3. Hinshaw HC, McDermott W (1950) Thiosemicarbazone therapy of tuberculosis in humans. Am Rev Tuberc 61:145.

4. Schatz A, Bugie E, Waksman SA (1944) Streptomycin. Substance exhibiting activity against Gram-positive and Gram-negative bacteria. Proc Soc Exp Biol Med 55:66–69.

5. Medical Research Council (1948) Streptomycin treatment of pulmonary tuberculosis. Br Med J 2:769–782.

6. Fox W, Sutherland I, Daniels M (1954) A five-year assessment of patients in a controlled trial of streptomycin in pulmonary tuberculosis. Quart J Med 23:347–366.

7. Medical Research Council (1952) The prevention of streptomycin resistance by combined chemotherapy. Br Med J 1:1157–1162.

8. Daniels M, Bradford Hill A (1952) Chemotherapy of pulmonary tuberculosis in young adults. Br Med J 1:1162–1168.

9. Medical Research Council (1953) Isoniazid in combination with streptomycin or with PAS in the treatment of pulmonary tuberculosis. Br Med J 2:1005–1014.

10. Medical Research Council (1955) Various combinations of isoniazid with streptomycin or with PAS in the treatment of pulmonary tuberculosis. Br Med J 1:435–445.

11. Chaves AD, Robins AB, Abeles H, Peizer LR, Dangler G, Widelock D (1955) The prevalence of streptomycin- and isoniazid-resistant strains of *Mycobacterium tuberculosis* in patients with newly discovered and untreated active pulmonary tuberculosis. Am Rev Tuberc Chest Dis 72:143–150.

12. Fox W, Wiener A, Mitchison DA, Selkon JB, Sutherland I (1957) The prevalence of drug-resistant tubercle bacilli in untreated patients with pulmonary tuberculosis. A National Survey 1955–56. Tubercle 38:71–84.

13. Crofton JW (1959) Chemotherapy of pulmonary tuberculosis. Br Med J 1:1610–1614.

14. International Union Against Tuberculosis (1964) An international investigation of the efficacy of chemotherapy in previously untreated patients with pulmonary tuberculosis. Bull Int Union Tuberc Lung Dis 34:79–191.

15. Tuberculosis Chemotherapy Centre, Madras (1959) A concurrent comparison of home and sanatorium treatment of pulmonary tuberculosis in South India. Bull WHO 21:51–144.

16. Andrews RH, Devadatta S, Fox W, Radhakrishna S, Ramakrishnan CV, Velu S (1960) Prevalence of tuberculosis among close family contacts of tuberculous patients in South India, and influence of segregation of the patient on the early attack rate. Bull WHO 23:463–510.

17. East African Hospitals/British Medical Research Council (1960) Comparative trial of isoniazid in combination with thiacetazone or a substituted diphenylthiourea (SU1906) or PAS in the treatment of acute pulmonary tuberculosis in East Africans. Tubercle 41:399–423.

18. East African/British Medical Research Council (1963) Isoniazid with thiacetazone

in the treatment of pulmonary tuberculosis in East Africa—second investigation. Tubercle 44:301–333.

19. East African/British Medical Research Council (1966) Isoniazid with thiacetazone (thioacetazone) in the treatment of pulmonary tuberculosis in East Africa—third investigation. Tubercle 47:1–32.

20. Tuberculosis Chemotherapy Centre, Madras (1966) Isoniazid plus thiacetazone compared with two regimens of isoniazid plus PAS in the domiciliary treatment of pulmonary tuberculosis in South Indian patients. Bull WHO 34:483–515.

21. East African/British Medical Research Council (1966) Isoniazid with thiacetazone (thioacetazone) in the treatment of pulmonary tuberculosis in East Africa—fourth investigation: the effect of increasing the dosage of isoniazid. Tubercle 47:315–339.

22. East African/British Medical Research Council (1973) Isoniazid with thiacetazone (thioacetazone) in the treatment of pulmonary tuberculosis in East Africa—third report of fifth investigation. Tubercle 54:169–179.

23. Briggs IL, Rochester WR, Shennan DH, Riddell RW, Fox W, Heffernan JF, Miller AB, Nunn AJ, Stott H, Tall R (1968) Streptomycin plus thiacetazone (thioacetazone) compared with streptomycin plus PAS and with isoniazid plus thiacetazone in the treatment of pulmonary tuberculosis in Rhodesia. Tubercle 49:48–69.

24. Hong Kong Anti-tuberculosis Association and Government Tuberculosis Service/ British Medical Research Council (1968) A controlled comparison of thiacetazone (thioacetazone) plus isoniazid with PAS plus isoniazid in Hong Kong. Tubercle 49:243–280.

25. Singapore Tuberculosis Services/Brompton Hospital/British Medical Research Council (1974) A controlled clinical trial of the role of thiacetazone-containing regimens in the treatment of pulmonary tuberculosis in Singapore: second report. Tubercle 55:251–260.

26. Miller AB, Fox W, Tall R (1966) An international cooperative investigation into thiacetazone (thioacetazone) side effects. Tubercle 47:33–74.

27. Miller AB, Nunn AJ, Robinson DK, Fox W, Somasunderam PR, Tall R (1972) A second international cooperative investigation into thioacetazone side effects 2. Frequency and geographical distribution of side effects. Bull WHO 47:211–227.

28. Fox W (1962) Self-administration of medicaments. A review of published works and a study of the problem. Bull Int Union Tuberc Lung Dis 32:307–331.

29. McCune RM, Feldmann FM, Lambert HP, McDermott W (1966) Microbial persistence. I. The capacity of tubercle bacilli to survive sterilization in mouse tissues. J Exp Med 123:445–468.

30. McCune RM, Feldmann FM, McDermott W (1966) Microbial persistence. II. Characteristics of the sterile state of tubercle bacilli. J Exp Med 123:469–486.

31. Batten J (1969) New drugs against tuberculosis. Lancet 1:1214.

32. Grosset J (1978) The sterilizing value of rifampicin and pyrazinamide in experimental short-course chemotherapy. Tubercle 59:287–297.

33. East African/British Medical Research Council (1972) Controlled clinical trial of

short-course (6-month) regimens of chemotherapy for treatment of pulmonary tuberculosis. Lancet 1:1079–1085.

34. East African/British Medical Research Council (1973) Controlled clinical trial of four short-course (6-month) regimens of chemotherapy for treatment of pulmonary tuberculosis. Lancet 1:1331–1339.

35. East African/British Medical Research Council (1974) Controlled clinical trial of four short-course (6-month) regimens of chemotherapy for treatment of pulmonary tuberculosis. Third report. Lancet 2:237–240.

36. East African/British Medical Research Council (1977) Results at 5 years of a controlled comparison of a 6-month and a standard 18-month regimen of chemotherapy for pulmonary tuberculosis. Am Rev Respir Dis 116:3–8.

37. East African/British Medical Research Council (1974) Controlled clinical trial of four short-course (6-month) regimens of chemotherapy for the treatment of pulmonary tuberculosis. Lancet 2:1100–1106.

38. East African/British Medical Research Council (1976) Controlled clinical trial of four 6-month regimens of chemotherapy for pulmonary tuberculosis. Second report. Am Rev Respir Dis 114:471–475.

39. East African/British Medical Research Council (1978) Controlled clinical trial of four short-course regimens of chemotherapy for two durations in the treatment of pulmonary tuberculosis. First report. Am Rev Respir Dis 118:39–48.

40. East African/British Medical Research Council (1980) Controlled clinical trial of four short-course regimens of chemotherapy for two durations in the treatment of pulmonary tuberculosis: Second report. Tubercle 61:59–69.

41. East African/British Medical Research Council (1978) Controlled clinical trial of five short-course (4-month) chemotherapy regimens in pulmonary tuberculosis. First report. Lancet 2:334–338.

42. East African/British Medical Research Council (1981) Controlled clinical trial of five short-course (4-month) chemotherapy regimens in pulmonary tuberculosis. Second report of the 4th study. Am Rev Respir Dis 123:165–170.

43. East and Central African/British Medical Research Council (1983) Controlled clinical trial of 4 short-course regimens of chemotherapy (three 6-month and one 8-month) for pulmonary tuberculosis. Tubercle 64:153–166.

44. East and Central African/British Medical Research Council (1986) Controlled clinical trial of 4 short-course regimens of chemotherapy (three 6-month and one 8-month) for pulmonary tuberculosis: Final Report. Tubercle 67:5–15.

45. Tanzania/British Medical Research Council (1985) Controlled clinical trial of two 6-month regimens of chemotherapy in the treatment of pulmonary tuberculosis. Am Rev Respir Dis 131:727–731.

46. Hong Kong Tuberculosis Treatment Services/British Medical Research Council (1975) Controlled trial of 6- and 9-month regimens of daily and intermittent streptomycin plus isoniazid plus pyrazinamide for pulmonary tuberculosis in Hong Kong. Tubercle 56:81–96.

47. Hong Kong Chest Service/British Medical Research Council (1977) Controlled trial of 6-month and 9-month regimens of daily and intermittent streptomycin plus isoniazid plus pyrazinamide for pulmonary tuberculosis in Hong Kong. The results up to 30 months. Am Rev Respir Dis 115:727–735.

48. Hong Kong Chest Service/British Medical Research Council (1978) Controlled trial of 6-month and 8-month regimens in the treatment of pulmonary tuberculosis. Am Rev Respir Dis 118:219–227.

49. Hong Kong Chest Service/British Medical Research Council (1979) Controlled trial of 6-month and 8-month regimens in the treatment of pulmonary tuberculosis: the results up to 24 months. Tubercle 60:201–210.

50. Hong Kong Chest Service/British Medical Research Council (1981) Controlled trial of four thrice-weekly regimens and a daily regimen all given for 6 months for pulmonary tuberculosis. Lancet 1:171–174.

51. Hong Kong Chest Service/British Medical Research Council (1982) Controlled trial of 4 three-times weekly regimens and a daily regimen all given for 6 months for pulmonary tuberculosis. Second report: the results up to 24 months. Tubercle 63:89–98.

52. Hong Kong Chest Service/British Medical Research Council (1987) Five-year follow-up of a controlled trial of five 6-month regimens of chemotherapy for pulmonary tuberculosis. Am Rev Respir Dis 136:1339–1342.

53. Hong Kong Chest Service/British Medical Research Council (1991) Controlled trial of 2, 4 & 6 months of pyrazinamide in 6-month, 3 × weekly regimens for smear-positive pulmonary tuberculosis, including an assessment of a combined preparation of isoniazid, rifampin & pyrazinamide. Am Rev Respir Dis 143:700–706.

54. Singapore Tuberculosis Service/British Medical Research Council (1979) Clinical trial of six-month and four-month regimens of chemotherapy in the treatment of pulmonary tuberculosis. Am Rev Respir Dis 119:579–585.

55. Singapore Tuberculosis Service/British Medical Research Council (1981) Clinical trial of six-month and four-month regimens of chemotherapy in the treatment of pulmonary tuberculosis: The results up to 30 months. Tubercle 62:95–102.

56. Singapore Tuberculosis Service/British Medical Research Council (1986) Long-term follow-up of a clinical trial of six-month and four-month regimens of chemotherapy in the treatment of pulmonary tuberculosis. Am Rev Respir Dis 133:779–783.

57. Singapore Tuberculosis Service/British Medical Research Council (1985) Clinical trial of three 6-month regimens of chemotherapy given intermittently in the continuation phase in the treatment of pulmonary tuberculosis. Am Rev Respir Dis 132:374–378.

58. Singapore Tuberculosis Service/British Medical Research Council (1988) Five-year follow-up of a clinical trial of three 6-month regimens of chemotherapy given intermittently in the continuation phase in the treatment of pulmonary tuberculosis. Am Rev Respir Dis 137:1147–1150.

59. Singapore Tuberculosis Service/British Medical Research Council (1991) Assessment of a daily combined preparation of isoniazid, rifampin and pyrazinamide

in a controlled trial of three 6-month regimens for smear-positive pulmonary tuberculosis. Am Rev Respir Dis 143:707–712.

60. Nunn P, Kibuga D, Gathua S, Brindle R, Imalingat A, Wasuna K, et al. (1991) Cutaneous hypersensitivity reactions due to thiacetazone in HIV-1 seropositive patients treated for tuberculosis. Lancet 337:627–630.

61. Tuberculosis Research Centre, Madras (1983) Study of chemotherapy regimens of 5 and 7 months' duration and the role of corticosteroids in the treatment of sputum-positive patients with pulmonary tuberculosis in South India. Tubercle 64:73–91.

62. Santha T, Nazareth O, Krishnamurthy MS, Balasubramanian R, Vijayan VK, Janardhanam B, Venkataraman P, Tripathy SP, Prabhakar R (1989) Treatment of pulmonary tuberculosis with short course chemotherapy in South India—5 year follow up. Tubercle 70:229–234.

63. Tuberculosis Research Centre Madras/National Tuberculosis Institute Bangalore (1986) A controlled clinical trial of 3- and 5-month regimens in the treatment of sputum-positive pulmonary tuberculosis in South India. Am Rev Respir Dis 134:27–33.

64. Balasubramanian R (1991) Fully intermittent six month regimens for pulmonary tuberculosis in South India. Indian J Tuberc 38:51–53.

65. Mitchison DA, Nunn AJ (1985) Influence of initial drug resistance on the response to short-course chemotherapy of pulmonary tuberculosis. Am Rev Respir Dis 133:423–430.

66. Hong Kong Chest Service/Tuberculosis Research Centre, Madras/British Medical Research Council (1989) A controlled trial of 3-month, 4-month, and 6-month regimens of chemotherapy for sputum-smear-negative pulmonary tuberculosis. Results at 5 years. Am Rev Respir Dis 139:871–876.

67. Hong Kong Chest Service/Tuberculosis Research Centre, Madras/British Medical Research Council (1984) A controlled trial of 2-month, 3-month, and 12-month regimens of chemotherapy for sputum-smear-negative pulmonary tuberculosis. Results at 60 months. Am Rev Respir Dis 130:23–28.

68. Tuberculosis Chemotherapy Centre, Madras (1964) A concurrent comparison of intermittent (twice-weekly) isoniazid plus streptomycin and daily isoniazid plus PAS in the domiciliary treatment of pulmonary tuberculosis. Bull WHO 31:247–271.

69. Tuberculosis Chemotherapy Centre, Madras (1970) A controlled comparison of a twice-weekly and three once-weekly regimens in the initial treatment of pulmonary tuberculosis. Bull WHO 43:143–206.

70. Tuberculosis Chemotherapy Centre, Madras (1973) A controlled comparison of two fully supervised once-weekly regimens in the treatment of newly diagnosed pulmonary tuberculosis. Tubercle 54:23–45.

71. WHO Collaborating Centre for Tuberculosis Chemotherapy, Prague (1971) A comparative study of daily and twice-weekly continuation regimens of tuberculosis chemotherapy, including a comparison of two durations of sanatorium treatment. Bull WHO 45:573–592.

72. Tuberculosis Chemotherapy Centre, Madras (1973) Controlled comparison of oral twice-weekly and oral daily isoniazid plus PAS in newly diagnosed pulmonary tuberculosis. Br Med J 2:7–11.

73. East African/British Medical Research Council (1974) A pilot study of two regimens of intermittent thiacetazone plus isoniazid in the treatment of pulmonary tuberculosis in East Africa. Tubercle 55:211–221.

74. Singapore Tuberculosis Service/British Medical Research Council (1977) Controlled trial of intermittent regimens of rifampin plus isoniazid for pulmonary tuberculosis in Singapore. The results up to 30 months. Am Rev Respir Dis 116:807–820.

75. WHO Collaborating Centre for Tuberculosis Chemotherapy, Prague (1977) A study of two twice-weekly and a once-weekly continuation regimen of tuberculosis chemotherapy, including a comparison of two durations of treatment 2. Second report: the results at 36 months. Tubercle 58:129–136.

76. Tuberculosis Research Centre, Madras (1981) Ethambutol plus isoniazid for the treatment of pulmonary tuberculosis—a controlled trial of four regimens. Tubercle 62:13–29.

77. Dickinson JM, Mitchison DA (1966) In vitro studies on the choice of drugs for intermittent chemotherapy of tuberculosis. Tubercle 47:370–380.

78. Dickinson JM, Mitchison DA (1970) Suitability of rifampicin for intermittent administration in the treatment of tuberculosis. Tubercle 51:82–94.

79. Dickinson JM, Mitchison DA (1970) Observations in vitro on the suitability of pyrazinamide for intermittent chemotherapy of tuberculosis. Tubercle 51:389–396.

80. Dickinson JM, Ellard GA, Mitchison DA (1968) Suitability of isoniazid and ethambutol for intermittent administration in the treatment of tuberculosis. Tubercle 49:351–366.

81. Awaness AM, Mitchison DA (1973) Cumulative effects of pulsed exposures of *Mycobacterium tuberculosis* to isoniazid. Tubercle 54:153–158.

82. Sirgel FA, Botha FJH, Parkin DP, Van de Wal BW, Donald PR, Clark PK, Mitchison DA (1993) The early bactericidal activity of rifabutin in patients with pulmonary tuberculosis measured by sputum viable counts. A new method of drug assessment. J Antimicrob Chemother 32:867–875.

83. Bekierkunst A, Bricker A (1967) Studies on the mode of action of isoniazid on mycobacteria. Arch Biochem Biophys 122:385–392.

84. Seydel JK, Schaper KJ, Wempe E, Cordes HP (1976) Mode of action and quantitative-structure-activity correlations of tuberculostatic drugs of the isonicotinic acid hydrazide type. J Med Chem 19:483–492.

85. Winder FG, Collins PB (1970) Inhibition by isoniazid of synthesis of mycolic acid in *Mycobacterium tuberculosis*. J Gen Microbiol 63:41–48.

86. Mitchison DA, Dickinson JM (1971) Laboratory aspects of intermittent drug therapy. Postgrad Med J 47:737–741.

87. Dickinson JM, Aber VR, Mitchison DA (1973) Studies on the treatment of

experimental tuberculosis of the guinea pig with intermittent doses of isoniazid. Tubercle 54:211–224.

88. Evans DAP, Manley KA, McKusic VA (1960) Genetic control of isoniazid metabolism in man. Br Med J 2:485–491.

89. Ellard GA (1976) Variations between individuals and populations in the acetylation of isoniazid and its significance for the treatment of pulmonary tuberculosis. Clin Pharmacol Therapeut 19:610–625.

90. Mitchison DA (1973) Plasma concentrations of isoniazid in the treatment of tuberculosis. In: Davies DS, Prichard BNC, eds. Biological Effects of Drugs in Relation to their Plasma Concentrations, pp. 169–182. London: MacMillan.

91. Biehl JP, Vilter RW (1954) Effect of isoniazid on vitamin B_6 metabolism; its possible significance in producing isoniazid neuritis. Proc Soc Exper Biol Med 85:389–392.

92. Krishnamurthy DV, Selkon JB, Ramachandran K, Devadatta S, Mitchison DA, Radhakrishna S, Stott H (1967) Effect of pyridoxine on vitamin B6 concentrations and glutamic-oxaloacetic transaminase activity in whole blood of tuberculous patients receiving high-dosage isoniazid. Bull WHO 36:853–870.

93. Tuberculosis Chemotherapy Centre, Madras (1960) A concurrent comparison of isoniazid plus PAS with three regimens of isoniazid alone in the domiciliary treatment of pulmonary tuberculosis in South India. Bull WHO 23:535–585.

94. Tuberculosis Chemotherapy Centre, Madras (1963) The prevention and treatment of isoniazid toxicity in the therapy of pulmonary tuberculosis. 1. An assessment of two vitamin B preparations and glutamic acid. Bull WHO 28:455–475.

95. Tuberculosis Chemotherapy Centre, Madras (1963) The prevention and treatment of isoniazid toxicity in the therapy of pulmonary tuberculosis. 2. An assessment of the prophylactic effect of pyridoxine in low dosage. Bull WHO 29:457–481.

96. Gangadharam PRJ, Devadatta S, Fox W, Nair CN, Selkon JB (1961) Rate of inactivation of isoniazid in South Indian patients with pulmonary tuberculosis. 3. Serum concentrations of isoniazid produced by three regimens of isoniazid alone and one of isoniazid plus PAS. Bull WHO 25:793–806.

97. Selkon JB, Devadatta S, Kulkarni KG, Mitchison DA, Nair CN, Ramachandran K (1964) The emergence of isoniazid-resistant cultures in patients with pulmonary tuberculosis during treatment with isoniazid alone or isoniazid plus PAS. Bull WHO 31:273–294.

98. Jindani A, Aber VR, Edwards EA, Mitchison DA (1980) The early bactericidal activity of drugs in patients with pulmonary tuberculosis. Am Rev Respir Dis 121:939–949.

99. Chan SL, Yew WW, Ma WK, Girling DJ, Aber VR, Felmingham D, Allen BW, Mitchison DA (1992) The early bactericidal activity of rifabutin measured by sputum viable counts in Hong Kong patients with pulmonary tuberculosis. Tubercle Lung Dis 73:33–38.

100. Botha FJH, Sirgel FA, Parkin DP, Van del Wal BW, Donald PR, Mitchison DA (1996) The early bactericidal activity of ethambutol, pyrazinamide and the fixed

combination of isoniazid, rifampicin and pyrazinamide (Rifater®) in patients with pulmonary tuberculosis. South African Med J 86:155–158.

101. Donald PR, Sirgel FA, Botha FJ, Siefart HI, Parkin DP, Vandenplas ML, Van de Wal, BW, Maritz JS, Mitchison DA (in press). The early bactericidal activity of isoniazid related to its dose size in pulmonary tuberculosis. Am J Respir Crit Care Med.

102. Canetti G (1955) The tubercle bacillus in the pulmonary lesion of man. In: Histobacteriology and Its Bearing on the Therapy of Pulmonary Tuberculosis. New York: Springer Publishing Co. Inc.

103. Long MW, Snider DE, Farer LS (1973) U.S. Public Health Service cooperative trial of three rifampin-isoniazid regimens in treatment of pulmonary tuberculosis. Am Rev Respir Dis 119:879–894.

104. Melero C, Rey R, Ussctti P, Munoz L, Caballero J, Ortega A, Ramos A (1994) Rifabutin vs rifampicin in the treatment of initial pulmonary tuberculosis. Follow-up at 5 years. Tubercle Lung Dis 75 (suppl):16.

105. Gonzales-Montaner LJ, Natal S, Yongchaiyud P, Olliaro P, the Rifabutin Study Group (1994) Rifabutin for the treatment of newly-diagnosed pulmonary tuberculosis: a multinational, randomized, comparative study versus rifampicin. Tubercle Lung Dis 75:341–347.

106. Grumbach F, Canetti G, Le Lirzin M (1969) Rifampicin in daily and intermittent treatment of experimental murine tuberculosis, with emphasis on late results. Tubercle 50:280–293.

107. Dickinson JM, Mitchison DA (1976) Bactericidal activity in vitro and in the guinea-pig of isoniazid, rifampicin and ethambutol. Tubercle 57:251–258.

108. Mitchison DA (1985) The action of antituberculosis drugs in short-course chemotherapy. Tubercle 66:219–225.

109. Dickinson JM, Mitchison DA (1981) Experimental models to explain the high sterilizing activity of rifampin in the chemotherapy of tuberculosis. Am Rev Respir Dis 123:367–371.

110. Konno K, Feldmann FM, McDermott W (1967) Pyrazinamide susceptibility and amidase activity of tubercle bacilli. Am Rev Respir Dis 95:461–469.

111. Heifets LB, Flory MA, Lindholm-Levy PJ (1989) Does pyrazinoic acid as an active moiety of pyrazinamide have specific activity against *Mycobacterium tuberculosis?* Antimicrob Agents Chemother 33:1252–1254.

112. Salfinger M, Crowle AJ, Reller LB (1990) Pyrazinamide and pyrazinoic acid activity against tubercle bacilli in cultured human macrophages and in the Bactec system. J Infect Dis 162:201–207.

113. Rastogi N, Potar M-C, David HL (1988) Pyrazinamide is not effective against intracellularly growing *Mycobacterium tuberculosis.* Antimicrob Agents Chemother 32:287.

114. Heifets LB, Lindholm-Levy PJ (1990) Is pyrazinamide bactericidal against *Mycobacterium tuberculosis?* Am Rev Respir Dis 141:250–252.

115. Crowle AJ, Dahl R, Ross E, May MH (1991) Evidence that vesicles containing living,

virulent *Mycobacterium tuberculosis* or *Mycobacterium avium* in cultured human macrophages are not acidic. Infect Immun 59:1823–1831.

116. Heifets L, Lindholm-Levy P (1992) Pyrazinamide sterilizing activity in vitro against semidormant *Mycobacterium tuberculosis* bacterial populations. Am Rev Respir Dis 145:1223–1225.

117. British Thoracic Association (1981) A controlled trial of six months chemotherapy in pulmonary tuberculosis. First report: results during chemotherapy. Br J Dis Chest 75:141–153.

118. British Thoracic Association (1982) A controlled trial of six months chemotherapy in pulmonary tuberculosis. Second report: results during the 24 months after the end of chemotherapy. Am Rev Respir Dis 126:460–462.

119. East African Hospitals/Medical Research Council (1960) Comparative trial of isoniazid alone in low and high dosage and isoniazid plus PAS in the treatment of acute pulmonary tuberculosis in East Africans. Tubercle 41:83–102.

120. Tuberculosis Chemotherapy Centre, Madras (1960) A concurrent comparison of isoniazid plus PAS with three regimens of isoniazid alone in the domiciliary treatment of pulmonary tuberculosis in South India. Bull WHO 23:535–585.

121. Matthews JH (1960) Pyrazinamide and isoniazid used in the treatment of pulmonary tuberculosis. Am Rev Respir Dis 81:348–351.

122. Canetti G, Parrot R, Porven G, Le Lirzin M (1969) Les taux de rifampicine dans le poumon et dans les lesions tuberculeuses de l'homme. Acta Tuberc Belg 60:315–322.

123. Ellard GA, Humphries MJ, Allen BW (1993) Cerebrospinal fluid drug concentrations and the treatment of tuberculous meningitis. Am Rev Respir Dis 148:650–655.

124. Canetti G, Grumbach F (1953) Diffusion de la streptomycine dans les lesions caseuses des tuberculeux pulmonaires. Ann Institut Pasteur 85:380–386.

125. Debeaumont DA (1966) Bacteriologie de la tuberculose osteoarticulare sous chemotherapie. Adv Tuber Res 15:126–188.

126. British Thoracic and Tuberculosis Association (1976) Short course chemotherapy in pulmonary tuberculosis. Lancet 2:1102–1104.

3

Experimental Chemotherapy of Mycobacterial Diseases

Jacques Grosset and Baohong Ji

1. Introduction

Experimental chemotherapy of tuberculois provides the opportunity to assess, in vivo, the antimicrobial activity of a newly developed drug in comparison with that of existing drugs, its antagonistic, additional, or synergistic activity when it is given in combination with other drugs, its ability to prevent the selection of mutants resistant to the other drugs, and, finally, its ability to sterilize the lesions of the experimentally infected animals. However, to obtain reliable information, experimental chemotherapy should be performed in an adequate model. Because of its exquisite susceptibility to *Mycobacterium tuberculosis* infection, the guinea pig has been long been the animal of choice for detecting the presence of a tiny number of tubercle bacilli in a clinical specimen. It has also been used in experiments to assess the airborne transmission of tuberculosis and the impact of chemotherapy on the transmission (1), the comparative virulence of different strains of *M. tuberculosis* (2), the protective value of *M. bovis* Bacille–Calmette–Guérin (BCG) against a subsequent challenge with *M. tuberculosis* (3), and the comparative antituberculosis activity of several drugs and drug combinations given daily or intermittently (4,5). In spite of its numerous advantages, the guinea pig has not been used extensively as an animal model for the experimental chemotherapy of tuberculosis because its size makes it difficult to be used in large numbers, its metabolism is different from that of humans, and, finally, it has a high sensitivity to intercurrent infections.

Other species such as dog, rat, rabbit (6), or monkey (7) have been or still are being used. The first two are basic models for toxicologic investigations, and the third has been used extensively by Lurie for studying the mechanism of immunity in tuberculosis, the nature and relative efficacy of acquired and native resistance to tuberculosis (8). However, all four species, because of their size, costs (supply and maintenance), and ethical reasons at least for dog and monkey, are not routinely used for experimental chemotherapy of mycobacterial disease. The model of choice is the mouse (9–13) because of its ease of handling in terms of size, supply, maintenance, robustness, and reproducibility, although the mouse is far from being as sensitive as the guinea pig and the monkey to *M. tuberculosis*. In addition, the course of the disease that follows the experimental infection with *M. tuberculosis* or other *Mycobacterium* species is different from that of the disease in humans. For example, the caseation process is always limited in mice and does not result in cavity formation with large bacillary population as in humans. But if used with care, the mouse model is able to provide results that can be extrapolated to man.

The present chapter deals with the experimental chemotherapy of, first, tuberculosis and, second, the infection with *M. avium-intracellulare* (MAC). Because of the limited experience of the authors, the experimental chemotherapy of mycobacterial diseases due to *M. kansasii, M. marinum, M. xenopi,* or *M. ulcerans* will not be covered in this chapter, although they are of real importance.

2. Experimental Chemotherapy of Tuberculosis in Mice

Experimental chemotherapy of tuberculosis is considered under three main factors, i.e. the experimental mouse, the tubercle bacillus and the antimicrobial activity of the drug(s) to be evaluated. A reliable and reproducible experiment of which the results may be extrapolated to man requires a careful balance of each of these factors, whose respective importance is given in the following sections. Finally, the importance of measures ensuring the biological safety in the experimental tuberculosis laboratory should be remembered.

2.1. Basic Requirements

THE EXPERIMENTAL MOUSE

Among the different strains of mice that have been proposed for the experimental chemotherapy of tuberculosis, the most frequently used is the common laboratory outbred "Swiss" mouse. Because the aim of experimental chemotherapy is to obtain results that can be extrapolated to human beings, there is no absolute need to use an inbred mouse strain, for example, Balb C, C57 BI/6, or C3H, although the individual variations would be smaller. The differences in the immune status among outbred Swiss mice exist also among humans, and a drug or a drug regimen that is active against the mycobacteria in the "Swiss" mouse is likely to be active in humans despite the natural differences in response among individuals. However, to compensate for individual variations, a sufficient number of "Swiss" mice should be used in experimental chemotherapy. This does not mean that inbred mice should be used in the experimental chemotherapy of tuberculosis. On the contrary, inbred mice are of great interest for comparing the antituberculosis activities of several drugs. By reducing variation between individual mice, fewer animals need be used. In addition, inbred mice are needed for assessing the role of the immune background in the results of chemotherapy (14).

Finally, to test the role of immunodeficiency in the response of *M. tuberculosis* to chemotherapy and the consistency of the results, immunodeficient mice such as athymic nude mice can be used (15). The difficulty with these mice is that they should be kept in completely sterile conditions in negative-pressure isolators.

For chemotherapy studies, mice are usually infected by the intravenous route with a standard amount of tubercle bacilli and the efficacy of a single drug or a drug combination is monitored by the survival/mortality rate, the evolution of body weight, the extent of gross lesions, and the enumerations of the colony-forming units (CFU) in the organs (spleen, lungs, liver) before, during, and after the course of treatment (16). Usually the CFU counts are performed in spleens (17,18), or lungs (19), or both (20,21). When results of spleen and lung cultures are compared (21), the overall results are similar. Therefore, either organ is appropriate for the assessment of bactericidal activity and, to simplify, there is no need to enumerate the CFU in both organs. In order to have reproducible results, mice should preferably be infected with 0.5 ml of the bacillary suspension when they are still young, 4 weeks old. For long-term experiments, it is preferable to use only female mice because males are frequently fighting and sometimes killing each other.

THE TUBERCLE BACILLUS

The strain of tubercle bacilli to be used should be well characterized, with standard virulence and drug susceptibility. The H37Rv strain of *M. tuberculosis,* whose virulence is maintained through regular passages in the mouse, is the strain of choice. Before infection, the strain is subcultured in Tween-80-containing media such as Dubos Tween Albumin or Middlebrook 7H9 broths (22) in order to disperse the bacilli as much as possible and have a well-calibrated inoculum. After intravenous infection with 0.1 mg wet weight (about 5×10^6 CFU) of *M. tuberculosis* H37Rv per mouse, up to 90% of mice die from overwhelming tuberculosis infection, with more than 10^8 CFU in the spleen or lung, within the first month after infection provided no active drug is administered (Table 3.1). When the inoculum is small (i.e., 10^4 CFU or less), mice are able to contain and control the infection after an initial multiplication of the organisms. The disease that follows

Table 3.1 Survival rate of 4-week old Swiss mice by 4 weeks after intravenous infection with 0.1 mg of *M. tuberculosis* H37Rv in five different experiments[a]

Exp. 1	Exp. 2	Exp. 3	Exp. 4	Exp. 5
2/30	5/30	0/30	12/30	4/30

[a]*Source:* references (Truffot-Pernot et al., 1991; Ji et al., 1991 [18 and 27])

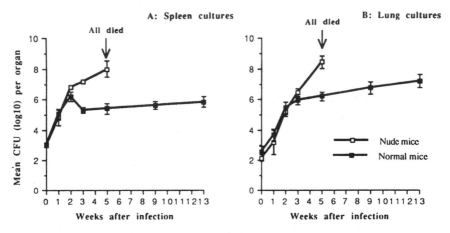

Figure 3.1. Enumeration of CFU ($\log_{10}$) in spleens (A) and lungs (B) of untreated mice after intravenous infection with 3.5×10^3 CFU of *M. tuberculosis*. Each point represents the mean of five mice; the error bars represent the standard deviations.

such limited infection remains chronic (23) and nonfatal, and, as shown in Figure 3.1, the bacilli population in both organs remains about 10^6 CFU or are only slowly increasing (15). To the contrary, athymic nude mice that are deprived of activated T lymphocytes cannot control the infection and die rapidly of overwhelming tuberculosis (Fig. 3.1).

To obtain a still more limited population of *M. tuberculosis,* mice can be vaccinated with *M. bovis* BCG and a month later, be infected with a small number, about 5×10^3, of virulent *M. tuberculosis*. In that case (24), the population of *M. tuberculosis* in the spleen increases by 2 logs during the first 2 weeks after infection, then remained stable about 10^5 CFU (Fig. 3.2).

DRUG ADMINISTRATION AND DOSAGE

Except for aminoglycosides, which are given subcutaneously in the upper part of the mouse back, all drugs are given orally with an esophageal cannula (gavage)

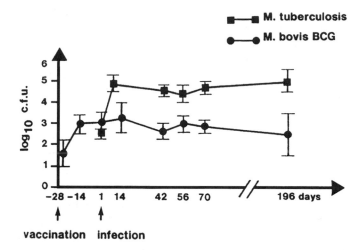

Figure 3.2. Enumeration of CFU in the spleens of *M. bovis* BCG-vaccinated and *M. tuberculosis*-infected mice (mean and SD from 10 mice at each point). BCG vaccination: intravenous inoculation of 2.74 $\log_{10}$ CFU; *M. tuberculosis* infection: intravenous inoculation of 3.38 $\log_{10}$ CFU.

at a volume of 0.2–0.3 ml per mouse according to their body weight, either daily (very often six times a week) or intermittently, depending on the objective of the experiment. For oral administration, the drugs are usually dissolved or suspended, and then diluted to the requested concentration in 0.05% agar-containing sterile distilled water. The drug solutions are prepared weekly and stored at 4°C.

The properties of absorption, distribution, metabolism, and excretion of a given drug in an experimental animal are almost always different from those in man. Therefore, extrapolating the activity of a drug from animal to man requires taking into consideration the pharmacokinetic differences of the drug between the species.

In general, effective dosages of drugs in the laboratory animal are larger than those in man. Drug dosages active in man are often too small to be active in the animal. Although there are important exceptions to the rule, it appears that the smaller the size of animal species, the larger the dosage required for activity. One possible reason for this is that drug metabolism correlates better with body surface area than body weight (25); and as shown in Table 3.2, the ratio between body surface area and body weight decreases sharply with increasing body weight. The usual dosages (mg per kg body weight) of a drug that provide an equally potent effect among different species are shown in Table 3.3. From this table, it appears that for most drugs, the equipotent dosage is 12 times larger in the mouse than in man. As a result, for a drug given to the mouse at a dosage equipotent to that in man, usually the peak serum level (C_{max}) is much higher and is achieved earlier (T_{max}), but the half-life of disappearance of the drug from the blood ($T_{1/2}$) is shorter in the mouse than in man. Actually, at equipotent drug dosages, what should be similar in the mouse and in man is the area, in units of drug concentration $\times$ time, under a curve (AUC) representing the concentration of the drug in the plasma or serum at different intervals after administration. At the least, a rough estimate of AUC will assist one to choose the relevant dosages for a valid assessment of the antimicrobial activity of a drug in the mouse. The data obtained from studies of the three main antituberculosis drugs (isoniazid, rifampin, and pyrazinamide), and the new fluoroquinolones are given as examples of the importance of taking into account the pharmacokinetics differences of drugs in man and in the mouse.

Comparative data on the main pharmacokinetics parameters of isoniazid, rifampin, and pyrazinamide in the mouse and in man are given in Table 3.4. From that table, it can be seen that the commonly used dosages give AUCs more favorable in the mouse than in man for rifampin, as favorable in the mouse as in the slow

Table 3.2 Comparison of body weight and body surface area for several species

Species	Mean body weight (g) (range)	Range of body surface area (cm²)	Ratio of surface area (cm²) on body weight (g)
Mouse	22 (18–25)	65–70	2.9–3.1
Rat	250 (200–350)	350–400	1.4–1.6
Rabbit	2,500 (1,800–2,600)	1,600–1,900	0.6–0.7
Dog	12,000 (10,000–15,000)	5,600–6,500	0.4–0.5
Man	66,000 (64,000–75,000)	16,000–18,000	0.2–0.3

Table 3.3 Usual equivalent dosages in mg per kg body weight for equipotent doses in several species

Species	Dosage (mg/kg)						Species ratio[a]
Mouse	50	100	200	400	800	1600	12
Rat	25	50	100	200	400	800	6
Rabbit	12.5	25	50	100	200	400	3
Dog	7.5	15	30	60	120	240	1.8
Man	4–4.2	8–8.5	17 (15–20)	34 (30–35)	67 (65–70)	133 (130–135)	1.0

[a]Ratio between dosage in experimental species and equipotent dosage in man.

Table 3.4 Major pharmacokinetics parameters of isoniazid, rifampin, and pyrazinamide in mice and humans after administration of a single oral dose

Test species	Drug dosage (mg/kg)	C_{max} (μg/ml)	T_{max} (h)	$t_{1/2}$ (h)	AUC (mg · h/liter)
	Isoniazid				
Mouse	25 mg/kg	28.20 ± 3.8	0.25 ± 0	1.7 ± 0.17	52.2 ± 2.2
Man[a]	6.2 ± 6 mg/kg				
	Rapid acetylators	5.4 ± 20	1.1 ± 0.5	1.54 ± 0.31	19.9 ± 6.1
	Solid acetylators	7.1 ± 1.9	1.1 ± 0.6	3.68 ± 0.59	48.2 ± 1.5
	Rifampin				
Mouse	10 mg/kg	10.58 ± 0.28	1.33 ± 0.58	7.61 ± 1.32	139.7 ± 10.7
Man[b]	10–15 mg/kg	14.91	2.84	2.46	117.93
	Pyrazinamide				
Mouse	150 mg/kg	146.1 ± 13.0	0.42 ± 0.2	1.05 ± 0.14	303.8 ± 17.9
Man[c]	27 ± 4 mg/kg	38.7 ± 5.9	1 ± 0	9.6 ± 1.8	520 ± 101

Note: C_{max}, peak level of drug in plasma; T_{max}, time to peak level of drug in plasma; $T_{1/2}$, half-life of elimination; AUC, area under the serum concentration–time curve.
[a]Data from Kim YG, Shin JG, Shin SG, Jang IJ, Kim S, Lee JS, Han JS, Cha YN (1993) Decreased acetylation of isoniazid in chronic renal failure. Clin Pharmacol Therapeut 54:612–620.
[b]Data from Kenny MT, Strates B (1981) Metabolism and pharmacokinetics of the antibiotic rifampin. Drug Metab Rev 12:159–218.
[c]Data from Lacroix C, Phan Hoang T, Nouveau J, Guyonnaud C, Laine G, Duwoos H, Lafont O (1989) Pharmacokinetics of pyrazinamide and its metabolites in healthy subjects. Eur J Clin Pharmacol 36:395–400.

Table 3.5 Comparative major pharmacokinetic parameters in mice and humans after administration of a single dose of sparfloxacin

Test species and dosage	C_{max} (μg/ml)	T_{max} (h)	$t_{1/2}$ (h)	AUC_{0-x} (mg h/L)
Mice				
5 mg/kg	0.25	0.3	5.0	0.74
50 mg/kg	2.80	0.75	5.0	15.18
100 mg/kg	8.54	0.75	5.5	47.94
Humans				
200 mg	0.70	4	20.8	18.75
400 mg	1.18	5	18.2	32.73

Note: C_{max}, peak level of drug in plasma; T_{max}, time to peak level of drug in plasma; $t_{1/2}$, terminal half-life; AUC_{0-x}, area under the plasma concentration–time curve from time zero infinity.

acetylator man for isoniazid, and less favorable in the mouse than in man for pyrazinamide.

Norfloxacin, ciprofloxacin, pefloxacin, ofloxacin, and sparfloxacin are the five major quinolones commercially available at present. Among them, ciprofloxacin is not completely absorbed from the gastrointestinal tract and, therefore, gives lower C_{max} levels than norfloxacin, pefloxacin, and ofloxacin after oral administration (26). Pefloxacin and ofloxacin are better absorbed, have much longer half-lives, and are more active against *M. tuberculosis* in the mouse and in man. The AUC (48 mg h/L) in mice treated with ofloxacin 150 mg/kg once daily is close to the AUC of 48.1 mg h/L in man treated with multiple doses of ofloxacin 400 mg every 12 h, i.e., 13 mg/kg (16,18). Thus, equipotent dosages of ofloxacin are about 12 times higher in mice than in humans. Similar conclusions can be drawn from the comparison of AUC in mice and human after administration of a single dose of sparfloxacin, the most recently developed fluoroquinolone. As shown in Table 3.5, 50 mg/kg of sparfloxacin in the mouse is equipotent to 200 mg daily in man, i.e., 4 mg/kg (21).

In conclusion, of the antituberculosis drugs, only rifampin shows similar C_{max} in mice and in humans after administration of the same dosage, 10 mg/kg, and its $T_{1/2}$ in mice is slightly longer than in humans (19,27). This unique characteristic explains why the rule that equipotent dosages are on average 12 times higher in mice than in humans is not always applicable, and main pharmacokinetic parameters of a new antituberculosis drug should be established before assessing the activity of that drug in the mouse.

SAFETY IN THE EXPERIMENTAL LABORATORY

Experiments with highly virulent tubercle bacilli require appropriate safety practices for the control of biological hazards, including applying all biosafety level 3 measures in the laboratory as well as in the animal room for protecting against infection by the inhalation route. As recently emphasized for the experimental tuberculosis laboratory (28), "only those who have mastered basic safe practices" should perform the intravenous injection of *M. tuberculosis,* maintenance, treatment, and autopsy of the infected mice, and the CFU counts of the infected organs.

2.2. Experimental Activity of a Single Drug

The first steps in evaluating the activity of a drug against *M. tuberculosis* in the mouse are to determine whether the drug is active or not; then, if the response is positive, to measure its minimal effective dose (MED) in terms of survival rate and of prevention of organ lesions; and finally, to determine whether the activity is of the bacteriostatic or bactericidal type.

DETERMINATION OF THE MINIMAL EFFECTIVE DOSAGE

When the tested dosages of the drug have been decided, the simplest and quick-est way to test the activity against *M. tuberculosis* in the mouse is to perform an experiment of preventive therapy, aiming to determine whether or not the drug is able to prevent the death of infected mice. For that purpose, a series of mice are each infected with 0.1 mg (about 5×10^6 CFU) of *M. tuberculosis* H37Rv strain by the intravenous route and treated daily for 4 weeks with various drug dosages, the treatment beginning the following day after infection. At the end of 4 weeks, all surviving mice are sacrificed and autopsied. Two control groups of mice should be used: (i) a negative control group of untreated mice of which a great majority will die by the 28th day after infection, with numerous gross lung lesions and significant splenomegaly and (ii) a positive control group of mice treated with 25 mg/kg isoniazid (INH) daily that should be all alive and without any organ lesions by the 28th day after infection. To facilitate the statistical analysis, ideally each group should consist of 30 animals. The activity of the tested drug can be mea-sured by several parameters: increase of body weight during treatment; mortality rate during treatment or survival rate at the end of treatment; and gross lung lesions and spleen enlargement of the surviving mice that are all sacrificed 28 days after infection. The minimal effective dosage (MED) is defined as the smallest dosage of drug that significantly increases the survival rate and also reduces the devel-opment of organ lesions among the treated mice.

As an example, Table 3.6 gives the survival rates of mice by 28 days after intravenous infection with 0.1 mg *M. tuberculosis* and treated once daily for 28 days with ofloxacin (18). The great majority of infected but untreated mice (neg-ative controls) died, whereas all of those treated as positive controls with 25 mg/kg isoniazid (INH) survived. Mice treated with 50 or 100 mg/kg ofloxacin be-

Table 3.6 Number of surviving mice treated with ofloxacin at different dosages in two different experiments (30 animals per group)

| Treatment group | No. of surviving at 28 days in | |
	Exp. 1	Exp. 2
Negative control	2	5
INH 25 mg/kg	30	30
Ofloxacin 50 mg/kg	3	
Ofloxacin 100 mg/kg	7	
Ofloxacin 150 mg/kg	23[a]	28[a]
Ofloxacin 300 mg/kg		30[b]

[a]All with slight or marked lung lesions.
[b]Twenty-four with no lung lesions.

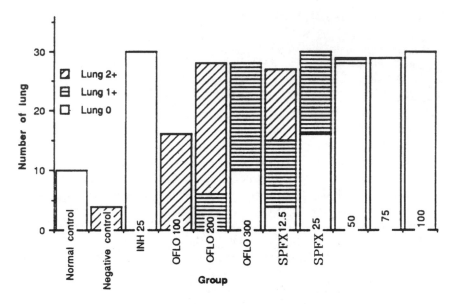

Figure 3.3. Lung lesions in mice at 28 days after infection with *M. tuberculosis.*

haved like negative controls. The majority of mice treated with 150 mg/kg survived, but all of the survivors had lung lesions. All 30 mice treated with 300 mg/
kg survived and only 6 of them had numerous lung lesions. Thus, in terms of
survival rate, the MED of ofloxacin against *M. tuberculosis* in mice was 150 mg/
kg; in terms of prevention of organ lesions, the MED was 300 mg/kg.

In another experiment (29), the activity of sparfloxacin given at different dosages was assessed comparative to that of ofloxacin. As an example, Fig. 3.3
illustrates the lung lesions in mice that survived 28 days after intravenous infection
with 0.1 mg *M. tuberculosis.* As measured by the ability of preventing lung lesions, sparfloxacin (SPFX) 12.5 mg/kg appeared more active than 200 mg/kg
ofloxacin (OFLO), SPFX 50 mg/kg appeared more active than OFLO 300 mg/kg
and SPFX 75 mg/kg as active as INH 25 mg/kg.

MEASURING THE BACTERICIDAL AND THE STERILIZING ACTIVITIES

By definition, a drug given at a dose that reduced the mortality and the development of organ lesions in treated mice was able to reduce or even to prevent the
multiplication of *M. tuberculosis* in the infected host and has at least some bacteriostatic activity. But the precise measurement of the antimicrobial activity relies
only on the evolution of CFU counts in the organs of mice before, during, and

after drug administration. That measurement should be made in well-defined conditions. If it is attempted in infected mice having much more than 10^6 CFU of *M. tuberculosis* in their lung or spleen, drug-resistant mutants will be present at the start of treatment (30) and selected during treatment. As a consequence, there will be interference between the decrease in the number of drug-susceptible CFU and the concurrent increase in the number of drug-resistant CFU, and the CFU counts might not provide a precise measurement of the antimicrobial activity of the tested drug. Of course, one can prevent that by performing total CFU counts and drug-resistant CFU counts on plain culture medium and drug-containing medium, respectively. An example of this is given in Figure 3.4, that shows the fate of INH-susceptible and INH-resistant organisms in the lungs of mice treated with isoniazid alone, the treatment being started when the lungs contained nearly 10^7 CFU of *M. tuberculosis* (31).

The easiest method is to begin treatment when mice have a bacillary population not larger than 10^6 CFU in their organs. In that case, no drug-resistant mutants

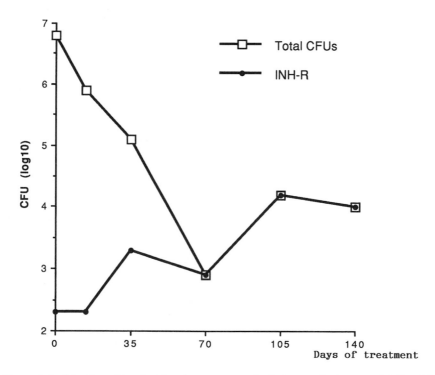

Figure 3.4. Selection of isoniazid-resistant mutants in the lungs of mice treated daily with 50 mg/kg INH.

will be present and possibly selected during treatment, and performing only the CFU counts on plain culture medium will, in fact, measure the fate of drug-susceptible CFU during treatment. In practice, there are two possibilities: either mice are infected with 0.1 mg wet weight *M. tuberculosis* (5×10^6 CFU) and started with the tested drug from the day after infection, or mice are infected with 0.0001 mg *M. tuberculosis* (5×10^3 CFU) and treated 14 days later when the bacillary population in their lung and spleen has reached 10^5–10^6 CFU. The crucial point is that, in both cases, the drug administration should be initiated when the bacillary population in lung and spleen is not larger than 10^6 CFU.

Treatment Without Delay after Infection with a Large Inoculum.   The technique to be followed is similar to that used for determining the MED, except that the CFU in the organs, lungs, and spleen, are enumerated at the beginning of treatment, which is the following day after infection with 0.1 mg (5×10^6 CFU) of *M. tuberculosis,* and at different intervals during treatment. The decrease in the CFU counts during treatment measures the bactericidal activity of the tested drug. For example, it has been used for assessing the activity of levofloxacin (LVFX), a levogyre isomer of ofloxacin (OFLO), against *M. tuberculosis* in the mouse compared to that of the parent compound ofloxacin and the new fluoroquinolone sparfloxacin (SPFX). As shown in Figure 3.5, by the 30th day after intravenous infection with 1.74×10^6 *M. tuberculosis* CFU, the great majority of untreated controls have died, as well as mice treated with 50 mg/kg LVFX. About 70% of those treated with OFLO 150 mg/kg survived, whereas all or almost all other

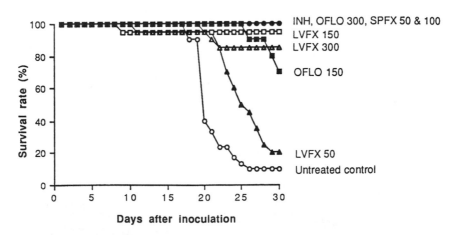

Figure 3.5. Survival rates of mice within 30 days after infection intravenously with 1.74 $\times$ 10^6 CFU of *M. tuberculosis* H37Rv. At the time (D1) the treatments were begun, there were 30 mice in the control group and 20 mice in each treated group.

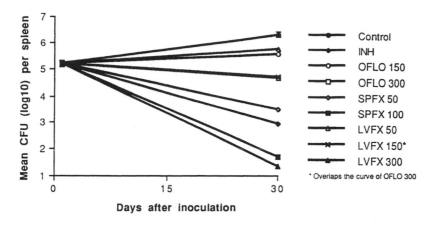

Figure 3.6. Enumerations of CFU in the spleens of mice. Mice were inoculated intravenously with 1.74×10^6 CFU of *M. tuberculosis* H37Rv, and treatments were begun the
following day after inoculation. Drugs were administered by gavage six times weekly for
4 weeks. Each point represents the mean number of CFU for 3–20 mice. Error bar represents the standard deviation of the control mice.

treated mice survived. The CFU counts in the spleen of mice performed the day
after infection and after 4 weeks in the spleen of those surviving provided a
comparative appraisal of the antimicrobial activity of the drugs (Fig. 3.6). As
expected, LVFX 150 mg/kg given 6 days a week had an activity similar to that
of OFLO 300 mg/kg, and LVX 300 mg/kg, perhaps equipotent to 800 mg daily
in man (32,33), was as bactericidal as SPFX 100 mg/kg and more bactericidal
than INH 25 mg/kg. Therefore, levofloxacin is potentially a "first-line" drug for
the treatment of multidrug-resistant tuberculosis.

Delayed Treatment After Infection with a Small Inoculum.   Mice are infected
by the intravenous route with 5×10^3 *M. tuberculosis* and treatment begins 14
days later when the CFU counts are stable or slowly increasing in the lungs and
spleen after an initial rapid increase. At that stage, the size of the bacillary population in the organs is between 10^5 and 10^6 CFU, thus insufficient to contain
drug-resistant mutants, and the activity of an antituberculous drug given alone
can be easily assessed without interference by the selection of resistant mutants.
For example, in a recent experiment (21), the bactericidal activity of 50 and 100
mg/kg daily sparfloxacin has been assessed in comparison to that of ofloxacin
and major drugs, isoniazid, rifampicin, and pyrazinamide. As illustrated in Table
3.7, after 8 weeks of treatment, sparfloxacin 100 mg/kg was as bactericidal as
rifampin 10 mg/kg, whereas sparfloxacin 50 mg/kg was as bactericidal as iso-

Table 3.7. Number of CFU of *M. tuberculosis* in the spleens of mice treated with single drugs

Treatment*	Mean no. of CFU ($\log_{10}$) $\pm$ S.D. at the following times (weeks) during treatment[a,b]			
	0 (week 2)	2 (week 4)	4 (week 6)	8 (week 10)
Untreated control	6.33 $\pm$ 0.18	5.70 $\pm$ 0.34	6.03 $\pm$ 0.33	6.13 $\pm$ 0.46
INH 25 mg/kg		4.57 $\pm$ 0.28	4.11 $\pm$ 0.21	3.47 $\pm$ 0.29
RMP 10 mg/kg		5.29 $\pm$ 0.17	4.12 $\pm$ 0.36	2.42 $\pm$ 0.55
PZA 150 mg/kg		5.20 $\pm$ 0.31	4.43 $\pm$ 0.36	4.17 $\pm$ 0.70
OFLO 300 mg/kg		5.70 $\pm$ 0.26	5.38 $\pm$ 0.22	5.31 $\pm$ 0.53
SPFX 50 mg/kg		5.35 $\pm$ 0.25	5.14 $\pm$ 0.21	3.65 = 0.54
SPFX 100 mg/kg		5.15 $\pm$ 0.20	4.25 $\pm$ 0.13	2.90 $\pm$ 0.58

[a]Mice were treated from 14 days (week 2) after inoculation with *M. tuberculosis*. Drugs were administered by gavage six times weekly.
[b]The results at each point represent the mean number of CFU in the spleens of 9–10 mice. The numbers in parentheses are weeks after inoculation.
S.D. = Standard deviation of the mean.

niazid 25 mg/kg. Ofloxacin 300 mg/kg and even pyrazinamide have a limited bactericidal activity. It is important to notice that during the initial 2 weeks of treatment, isoniazid was the most bactericidal of all antituberculosis drugs, and during the initial 8 weeks of treatment, none of the major existing antituberculosis drugs alone was able to kill 4 $\log_{10}$ *M. tuberculosis*.

Delayed Treatment in M. bovis BCG-Vaccinated Mice Superinfected with a Small Inoculum of M. tuberculosis.   In order to mimic the limited size and dormant status of tubercle bacilli supposedly present in subjects with latent tuberculosis infection, mice are first vaccinated with *M. bovis* BCG, then 1 month later superinfected with 5×10^3 *M. tuberculosis*. Because of the presence of protective immunity, the growth of *M. tuberculosis* is rapidly stopped and a chronic disease takes place with a rather stable and limited bacillary population, the size of which is around 10^5 CFU in the spleen and lung. Such a model has been used to test the sterilizing activity of single drugs or drug combinations for preventive therapy of tuberculosis (24). It was demonstrated that the combination rifampin–pyrazinamide given for 2 months was as active as rifampin alone given for 3 months and more active than isoniazid alone given for 6 months, as evidenced by the rate of spleen cultures conversion to negatively at the end of treatment and the re-growth of *M. tuberculosis* during the follow-up after stopping treatment (Fig. 3.7).

The same system has been used to compare the activities of three rifamycin derivatives, rifampin, rifabutin, and rifapentine, in the preventive therapy of tuberculosis (27). The drugs were given at a dosage of 10 mg/kg either daily or

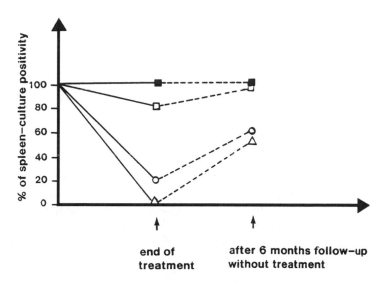

Figure 3.7. Proportion of mice with spleen culture-positive at the end of treatment (solid lines) for 2 months with RMP + PZA (open triangles) or RMP + PZA + INH (open squares), for 3 months with RMP alone (open circles), or for 6 months with INH alone (closed squares), and after a 6-month period of follow-up without treatment (dashed lines).

intermittently. The bactericidal activities of rifabutin given daily for 6 weeks or rifapentine given twice weekly for 12 weeks were comparable to that of rifampin given daily for 12 weeks. As the bactericidal activity of rifapentine given once weekly for 12 weeks was comparable to that of rifampin given daily for 6 weeks, rifapentine appeared suitable for the fully supervised once-weekly preventive therapy of tuberculosis. In another experiment (15), it was demonstrated that the bactericidal activity of once-weekly rifapentine was significantly increased by the addition of once-weekly isoniazid 75 mg/kg, suggesting that the combination of rifapentine plus isoniazid once weekly might become the combination of choice for fully supervised preventive therapy of tuberculosis or even for the curative therapy of active tuberculosis, especially during the secondary phase (continuation phase) of treatment after the initial 2-month intensive phase of daily treatment.

Antimicrobial Activity of Drug(s) in the Late Phase of Treatment.   The major objective of chemotherapy for tuberculosis in humans is to reduce the bacillary population in the lesions to such a limited size that no relapse would occur after stopping drug administration by remultiplication of the few bacilli called "persisters" (34) that have survived the treatment with active drugs. The ability of a given drug to eliminate persisters, in other words to sterilize the lesions, is therefore of crucial importance for the success of treatment.

The sterilizing activity of a given drug can be best measured during the continuation phase of chemotherapy when only a few hundreds or thousands CFU remained in the organs of mice after the initial, usually 2-month, intensive phase of therapy. At that stage, single drugs can be given without risk of selecting drug-resistant mutants and their respective ability to sterilize the lesions can be easily measured. As assessed by the rate of mice that became culture negative at the end of treatment (21), rifampin was the most sterilizing drug, followed closely by a new fluoroquinolone, sparfloxacin, and isoniazid, whereas pyrazinamide at that stage only displayed a bacteriostatic activity (Fig. 3.8). Using the same model, it was also possible to study whether or not the animals will remain culture negative after stopping administration of a particular drug or drug combination. As emphasized by McCune et al. (35,36), all mice that had become culture negative did not get rid of viable *M. tuberculosis* completely. A number of them converted their infection to a latent state characterized by negative cultures of spleen and lungs at the end of drug administration, but followed by positive cultures some time after stopping treatment. The drugs or drug combinations that can best ensure the eradication of the latent *M. tuberculosis* (i.e., the true sterilization of the mouse tissues) are those able to prevent relapses after chemotherapy.

In this respect, rifampin is the key sterilizing drug. Mice treated with the combination isoniazid–rifampin are all culture negative by the sixth month of treatment, although 50% of them become culture positive again after a follow-up of

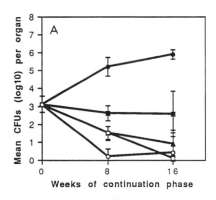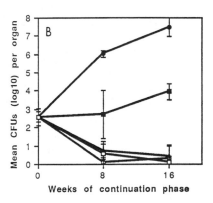

Figure 3.8. Spleen (A) and lung (B) cultures during the continuation phase of chemotherapy. Mice were inoculated intravenously with 0.1 mg of *M. tuberculosis* H37Rv and were initially treated with INH–RMP–PZA for 8 weeks. Then the mice were randomly allocated to an untreated control group (●) or were treated with INH (25 mg/kg) (▲), RMP (10 mg/kg) (○), PZA (150 mg/kg) (■), or SPFX (5100 mg/kg) (□) six times a week for 16 weeks. Each point represents the mean number of CFU for 10 mice. Error bars represent standard deviations.

6 months without treatment. If the treatment with rifampin is continued up to 9 months, all mice remain culture negative in spite of a follow-up of 6 months without treatment, a result that cannot be obtained by other regimens without rifampin (37).

An elegant mouse model to test the sterilizing properties of antituberculous drugs would be the model developed by McCune et al. at Cornell University (36): Mice infected intravenously with 0.1 mg *M. tuberculosis* and treated for 100 days with daily dosage equivalent to 18–20 mg/kg INH and 3500 mg/kg PZA given in the diet were all culture negative at the end of treatment, but a percentage of them relapsed with spleen and lungs culture positive after a few months without treatment. Using that model, it is possible to determine whether or not a drug given at the end of the first 100 days of treatment with INH + PZA is able to prevent relapses; in other words, whether or not the drug has a sterilizing activity.

2.3. The Mouse Model for Testing the Experimental Activity of Drug Combination

The main lesion of pulmonary tuberculosis in humans is the tuberculosis cavity that contains a large population of about 10^8 actively multiplying tubercle bacilli and drug-resistant mutants in a proportion of 10^{-6} to 10^{-7} (30). These mutants are selected by monotherapy (38), but their selection is prevented by combined therapy (39). An experimental model aiming at reproducing the bacteriologic condition of human pulmonary tuberculosis should also provide a large population of actively multiplying tubercle bacilli with drug-resistant mutants. Such a model is obtained in mice infected by the intravenous route with 0.1 mg wet weight *M. tuberculosis* (i.e., about 5×10^6 CFU) and kept without treatment for 14 days. During the first 14 days following infection, the bacilli actively multiply to reach 10^7–10^8 in the lung and spleen, and the infection would be fatal by the 21th–28th day if the mice were not treated effectively. As already illustrated by Figure 3.4, monotherapy begun by the 14th day after infection selects resistant mutants. Combined drug therapy with at least two active drugs prevents the selection of drug-resistant mutants, and the CFU counts in the lung and spleen decrease progressively as a function of the bactericidal activity of the drug combination. If the drug combination has a sterilizing activity and treatment is continued for enough duration, the lungs and spleen eventually become culture negative, and the mice are considered to be cured. Thus, in the continuation phase of chemotherapy also, the fate of *M. tuberculosis* in the tissues of the infected host has numerous points of similarity in mice and human beings, and the information derived from the chemotherapy of experimental murine tuberculosis is highly suggestive of what can be expected from the treatment with the same drug combination in humans.

Because it is not possible to analyze in detail the results of all experiments that had been performed during the last 50 years and because chemotherapy for tuberculosis has considerably improved since the introduction of streptomycin in 1944 (40), para-aminosalicylic acid in 1946 (41) isoniazid in 1952 (42,43), and rifampin in 1965 (44), the present section is subdivided in three parts: the activity of standard chemotherapy with isoniazid and streptomycin; the short-course chemotherapy with rifampin and pyrazinamide; and the issues that experimental chemotherapy might address in the near future.

STANDARD CHEMOTHERAPY FOR TUBERCULOSIS

In mice infected with 0.1 mg *M. tuberculosis* and treated 14 days later when the lung and spleen contain more than 10^7 CFU, daily treatment with either 200 mg/kg streptomycin alone or 25 mg/kg isoniazid alone resulted in the selection of drug-resistant mutants (12,31). When similarly infected mice were treated daily with the combination of both drugs given at the same dosages as in monotherapy, the selection of drug-resistant mutants was prevented and drug-susceptible organisms were progressively killed (12): By the 3rd month of treatment, 1000 CFU were isolated from the lung, by the 6th month only 100 CFU, and none at all or very few CFU by the 12th month (Fig. 3.9). Thus, the combination isoniazid–streptomycin was able to cure mice as it was able to cure infected humans patients.

However, the results of chemotherapy of experimental tuberculosis in mice with the combination isoniazid and streptomycin were suggesting that the combination was far from optimal for the treatment of tuberculosis. First, even though isoniazid and streptomycin were given daily up to 15 or even 18 months, as many as 35% of the animals remained culture positive with a few colonies of drug-susceptible organisms (45). In addition, after a follow-up of 3–6 months without treatment, 75% of all treated mice became culture positive again, indicating that the sterilizing activity of the combination is limited. Similar findings were observed in humans, and a certain proportion, at least 10%, of patients treated daily with the combination isoniazid, streptomycin, and para-aminosalicylic acid had relapsed despite prolonged therapy (46).

Second, even though isoniazid and streptomycin were administered daily at a high dose, 25 mg/kg and 200 mg/kg, respectively, the combination was never able to prevent the selection of isoniazid-resistant mutants in 100% of the treated mice: On average, that selection occurred in 10% of the mice (12).

Third, when the daily dose of isoniazid was reduced from 25 mg/kg to 5 mg/kg or when streptomycin was not given daily but every other day, the decline in the CFU counts was slowed down and the selection of drug-resistant mutants, to isoniazid or to streptomycin, was favored in mice (12,47) as in humans (48). However, when streptomycin was stopped after 3 months of daily combined administration with isoniazid, and the treatment was continued with isoniazid alone up to 12 months, the antimicrobial activity remained similar to that of the com-

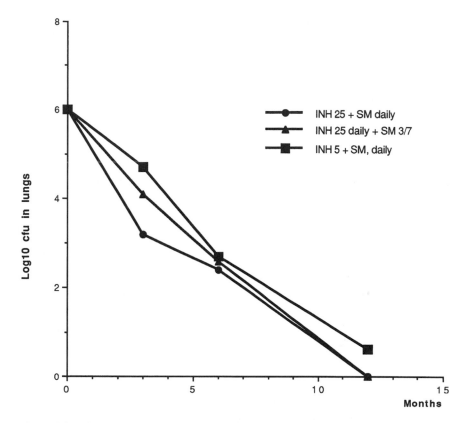

Figure 3.9. Median CFU counts in the lungs of mice treated with isoniazid (INH) and streptomycin (SM). SM given daily or three times weekly (³⁄₇) at a daily dose of 200 mg/ kg; INH given daily either at 25 mg/kg or 5 mg/kg.

bination isoniazid plus streptomycin, in terms of reduction of the CFU counts, of the proportion of mice with isoniazid resistant strains, and of the number of colonies persisting at the end of treatment (47,49). Such findings confirmed that streptomycin has no sterilizing activity, and combined therapy is not required when the size of the bacillary population is limited.

SHORT-COURSE CHEMOTHERAPY

In 1956, McCune et al. (36) were able to render the organs of mice culture negative after 3 months of daily treatment with the combination isoniazid (18– 20 mg/kg) and high-dosage pyrazinamide (3000 mg/kg), both drugs being administered in the mouse diet. Such a result had never been achieved before with any of the then existing drug regimens. However, after 3 months of follow-up without treatment, one-third of the animals relapsed with fully drug-susceptible organisms, indicating that tubercle bacilli were still persisting in a latent state among culture-negative animals. The importance of these findings for the future short-course chemotherapy became apparent only later, when rifampin was available.

In 1967, Grumbach and Rist (50) were also able to render mice culture negative after a treatment for 4 months with the daily combination of 25 mg/kg isoniazid and 25 mg/kg rifampin. Mice treated with isoniazid 25 mg/kg alone or rifampin 25 mg/kg alone remained culture positive, and drug-resistant organisms were

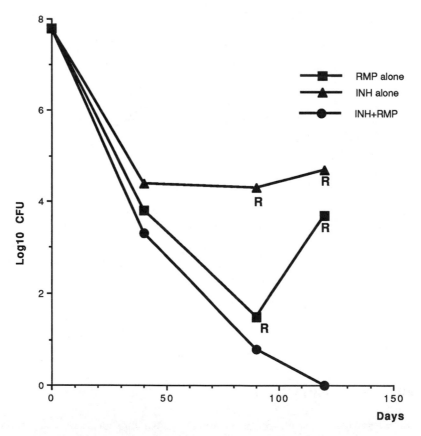

Figure 3.10. CFU counts in the lungs of mice treated with isoniazid alone, rifampin alone, or the combination isoniazid plus rifampin. R = drug-resistant mutants.

selected (Fig. 3.10). The results of treatment with the combination isoniazid-rifampin were so good and obtained with so realistic drug dosages, that they had ushered numerous clinical trials and animal experiments, which had eventually led to the development of modern short-course chemotherapy.

Short-Course Chemotherapy with Rifampin-Containing Regimens.   First, the very potent activity of rifampin given alone or in combination with isoniazid has been confirmed by numerous workers (51–53). Then, activities of rifampin in combination with drugs other than isoniazid were investigated. As shown in Table 3.8, with the exception of isoniazid, none of the drugs given in combination with rifampin was capable of rendering culture negative all infected and treated mice within 6 months of treatment; for instance, no more than 75% of mice were culture negative after 6 months of daily treatment with the combination ethionamide 25 mg/kg plus rifampin, or ethambutol 100 mg/kg plus rifampin (54,55). All of the remaining 25% that were still culture positive had organisms resistant to rifampin. Therefore, at the tested dosages, ethambutol and ethionamide contributed little to the effectiveness of rifampin in the mouse and were not capable of preventing completely the selection of rifampin-resistant mutants. The additive effect of streptomycin to the combination isoniazid–rifampin was also limited: Streptomycin contributed only to speeding up the reduction of bacterial counts during the first month of treatment but did not change the final results (54).

To facilitate the supervision of treatment and reduce its costs, intermittent (e.g., once, twice, or three times weekly) drug administration has been tested. Rifampin is not a drug very suitable for intermittent use for the following two reasons. First, its bactericidal activity is reduced by intermittent administration. As shown in

Table 3.8. Short-course chemotherapy of experimental murine tuberculosis; percentage of lung and spleen culture negative after 6 months of daily treatment with various regimens

Drug regimen[a] (mg/kg)	% Cultures negative at 6 months
INH 25 alone	0[b]
INH 25 + SM 200	10 (79% after 18 months)
RMP 25 alone	50–90[b]
RMP 25 + INH 25	100
RMP 25 + ETHIO 25	75*
RMP 25 + EMB 100	75*
RMP 25 + INH 25 + SM	100

[a]INH = isoniazid; SM = streptomycin; RMP = rifampin; ETHIO = ethionamide; EMB = ethambutol.

[b]All positive cultures yielded organisms resistant to the main drug used.

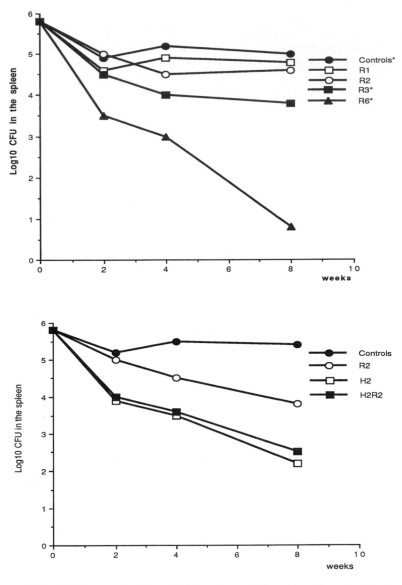

Figure 3.11. (A) Comparative bactericidal activity of 10 mg/kg rifampin (R) given at different intervals in mice infected intravenously with 1.3×10^3 *M. tuberculosis* CFU and treated 14 days later. Each point represents the mean number of CFU of 10 mice. R1, 2, 3, 6: once, twice, three times, and six times, respectively, weekly administration of R. *p < .05. (B) Comparative bactericidal activity of 10 mg/kg rifampin and 50 mg/kg isoniazid given twice a week either alone or in combination. Mice were infected intravenously with 2.4×10^3 *M. tuberculosis* CFUs and treated 14 days later. Each point represents the mean number of CFUs of 10 mice. R = rifampin; H = isoniazid.

Figure 3.11A, the reduction of the *M. tuberculosis* CFU counts resulting from 8 weeks of treatment with rifampin 10 mg/kg was more than 4 $\log_{10}$ CFU when the drug was given six times weeky, only 2 $\log_{10}$ when the drug was given three times weekly ($p < .05$), and almost nil when the drug was given twice weekly (56). It is, therefore, not surprising that in mice treated twice weekly, the killing activity of the combination isoniazid 50 mg/kg and rifampin 10 mg/kg was similar to that of isoniazid alone (Fig. 3.11B), and, in addition, only a limited percentage of mice became culture negative by the sixth month of treatment (54). Second, the reduced activity of rifampin given intermittently can be compensated by an increase in the dosage of rifampin in the mouse (54) and in the guinea pigs (4). But, in man, the intermittent administration of rifampin at dosages higher than 10 mg/kg is prohibited because of the risk of serious side effects.

An amazing property of rifampin is its apparently delayed activity when given for only a short initial period of time. That property was well demonstrated in a mouse experiment (54,57). Two randomized groups of mice were infected with 5×10^6 *M. tuberculosis* and treated 14 days later for 3 months with the combination isoniazid plus streptomycin daily. The first group of mice received only the two drugs and the second group received a supplement of daily rifampin during the first month. At the end of that month of treatment, both groups of mice, regardless of whether they had received rifampin, had almost a similar number of CFU in their lungs. But after 2 additional months during which both groups received the same treatment—isoniazid plus streptomycin—the mean number of CFU was far lower in those animals that had received rifampin than in those that had not. Similar observation was made in guinea pigs (5). As pointed out by Mitchison (58), the most likely explanation of these findings is that rifampin was specifically acting against the nonmultiplying bacillary subpopulation, which otherwise escaped the killing of isoniazid and streptomycin and persisted in the tissues.

The sterilizing activity of rifampin was also demonstrated by numerous experiments to determine the relapse rate after treatment with rifampin for different lengths of time (19). Table 3.9 gives a summary of the experiments, performed since 1969, on relapse rates after treatment with various lengths of isoniazid plus rifampin-containing regimens and with isoniazid plus streptomycin-containing regimens. After 6 months' treatment with isoniazid plus rifampin, the relapse rate is 20%, a fairly high rate but quite low in comparison with the relapse rate of 75% after an 18-month treatment with the combination isoniazid plus streptomycin. Therefore, the results of mouse experiments had suggested, as early as 1969, that isoniazid plus rifampin for 6 months might be more effective than the standard regimen of isoniazid plus streptomycin for 18 months, which was once regarded as very effective. Further experiments in mice (37,59) were performed to determine the necessary length of treatment resulting in no relapses at all. It appears that no relapse occurs after 9 months of treatment with the two-drug

Table 3.9. Percentage of mice culture positive on completion of treatment with isoniazid (INH) plus streptomycin (SM) or INH plus rifampin (RMP) and after follow-up without treatment

Drug regimen (mg/kg)	Duration of chemotherapy (months)	Percentage of mice culture positive[a]		
		On completion of treatment	After a follow-up of	
			3 months	6 months
INH (25 + SM 200)	6	100	100	
INH (25 + SM 200)	18	35	75	
INH (25 + RMP 25)	6	0		20
INH (25 + RMP 25)	9	0		0
INH (25 + RMP 25)	6			
Followed by INH alone	+3	0		20

Note: Mice infected intravenously with 5×10^6 *M. tuberculosis* CFU and treatment started 14 days later.
[a]Lungs.

Table 3.10. Reactivation of experimental tuberculosis in mice by daily cortisone administration after completion of treatment

Treatment		Duration of administration of 20 mg/kg cortisone	% Relapses
Drug	Duration		
INH + PZA	3 months	0	75
		1 month	100
INH + PZA	6 months	0	20
INH + RMP	6 months	0	20
		1 month	30
INH + RMP	9 months	0	0
		2 months	60
INH + RMP	12 months	0	0
		2 months	60

regimen isoniazid plus rifampin, with the condition that rifampin be given throughout the whole period of treatment. If rifampin is withdrawn after the sixth month and the treatment continued with isoniazid alone for an additional 3 months, the relapse rate is 20%. Thus, giving INH alone between the sixth and the ninth month does not improve the results achieved by 6 months of treatment with the two-drug regimen isoniazid plus rifampin. The results underline the sterilizing activity of rifampin and, comparatively, the poor sterilizing activity of isoniazid.

As early as in 1956, McCune et al. (36) questioned whether all negative cultures after effective long-duration chemotherapy really represent sterilization or only a latent state of infection. From experiments in mice, as shown in Table 3.10, it appears that in spite of 1 year of chemotherapy with isoniazid plus rifampin, it is possible to induce a relapse rate of 60% by giving 2 months of high-dosage cortisone after the end of treatment (59). Thus, in the mouse, very effective chemotherapy is more likely to achieve a latent state of infection than to achieve a real sterilization. Whatever the theoretical implications of these findings may have, they mean that cure of tuberculosis in the immunocompetent host does not necessarily require a real sterilization, in the mouse at any rate.

Short-Course Chemotherapy with Pyrazinamide-Containing Regimens.   As already emphasized (36), pyrazinamide given at a daily dose of 3000 mg/kg in combination with 18–20 mg/kg isoniazid exhibited a powerful sterilizing activity. Even when given at a dosage comparable to that given in humans, pyrazinamide appeared as a sterilizing drug. For example, after 2 months of treatment with either isoniazid + streptomycin or isoniazid + streptomycin + pyrazinamide 150 mg/kg daily (Table 3.11), the results of organ cultures were exactly the same whether mice had received or not pyrazinamide; but after 6 months of treatment, the cultures from mice having received pyrazinamide yielded only one or two colonies, whereas those from mice not having received pyrazinamide yielded 100 times more colonies; after 9 and 12 months, the culture-positive rate among the animals having received pyrazinamide was only 10% and 0%, respectively, but was 80% and 70%, respectively, among the animals not having received pyrazinamide (19).

Table 3.11. Percentage of culture-positive mice (lung and/or spleen) on treatment with either isoniazid (INH) plus streptomycin (SM) or INH + SM + pyrazinamide (PZA)

Months of treatment	% Mice culture positive on treatment with	
	INH + SM	INH + SM + PZA
0	100	100
2	100	100
6	100	90[a]
9	80	10[b]
12	70	0

Note: Mice were infected with 5×10^6 *M. tuberculosis* CFU and treatment was started 14 days later with INH 25 mg/kg, SM 200 mg/kg, and pyrazinamide 150 mg/kg. Each point corresponds to 10 mice.
[a]Only one or two colonies in lung and/or spleen cultures.
[b]One mouse with only one colony in spleen culture.

Short-Course Chemotherapy with Rifampin and Pyrazinamide-Containing Regimens.   Because of their common sterilizing activity against *M. tuberculosis,* rifampin and pyrazinamide drugs are now combined with isoniazid in the short-course (6-month) chemotherapy of tuberculosis. The additive effect of combining rifampin and pyrazinamide was easily demonstrated in the mouse (60). Mice infected intravenously with 3.7×10^6 CFU of *M. tuberculosis* were treated 14 days later for 6 months with either 25 mg/kg isoniazid plus 10 mg/kg rifampin or isoniazid plus rifampin plus an initial 2-month supplement of 150 mg/kg pyrazinamide, all drugs being given daily. At the end of treatment, mice were kept without treatment for an additional 6-month period and then killed to determine the organ culture-positive rates. As shown in Table 3.12, 10% of those that received the initial supplement of pyrazinamide were organ (spleen and/or lung) culture positive versus 36% of those that did not!

As in humans, the additive effect of pyrazinamide was not demonstrated when the drug was given only during the continuation phase of chemotherapy. In an experiment (60), mice were infected intravenously with 1.1×10^6 *M. tuberculosis,* and the treatment began 14 days later when the mean number of CFU was 1.1×10^7 and 6.8×10^7 in the spleens and lungs, respectively. All mice were treated daily with isoniazid 25 mg/kg plus rifampin 10 mg/kg for the first 3 months. Then, they were subdivided into three groups and treated daily for 3 additional months with either rifampin alone, or rifampin plus isoniazid, or rifampin plus isoniazid plus pyrazinamide. As measured by the culture-positive rate of lungs and spleens on completion of treatment and after 6 months of follow-up without treatment, the sterilizing activity of the three drug regimens did not differ significantly, suggesting that neither pyrazinamide nor isoniazid contributed during the continuation phase to the overall sterilizing activity of the treatment (Table 3.13).

Table 3.12. Experimental activity of pyrazinamide (PZA) in combination with isoniazid (INH) and rifampin (RMP) during the initial phase of chemotherapy

| | | After 6-month follow-up without treatment | |
| | | Culture positive | |
Drug regimen[a] for 6 months	Total mice	No.	%
INH + RMP	52	19	36.5
INH + RMP + PZA during initial 2 months	47	5	10.7

Note: Mice were infected intravenously with 3.7×10^6 *M. tuberculosis* CFU. At the start of treatment and 14 days later, there were 1×10^7 and 4.5×10^7 CFU in spleens and lungs, respectively.
[a]INH: 25 mg/kg; RMP: 10 mg/kg; PZA: 150 mg/kg.

Table 3.13. Activity of pyrazinamide and isoniazid in combination with rifampin during the continuation phase of chemotherapy

Drug regimen[a]		No. of mice culture positive		
Initial 3 month	last 3-month	On completion of treatment	after 6-month follow-up	(%)
RMP + INH	RMP	1/10[b]	35/53	(66)
RMP + INH	RMP + INH	1/10[c]	31/52	(60)
RMP + INH	RMP + INH + PZA	0/10	32/55	(58)

Note: Mice were infected intravenously with 1.1×10^6 *M. tuberculosis* CFU. At the start of treatment and 14 days later, there were 1.1×10^7 and 6.8×10^7 CFU in spleens and lungs, respectively. After 3 months treatment with RMP + INH, there were 11 CFU in the spleen and 140 in lungs.
[a]INH: 25 mg/kg; RMP: 10 mg/kg; PZA: 150 mg/kg.
[b]One mouse with 34 CFU in lung culture and 1 CFU in spleen culture.
[c]One mouse with 1 CFU in lung culture.

Although pyrazinamide enhances the sterilizing activity of the combination isoniazid–rifampin in mice as well as in humans, the response of tubercle bacilli in mice to the combination of the major drugs isoniazid, rifampin, and pyrazinamide might not be totally similar to the response in humans. Actually, in two experiments (17,24), it was demonstrated that the daily combination rifampin + pyrazinamide was more bactericidal in terms of culture-negative rate on completion of the 6-month treatment, and more sterilizing, in terms of prevention of relapses after stopping chemotherapy, than the combination isoniazid + rifampin + pyrazinamide. Such results suggested the existence of some antagonism between isoniazid and the combination rifampin + pyrazinamide.

The nature of antagonism between isoniazid and the combination rifampin + pyrazinamide in mice could be pharmacological, microbiological, or both. Pharmacokinetic analysis demonstrated a negative interaction between isoniazid and rifampin because concomitant treatment with isoniazid significantly reduced the C_{max} of RMP from 14 to 8 μg/ml and the AUC of RMP from 160 to 97 mg h/L. The findings in mice are possibly limited to that species because rifampin is given at the same dosage in mice and in humans, whereas the two other drugs, isoniazid and pyrazinamide, are given at a significantly higher dosage in mice than in man.

THE ISSUES TO BE ADDRESSED BY EXPERIMENTAL CHEMOTHERAPY IN THE NEAR FUTURE

Among all important issues in research for the therapy of tuberculosis in the 1990s, at least three might be addressed by experimental chemotherapy: (i) the improvement of short-course chemotherapy; (ii) the development of drug regimens that are active against multidrug (isoniazid plus rifampin)-resistant tuberculosis; and (iii) the preventive and curative treatment of tuberculosis in HIV-infected subjects.

Improvement of Short-Course Chemotherapy.   In comparison with the 18-month duration of standard chemotherapy of tuberculosis with isoniazid plus streptomycin and para-aminosalicylic acid (PAS), the 6-month duration of short-course chemotherapy with the three-drug combination, isoniazid, rifampin, and pyrazinamide, is a dramatic progress. However, from the viewpoint of patients and health care staff, the so-called short-course chemotherapy is still a treatment of long duration. Further shortening of the chemotherapy for tuberculosis could, in theory, be obtained from a better use of existing antituberculosis drugs or by supplementing the present drug combinations with new antimicrobial drugs or immunodulators. As none of the multiple combinations of the presently available drugs has demonstrated a better activity than the daily administration of all three drugs, an additional shortening of therapy is more likely to result from adding a new drug to the main three drugs than from a new combination of those already existing drugs. Among the newly developed antituberculosis drugs, fluoroquinolones are, at present, the most promising. Although one of the most recent and most active fluoroquinolones against *M. tuberculosis,* sparfloxacin (29), failed to enhance the bactericidal activities of the three main drug combination during the initial phase of experimental chemotherapy (21), its role remains to be tested in the continuation phase at different dosages and rhythms of administration.

Until now, no immunomodulator has proved to be effective in combination with chemotherapy. However it is conceivable that an improvement of the immune response of the infected host could speed up the clearing of viable *M. tuberculosis* during chemotherapy and prevent relapses after stopping chemotherapy. After having been infected with *M. tuberculosis* and treated with the same drug regimen, strains of mice with different genetic background relapse at a different frequency after stopping treatment, suggesting that the immune response is playing a role in the cure of tuberculosis. If the immune response of the host could be improved by the use of immunomodulators in the initial or the continuation phase of chemotherapy, one might expect the results of chemotherapy to be improved.

Chemotherapy for Multidrug-Resistant Tuberculosis.   Mycobacterium tuberculosis isolates that are resistant to multiple drugs, including rifampin and isoniazid, have become a serious problem in many parts of the world (62–63). At present, only a limited number of alternative drug regimens are available, none of them is very effective, and the mortality from multidrug-resistant tuberculosis is high (61,63). Therefore, effective new antituberculosis drugs with bactericidal mechanisms different from those of the presently available drugs are urgently needed to treat multidrug-resistant tuberculosis as well as to improve the antibacterial activity of chemotherapy against fully drug-susceptible organisms. In this respect, all new antimicrobial drugs with antituberculosis activity should be tested in combination with the "reserve" drugs to which multidrug-resistant *M. tuberculosis* are most likely to be still susceptible, such as cycloserin, ethambutol, kanamycin, and amikacin, as well as the "first-line" drug, pyrazinamide, to which the multidrug-resistant strains remain sometimes susceptible.

As an example, the antituberculosis activity of sparfloxacin has been tested in combination with "reserve" drugs in the initial phase of chemotherapy. The combination of sparfloxacin with kanamycin and pyrazinamide was less actively bactericidal than the combination of sparfloxacin with streptomycin and pyrazinamide but was as bactericidal as the combination of isoniazid, rifampin, and pyrazinamide (21).

A technical point concerning the assessment of new drugs for the treatment of multidrug-resistant tuberculosis should be emphasized. We do not recommend testing their activity directly against multidrug-resistant strains of *M. tuberculosis* in the mouse for the following two reasons: (i) because of their usually high resistance to isoniazid, multidrug-resistant strains are potentially of less virulence for the guinea pig (64) and the mouse, and, consequently, the disease that follows the infection may be difficult to standardize; (ii) it is preferable to avoid the biohazards related to the manipulation of multidrug-resistant strains of *M. tuberculosis*. Actually, the most simple and suitable mouse model for testing new drugs or drug combination active against multidrug-resistant strains is the mouse infected with the highly virulent, fully drug-susceptible *M. tuberculosis* H37Rv strain and not treated with isoniazid and rifampin!

Chemotherapy of Tuberculosis in HIV-Infected Subjects. Subjects coinfected with *M. tuberculosis* and the human immunodeficiency virus (HIV) are at much higher risk of developing clinically active tuberculosis than HIV-negative subjects infected with *M. tuberculosis* (65–67). For this reason, preventive therapy of tuberculosis in HIV-positive subjects has often been advocated. The efficacy of such therapy can be tested in the mouse model on the condition that the mouse is suffering from some immunodeficiency. Because of their lack of cell-mediated immunity, athymic nude mice (15) have been used for that purpose despite the fact that they can mimic the most severe immunodeficient status of patients with advanced acquired immunodeficiency that occurs during the natural course of HIV infection. In these mice, the response to treatment with some drug regimens (e.g., rifampin plus pyrazinamide daily or rifapentine alone given every 2 weeks) was less favorable than in normal mice. With other regimens, (e.g., isoniazid alone), the response was at least as favorable as in normal mice. Therefore, the immunologic status of the host might interfere with the response of tuberculosis infection to chemotherapy. In addition, after stopping therapy, virtually all treated nude mice relapsed, suggesting that preventive therapy for tuberculosis may have to be given on a lifelong basis to immunodeficient persons.

Until now, no long-term experiment of tuberculosis curative therapy has been undertaken in immunodeficient mice, although it would be of great interest from the public health point of view to determine the optimal length of time during which the optimal drug combination should be given to prevent relapse after stopping treatment in the immunodeficient host.

3. Experimental Chemotherapy of Disseminated Mycobacterium avium-Complex Infection in Mice

Only recently, the disseminated *M. avium*-complex (MAC) infection has become an important subject for chemotherapy research due to the epidemic of AIDS, because MAC infection is the most common disseminated bacterial infection in patients during the later stage of AIDS. For experimental chemotherapy research of MAC infection, because of its easiness of handling, especially when relatively large sample size is needed, the mouse is the animal of choice; in fact, very few other species of animal have been selected for this purpose (68).

3.1. Factors That Interfere with the Results of Experimental Chemotherapy of MAC Infection

As already described earlier in the section two on tuberculosis, results of experimental chemotherapy of MAC infection are also mediated by three factors: the immunological status of the host, the strains of MAC, and the antimicrobial activities of the drug to be tested.

THE MOUSE

To establish the experimental infection of MAC infection, different kinds of mice have been used: immunocompetent (normal) mice (69–73), thymectomized and CD4 T-cell-deficient C57Bl/6 (T $\times$ CD4$^-$) mice (74), C57Bl/6J bg^j/bg^j (beige) (75–77), and congenital athymic (nude) mice (69,73).

The MAC may multiply in normal mice (69–73), and the available results indicate that their responses to chemotherapy (70–73) are basically in agreement with those obtained from other mouse models, including the beige mouse model (77). However, unlike the disseminated MAC infection in patients with AIDS, the multiplication of MAC in normal mice is self-limited, the CFU ($\log_{10}$) per organ rarely reach 10^8 even 6 months after intravenous inoculation with 1.5×10^7 CFU per mouse (72), and only results in a chronic and nonfatal infection (69–72). Susceptibility to the infection of MAC varied among the strains of mice; based on CFU counts, the C57Bl/6 and BALB/c mice are more susceptible than other strains (69,78).

T $\times$ CD4$^-$ mice are C57Bl/6 mice that had been thymectomized at 4 weeks of age, followed 1 week later by intravenous infusion with monoclonal antibody to selectively deplete CD4$^+$ T cells (74). In terms of CFU counts, it was found that both beige and T $\times$ CD4$^-$ mice were more susceptible to the MAC infection than were immunocompetent C57Bl/6 mice (74). However, T $\times$ CD4$^-$ mice were less susceptible to MAC infection in the lungs than were beige mice (74), and it seems rather demanding to organize a moderate-scale experiment using T $\times$ CD4$^-$ mice because the manipulations to remove the thymus and to deplete CD4$^+$ T cells are laborious.

The nude mouse is susceptible to MAC infection (69,73), but, to our knowledge, its susceptibility to MAC has not been properly compared with that of the beige mouse. Therefore, whether or not the nude mouse has a role in experimental chemotherapy research of MAC infection remains unclear. The major disadvantage in using nude mice, as compared with other mouse models, is the difficulties in handling because they must be kept in a specific pathogen free (SPF) condition. The interest in using nude mice for experiments of MAC infection might be revived in the future if it turns out that the infection of the beige mouse model cannot satisfy all experimental purposes.

To date, the beige mouse is the most widely used animal model for experiments of MAC infection. It is more susceptible to MAC than is immunocompetent mouse (73,74,78), probably in relation to its deficiency in natural killer cells (79). After intravenous inoculation, MAC multiply more or less progressively in beige mice and result to a fatal infection in a significant proportion of animals. Another important feature of the beige mouse is that it can be kept in conventional conditions.

THE STRAIN OF MAC

It is well known that the virulence varied among the strains of MAC. For example, the following are claimed to be virulent strains by different investigators: *M. avium* 724 and *M. intracellulare* D673 (69), *M. intracellulare* 578-1 (75,78), and MAC strain 101 (9). However, in a study involving three strains of MAC (101, Lpr, and MO1), the mortality rate of untreated control mice did not differ significantly among animals inoculated with different strains but was correlated with the inoculum size and the duration of the experiment (77). Unlike the H37Rv strain for *M. tuberculosis,* there is no standard strain of MAC for animal experiment; nevertheless, to date, strain 101 is widely used in most laboratories.

In most laboratories, only transparent colonies are used for animal inoculation because they are known to be associated with higher virulence to experimental animals (80–82) and resistance to most antimicrobial agents and chemicals (83,84).

THE DRUGS

Different classes of antimicrobials, including almost all the antituberculosis drugs, have been tested for their activities against MAC in mice, especially in the beige mouse model (77,85–92). However, apart from clarithromycin, the list of drugs that show bactericidal effects against MAC in beige mice is rather short (77). The activities of individual drugs will be discussed in later sections.

3.2. Assessing the Efficacies of Antimicrobial Agents Against MAC in the Beige Mouse Model

BASIC METHODS

Mice. In general, 6–10-week-old beige mice, with body weights 18–20 g, are inoculated. We use female mice only, as in the experiments of *M. tuberculosis* or *M. leprae;* some investigators also use male beige mice, caged separately from the females. The number of mice required in the experiment depends on the number of groups and number of mice per group, and the latter depends on the number of sacrifices for each group. Whenever possible, at least eight mice are sacrificed at each time point per group. When calculating the required number of mice, the mortality of mice during the course of the experiment should always be taken into account. The mortality rates depend on the duration of experiment, and varies among the groups (77). The mortality rate in untreated controls and mice treated with effective regimen is estimated respectively to be 20–30% and 10% in a 4-week experiment, and increased to 50% and 15%, respectively, if the experiment lasts for 12–16 weeks; for those mice treated with regimens whose efficacy remains to be identified, their mortality should be estimated as that of untreated control mice.

Inoculation of Mice. To prepare the inoculum, we normally collect 300–400 colonies, predominantly smooth and transparent, of MAC from 10% oleic acid–albumin–dextrose–catalase (OADC)-enriched 7H11 agar medium and subculture them in 100–120 ml of Dubos broth at 37°C for 7 days. The turbidity of the resulting suspension is adjusted with normal saline to match that of a standard suspension of *M. bovis* BCG (1 mg/ml), and the suspension is further diluted five-fold for inoculation (77). The inoculum contains about 10^7 CFU/ml, as determined by plating five appropriate dilutions on 10% OADC-enriched agar medium. Each mouse is inoculated with 0.5 ml of MAC suspension via the tail vein.

Chemotherapy. Either the next day (D1) or 14 days (D14) after inoculation, 10 mice are sacrificed for establishing the baseline (pre-treatment) values of spleen weights and organ CFU counts. The remaining mice are allocated randomly to an untreated control group and various numbers of treated groups, and treatments are begun on the same day. In general, the drugs are administered six times weekly through an esophageal cannula, except that aminoglycosides are injected subcutaneously. The duration of treatment varied from 4 to 16 weeks (77,93,94), depending on the objectives of the experiment: If the purpose is only to determine the drug activity, 4 weeks of treatment is enough; whereas if the purpose is related to the selection of drug resistant mutants, treatment should be given for at least 12 weeks (93,94). To date, virtually no experiment was devoted to evaluate the sterilization effect of the treatment; with the currently available drugs (77), probably no sterilization will be observed, even though the duration of treatment was continued for 24 weeks. To avoid or reduce the carryover effects of the drugs on the number of CFU, treated mice are sacrificed 48–72 h after administration of the last dose of treatment.

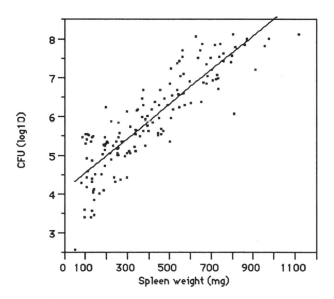

Figure 3.12. Scattergram of spleen CFU ($\log_{10}$) against spleen weight (mg) of 150 mice. After inoculation with $10^{6.70}$ CFU of MAC strain 101 per mouse, treatment was begun on D1 at a frequency of six times weekly and continued for 4 weeks. All mice were sacrificed 48 h after the last dose of treatment.

Assessment of the Results.   Mortality, spleen weights, and the number of CFU in the organs (spleens, lungs, and livers) are applied as parameters for assessing the severity of infection and the effectiveness of treatment. As shown in Figure 3.12, there is a linear correlation between the spleen weights and the spleen CFU counts; in other words, the greater the spleen weight, the higher the CFU count. The bactericidal effect of the treatment is defined as a significant decrease in the mean number of CFU in the treated group from the pretreatment value. For the experiments related to the selection of drug-resistant mutants, the CFU of drug-resistant mutants in the organs are also determined.

MAJOR FEATURES OF DISSEMINATED MAC INFECTION IN UNTREATED CONTROL MICE

Two weeks after inoculation, deaths begin to occur in untreated control mice, then their frequency gradually increases. As mentioned already, the mortality was about 10–30% by the end of 4 weeks, and more than 50% by the end of 14 weeks (77,94). The mortality data indicate that the nature of disseminated MAC infection in beige mice was chronic rather than acute, as was originally thought (75). From an experimental chemotherapy point of view, such a mortality rate is annoying because it is not high enough to easily make a distinction between an active treatment and an inactive one and it interferes significantly with the enumeration of CFU, by far the most important parameter in assessing the in vivo activity of the antimicrobial treatment.

The splenomegaly is observed the day after inoculation; the mean spleen weight increases tremendously during the initial 14 days, becoming about six to eight times greater than the normal value, slightly increases, and reaches the peak at 16 weeks (Fig. 3.13).

The CFU counts progressively increase in the spleens and lungs, often reach 10^{10} per spleen or lung at 16 weeks after inoculation, and do not increase further afterward (Fig. 3.14). The mean number of CFU increases by more than two orders of magnitude between D1 and 4 weeks and by two to three orders of magnitude between 4 and 16 weeks, suggesting that the multiplication of MAC is more rapid during the early stage of infection, slowing down after, perhaps as the result of an immune response.

DETERMINATION OF THE ANTI-MAC ACTIVITY OF A SINGLE DRUG

The anti-MAC activities of 4-week monotherapy with clarithromycin (CLARI), rifampicin (RMP), rifabutin (RBT), amikacin (AMIKA), ethambutol (EMB),

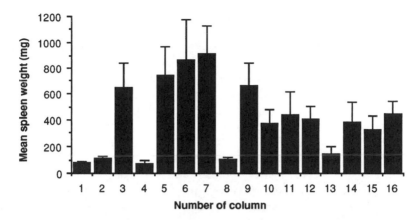

Figure 3.13. Mean spleen weights in various groups of beige mice. Treatments began 4 weeks after inoculation with *M. avium* strain 101 and continued up to the end of the 20th week. All drugs were given by gavage, except AMIKA was administered by injection, six times weekly at the following dosages: CLARI 200 mg/kg, AMIKA 100 mg/kg, RBT 10 mg/kg, and EMB 125 mg/kg.

Columns: 1—noninfected D1; 2—control D1; 3—control 2 weeks; 4—noninfected 4 weeks; 5—control 4 weeks; 6—control 11 weeks; 7—control 16 weeks; 8—noninfected 20 weeks; 9—control 20 weeks. Columns 10–16 are results from treated mice sacrificed at the end of the 20th week: 10—CLARI alone; 11—CLARI + (AMIKA 2 weeks); 12—CLARI + (AMIKA 4 weeks); 13—CLARI + (AMIKA 8 weeks); 14—CLARI + RBT; 15—CLARI + EMB; 16—CLARI + EMB + RBT.

sparfloxacin (SPFX), and clofazimine (CLO) were compared in beige mice. These compounds represent almost all the classes of antimicrobials with potential activity against MAC, and their activities were compared in a single experiment. The mean spleen weights and the, mean number of CFU in the spleens and lungs of each group of mice are presented in Table 3.14.

The results indicate that CLARI was active, as shown by the significantly lower spleen weights and the number of CFU in the organs of all three treated groups (50, 100, and 200 mg/kg) compared with those of untreated control mice sacrificed simultaneously ($p < .01$). The activity of CLARI was dose related because the spleen weights and CFU counts decreased progressively when the dosages of CLARI increased. At a daily dose of 50 mg/kg, CLARI displayed only a bacteriostatic effect, whereas at a dose of 200 mg/kg, it was bactericidal, and 4 weeks of treatment reduced the number of CFU by 1 $\log_{10}$ unit from the pretreatment values. Because CLARI at a dosage 200 mg/kg displayed similar bactericidal activity against all three strains of MAC tested in our laboratory, because the same dosage of CLARI was active against all four other strains of MAC tested by other investigators (96), and because all AIDS patients with disseminated MAC infection responded favorably to treatment with CLARI (30,31), it is reasonable to assume that if a drug was truely active against MAC, it should be expected to be active, as in the case of CLARI, against the majority of the MAC strains in beige mice as well as in patients with disseminated MAC infection.

Both RMP and RBT were inactive. AMIKA and EMB displayed modest degrees of bactericidal activity that were similar to that of CLARI at 100 mg/kg. SPFX showed a dose-related bacteriostatic effect, and the activity of SPFX 100 mg/kg was similar to that of CLARI 50 mg/kg. CLO displayed a very weak bacteriostatic effect that was similar to or weaker than that of SPFX 50 mg/kg. Although it was observed that the cultures were either negative or showed very low CFU counts in the organs from mice that had been treated with longer duration of CLO-containing regimens (77,85,86), it is concluded that the negative cultures or low CFU counts were due to the carryover of CLO in the medium (77).

DETERMINATION OF THE ANTI-MAC ACTIVITY OF COMBINED THERAPY

The activities of various multidrug regimens that did not include CLO were compared in beige mice. After 4 weeks of treatment, both the spleen weights and

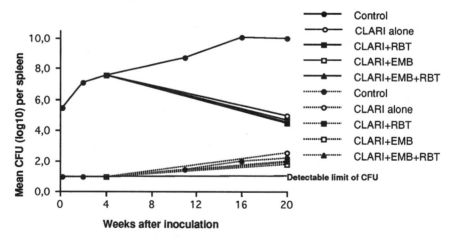

Figure 3.14. Number of CFU of total organisms and of CLARI-resistant mutants in spleens of untreated control and treated groups in which resistant mutants were isolated. Each mouse was inoculated intravenously with $10^{6.70}$ CFU of MAC strain 101. Treatment began at 4 weeks and continued up to 20 weeks. All drugs were given by gavage six times weekly at the following dosages (mg/kg): CLARI 200, RBT 10, and EMB 125. Solid lines represent the curves of total organisms; dotted lines represent the curves of CLARI-resistant mutants.

Table 3.14. Mean spleen weights and organ CFU counts in control groups and mice treated with various antimicrobials for four weeks[a]

Group [daily dose (mg/kg)]	Spleen weight (mg)	CFU ($\log_{10}$) per organ[b]	
		Spleen	Lung
Noninfected, D1	80 ± 12	ND[c]	ND
Infected and untreated (control), D1	111 ± 16	5.39 ± 0.29	4.12 ± 0.23
Control, D14	649 ± 188	7.04 ± 0.33	5.42 ± 0.87
CLARI 200 D14	137 ± 40	4.00 ± 0.52	2.75 ± 0.43
Noninfected, D28	75 ± 18	ND	ND
Control, D28	740 ± 230	7.57 ± 0.59	6.46 ± 0.62
CLARI (50)	340 ± 115	5.52 ± 0.51	3.78 ± 0.69
CLARI (100)	275 ± 102	5.17 ± 0.27	3.09 ± 0.89
CLARI (200)	137 ± 60	4.00 ± 0.54	2.48 ± 0.57
RBT (10)	748 ± 294	7.63 ± 0.37	6.15 ± 0.23
RMP (10)	782 ± 179	7.62 ± 0.24	6.51 ± 0.53
AMIKA (100)	221 ± 56	5.17 ± 0.17	2.15 ± 0.63
EMB (125)	294 ± 61	5.27 ± 0.43	3.33 ± 0.37
EMB (125) + RBT (10)	375 ± 97	5.53 ± 0.50	3.53 ± 0.48
SPFX (50)	472 ± 137	6.50 ± 0.42	4.89 ± 0.30
SPFX (100)	411 ± 189	5.89 ± 0.48	3.71 ± 0.43
CLO (20)	553 ± 102	6.52 ± 0.22	5.88 ± 0.25

[a]Treatment was begun on D1 and was continued to D28. Mice were sacrificed 48 h after the last dose of treatment. Each figure represents the mean of 10 mice.
[b]The CFU was enumerated on 10% OADC-enriched 7H11 agar medium.
[c]Not determined.

Table 3.15. Mean spleen weights and organ CFU in control groups and mice treated with various multidrug regimens 4 weeks[a]

Group	Mean spleen weight (mg)	Mean CFU ($\log_{10}$)[b] per		
		Spleen	Liver	Lung
Control (before treatment)	519 ± 86	7.93 ± 0.55	7.91 ± 0.55	6.33 ± 0.33
Control (end of treatment)	960 ± 166	8.75 ± 0.52	8.93 ± 0.37	8.63 ± 0.28
CLARI alone	736 ± 134	7.34 ± 0.90	7.42 ± 0.50	6.70 ± 0.37
CLARI + EMB + RPT	536 ± 112	6.58 ± 0.71	6.90 ± 0.25	5.14 ± 0.37
CLARI + EMB + RMP	556 ± 89	6.73 ± 0.57	6.96 ± 0.71	05.24 ± 0.19
CLARI + EMB + RBT	463 ± 102	6.87 ± 0.45	7.20 ± 0.46	5.01 ± 0.53
CLARI + EMB + SPFX	717 ± 192	7.14 ± 0.22	7.17 ± 0.23	5.49 ± 0.33

[a]Mice were sacrificed 48 h after administration of the last dose of treatment. Each value represents the mean of 7–10 mice.
[b]The CFU were enumerated on 5% OADC-enriched Mueller–Hinton agar medium.

the CFU counts of the treated groups were significantly smaller than those for the corresponding control mice (Table 3.15), indicating that all the treatments were active. The activity of the combination CLARI–EMB–SPFX did not differ significantly from or was only slightly better than that of CLARI alone; the activities of the three combinations consisting of CLARI + EMB, + RMP, or RBT, or RPT were virtually the same and were slightly greater than that of CLARI alone. Therefore, the combination of EMB with either SPFX, RMP, RBT, or RPT did not or only marginally increased the activity of CLARI, indicating that the in vitro synergistic effects of the triple-drug combinations containing both CLARI and EMB (95) were not confirmed in beige mice.

In other experiments, the activities of 16 weeks of treatment with various CLARI-containing multidrug regimens were compared with CLARI alone. Treatments began 4 weeks after inoculation with MAC strain 101 and were continued up to the end of the 20th week. As expected, by the end of treatment, the mean number of CFU in every treated group was significantly smaller ($p < .01$) than the pretreatment values (Table 3.16); thus, all the regimens studied displayed a certain degree of bactericidal activity against the MAC strain. Among the groups treated with CLARI-containing combined regimens, only those treated with 16 weeks of CLARI plus an initial 4 weeks of AMIKA, CLARI–AMIKA (4 weeks), or 16 weeks of CLARI plus an initial 8 weeks of AMIKA, CLARI-AMIKA (8 weeks), yielded mean CFU counts in both the spleens and lungs less than those

Table 3.16. Mean CFU counts in the spleens and lungs of beige mice

Time after inoculation	Group	n	Mean CFU ($\log_{10}$) per Spleen	Lung
D1	Control	10	5.39 ± 0.29	4.12 ± 0.23
2 weeks	Control	10	7.04 ± 0.33	5.42 ± 0.87
4 weeks	Control	10	7.57 ± 0.59	6.46 ± 0.62
11 weeks	Control	10	8.71 ± 1.05	8.42 ± 0.63
16 weeks	Control	10	10.09 ± 0.62	9.60 ± 0.40
20 weeks	Control	22	10.01 ± 0.75	9.76 ± 0.70
20 weeks	CLARI alone	8	4.92 ± 0.48	4.55 ± 0.58
	CLARI + (AMIKA 2 weeks)	10	$4.21 \pm 1.08 \ [0.105]^a$	$3.59 \pm 1.12 \ [0.044]$
	CLARI + (AMIKA 4 weeks)	10	$3.43 \pm 0.55 \ [<0.01]$	$3.38 \pm 0.75 \ [<0.01]$
	CLARI + (AMIKA 8 weeks)	9	$2.59 \pm 0.67 \ [<0.01]$	$3.26 \pm 0.42 \ [<0.01]$
	CLARI + RBT	10	$4.69 \pm 0.47 \ [0.311]$	$4.15 \pm 1.05 \ [0.356]$
	CLARI + EMB	9	$4.43 \pm 0.77 \ [0.141]$	$3.64 \pm 1.05 \ [0.047]$
	CLARI + EMB + RBT	9	$4.50 \pm 0.40 \ [0.07]$	$3.42 \pm 0.87 \ [<0.01]$

^{a}Value in brackets are p-values in comparison with the counts from mice treated with CLARI alone.

among the mice treated with CLARI alone. In addition, among the groups treated with CLARI plus various durations of AMIKA, the mean number of CFU were inversely related to the duration of initial treatment with AMIKA. These results indicate that among the regimens tested, only the combination CLARI–AMIKA (4 weeks) and CLARI–AMIKA (8 weeks) displayed significantly greater bactericidal activity in both spleens and lungs than CLARI alone; in other words, even CLARI in combination with an initial 4 weeks of AMIKA made a difference in the bactericidal activity of CLARI.

In separate experiments, the spleen weights and CFU counts in mice treated with 4 weeks of EMB–RBT did not differ significantly from those of mice treated with EMB alone (Table 3.14); even though the treatment continued for 12 weeks, the combination merely displayed a bacteriostatic effect (77). Therefore, the combination EMB–RBT did not show a synergistic effect in the beige mouse model.

SELECTION OF DRUG-RESISTANT MUTANTS AND PREVENTING THE SELECTION BY EFFECTIVE MULTIDRUG THERAPY

Although CLARI is by far the most important antimicrobial agent for the treatment of disseminated MAC infection in AIDS patients, CLARI-resistance developed rapid in patients (97,98). To reproduce, experimentally, the emergence of drug-resistance during treatment of MAC infection, we have successfully selected the CLARI-resistant mutants and also prevented the selection by effective multidrug regimens in beige mice (77,93,94). The CLARI-resistant mutants are defined as those who multiplied on 10% OADC-enriched agar (7H11 or Mueller–Hinton) medium containing CLARI at concentration four times of the MIC.

Our repeated experiments demonstrated that the frequency or the proportion of CLARI-resistant mutants in an untreated bacterial population (i.e., the ratio between the number of CFU of CLARI-resistant organisms and the total CFU) is basically constant, about 1 per 10^8 (77,93,94). To increase the opportunity in selecting the CLARI-resistant mutants, treatments should begin when the CFU was far greater than 10^7 per spleen, normally 4 weeks after intravenous inoculation with 10^6–10^7 CFU/mouse (77,93).

In untreated control mice, a small number of CLARI-resistant mutants began to be isolated from some of the spleens at the end of 6 weeks (77) and were isolated from the great majority of spleens around 10 weeks when the total CFU was greater than 10^8 per spleen (77,94). Although the number of resistant mutants was slightly increased afterward, the proportion of resistant mutants remained constant (Fig. 3.14). Preliminary results indicate that the mutants resistant to drugs other than CLARI may also be isolated from untreated control mice when the bacterial population has reached a certain size.

In mice treated with CLARI 200 mg/kg alone daily, in spite of a significant decrease of the total CFU in the spleens, a small number of CLARI-resistant mutants were isolated from majority of the spleens after 4 weeks of treatment and were isolated from virtually all CLARI-treated mice after 8 weeks of treatment (95). Besides the progressively increase of the number of resistant mutants, the proportion of resistant mutants also increased significantly, which may be as high as 1 per 10^1 to 1 per 10^2 at the end of 12–16 weeks of treatment, indicating that the mutants were selected during the course of monotherapy with CLARI. It seems mutants resistant to other drugs are also likely to be selected by a similar mechanism, as long as the monotherapy displays enough bactericidal effect.

CLARI-resistant mutants were also isolated from mice treated with 12 weeks of MINOCYCLINE (CLARI–MINO) (10), or 16 weeks of CLARI–EMB, CLARI–RBT, or CLARI–EMB–RBT (94) (Fig. 3.14). The CFU counts and the proportion of CLARI-resistant mutants isolated from these groups did not differ significantly from those of mice treated with CLARI alone, indicating that MINO, EMB, and RBT given alone or in combination could not prevent the selection of mutants resistant to CLARI. The activities of 12–16 weeks of treatment with MINO, EMB, RBT, or EMB–RBT were so weak that they were unable to contain the multiplication of the tiny number of CLARI-resistant mutants, estimated to be less than $10^{0.94}$ CFU per spleen at the beginning of treatment (77,94).

However, no CLARI-resistant mutants were isolated from mice that had been treated with 12 weeks of CLARI–AMIKA (77). We further demonstrated that multiplication of CLARI-resistant mutants was significantly reduced in mice treated with 16 weeks of CLARI plus an initial 2 weeks of AMIKA [CLARI–AMIKA (2 weeks)], and their selection was prevented by treatment with CLARI–AMIKA (4 weeks) or CLARI–AMIKA (8 weeks) (94), suggesting that 2 weeks of AMIKA killed the majority of CLARI-resistant mutants and 4–8 weeks of AMIKA completely killed the mutants. This is the first demonstration that a companion drug has successfully reduced or prevented the selection of CLARI-resistant mutants. However, because the concentration of MAC in AIDS patients with disseminated MAC infection may be hundred times greater than that in beige mice at the time when the treatment was begun (94), the number of CLARI-resistant mutants in AIDS patients might be more than two orders of magnitude greater than that in the beige mice. Therefore, with respect to the optimal duration of treatment with AMIKA, one should be extremely cautious in extrapolating the results from beige mice to AIDS patients.

In summary, only clarithromycin, azithromycin (90,92), and amikacin displayed in the beige mouse model definite bactericidal activities against MAC; ethambutol showed marginal bactericidal effect; sparfloxacin and clofazimine displayed weak bacteriostatic activities; and rifamycin derivatives (rifampicin, rifabutine, rifapentine, and possibly others) were virtually inactive. The in vitro synergistic effects against MAC which have frequently been observed in various ethambutol-containing combined regimens have not been confirmed in beige mice. Clarithromycin-resistant mutants of MAC have been selected in beige mice during the course of monotherapy with clarithromycin. Among the combined regimens, only treatment with clarithromycin combined with amikacin could prevent the selection of clarithromycin-resistant mutants. Because the bactericidal activity of clarithromycin, azithromycin, and amikacin are modest to moderate, it is unlikely that combinations of these drugs would be capable of sterilizing the disseminated MAC infection in AIDS patients. Therefore, additional drugs, especially with bactericidal mechanisms entirely different from those of available agents, are needed; and the most realistic and economic approach is to continue the screening of anti-MAC activities among antimicrobial agents that display potent activities against gram-positive microorganisms.

References

1. O'Grady F, Riley RL (1963) Experimental airborne tuberculosis. Adv Tuberc Res 12:150–190.

2. Mitchison DA, Wallace JG, Bhatia AL, Selkon JB, Subbaiah TV, Lancaster MC (1960) A comparison of the virulence in guinea-pigs of South Indian and British tubercle bacilli. Tubercle 41:1–22.

3. Smith DW (1985) Protective effect of BCG in experimental tuberculosis. Adv Tuberc Res 22:1–99.

4. Dickinson JM, Mitchison DA (1970) Suitability of rifampicin for intermittent administration in the treatment of tuberculosis. Tubercle 51:82–94.

5. Dickinson JM, Mitchison DA (1976) Bactericidal activity *in vitro* and in the guinea-pig of isoniazid, rifampicin and ethambutol. Tubercle 57:251–258.

6. Dannenberg AM Jr. (1994) Rabbit model of tuberculosis. In: Bloom BR, ed. Tuberculosis: Pathogenesis, Protection, and Control, pp. 149–156. Washington, DC: American Society for Microbiology.

7. Good RC (1994) Diseases in nonhuman primates. In: Kubica GP, Wayne LG, eds. The Mycobacteria: A Source Book, pp. 903–924. New York: Marcel Dekker, Inc.

8. Lurie MB (1964) Resistance to Tuberculosis: Experimental Studies in Native and Acquired Defensive Mechanisms. Cambridge, MA: Harvard University Press.

9. Raleigh GW, Youmans GP (1948) The use of mice in experimental chemotherapy of tuberculosis. I. Rationale and review of literature. J Infect Dis 82:197–204.

10. Pierce SH, Dubos RJ, Schaefer WB (1953) Multiplication and survival of tubercle bacilli in the organs of mice. J Exp Med 97:189–205.

11. Youmans GP, Youmans AS (1964) Experimental chemotherapy of tuberculosis and other mycobacterial infections. In: Schnitzer RJ, Hawking F, eds. Experimental Chemotherapy. Vol. II, pp. 393–499. New York: Academic Press.

12. Grumbach F (1965) Etudes chimiothérapiques sur la tuberculose avancée de la souris. Adv Tuberc Res 14:31–96.

13. Lefford MJ (1984) Diseases in mice and rats. In: Kubica GP, Wayne LG, eds. The Mycobacteria: A Source Book, pp. 947–977. New York: Marcel Dekker, Inc.

14. Lecoeur HF, Lagrange Ph, Truffot-Pernot Ch, Gheorghiu M, Grosset J (1989) Relapses after stopping chemotherapy for experimental tuberculosis in genetically resistant and susceptible strains of mice. Clin Exp Immunol 76:458–462.

15. Chapuis L, Ji B, Truffot-Pernot C, O'Brien RJ, Raviglione MC, Grosset JH (1994) Preventive therapy of tuberculosis with rifapentine in immunocompetent and nude mice. Am J Respir Unit Care Med 150:1355–1362.

16. Fenner F (1951) The enumeration of viable tubercle bacilli by surface plate counts. Am Rev Tuberc 64:353–380.

17. Grosset J, Truffot-Pernot C, Lacroix C, Ji B (1992) Antagonism between isoniazid and the combination pyrazinamide–rifampin against tuberculosis infection in mice. Antimicrob Agents Chemother 36:548–551.

18. Truffot-Pernot C, Ji B, Grosset J (1991) Activities of pefloxacin and ofloxacin against Mycobacteria: *in vivo* and mouse experiments. Tubercle 72:57–64.

19. Grosset J (1978) The sterilizing value of rifampicin and pyrazinamide in experimental short-course chemotherapy. Bull Int Union Tuberc 53:5–12.

20. Dickinson JM, Mitchison DA (1991) Efficacy of intermittent pyrazinamide in experimental murine tuberculosis. Tubercle 72:110–114.

21. Lalande V, Truffot-Pernot C, Paccaly-Moulin A, Grosset J, Ji B (1993) Powerful bactericidal activity of sparfloxacin (AT-4140) against *M. tuberculosis* in mice. Antimicrob Agents Chemother 37:407–413.

22. Dubos RJ, Middlebrook G (1947) Media for tubercle bacilli. Am Rev Tuberc 56:334–345.

23. Orme IM, Collins FM (1994) Mouse model of tuberculosis. In: Bloom BR, ed., Tuberculosis: Pathogenesis, Protection, and Control, pp. 113–134. Washington, DC: American Society for Microbiology.

24. Lecoeur HF, Truffot-Pernot C, Grosset JH (1989) Experimental short-course preventive therapy of tuberculosis with rifampin and pyrazinamide. Am Rev Respir Dis 140:1189–1193.

25. Rowland M, Tozer TN (1980) Clinical pharmacokinetics, concepts and applications. Philadelphia: Lea & Febriger.

26. Hooper DC, Wolfson JS (1985) The fluoroquinolones: pharmacology, clinical uses, and toxicities. Antimicrob Agents Chemother 28:716–721.

27. Ji B, Truffot-Pernot C, Lacroix C, Raviglione MC, O'Brien RJ, Olliaro P, Roscigno G, Grosset J (1993) Effectiveness of rifampin, rifabutin, and rifapentine for preventive therapy of tuberculosis in mice. Am Rev Respir Dis 148:1541–1546.

28. Barkley WE, Kubica GP (1994) Biological safety in the experimental tuberculosis laboratory. In: Bloom BR, ed. Tuberculosis: Pathogenesis, Protection and Control, pp. 61–71. Washington, DC: American Society for Microbiology.

29. Ji B, Truffot-Pernot C, Grosset J (1991) *In vitro* and *in vivo* activities of sparfoxacin (AT-4140) against *Mycobacterium tuberculosis*. Tubercle 72:181–186.

30. Grosset J (1980) Bacteriologic basis of short-course chemotherapy for tuberculosis Clinics. Chest Med 1:231–141.

31. Canetti G, Grumbach F, Grosset J (1960) Studies of bacillary populations in experimental tuberculosis of mice treated by isoniazid. Am Rev Respir Dis 82:295–313.

32. Flor SC, Rogge MC, Chow AT (1993) Bioequivalence of oral and intravenous ofloxacin after multiple-dose administration to healthy male volunteers. Antimicrob Agents Chemother 37:1468–1472.

33. Israel D, Gillum JG, Turik M, Harvey K, Ford J, Dalton H, Towle M, Echols R, Heller A, Polk R (1993) Pharmacokinetics and serum bactericidal titers of ciprofloxacin and ofloxacin following multiple oral doses in healthy volunteers. Antimicrob Agents chemother 37:2193–2199.

34. McDermott W (1958) Microbial persistence. Yale J Biol Med 30:257–291.

35. McCune RM Jr, Tompsett R (1956) Fate of *Mycobacterium tuberculosis* in mouse tissues as determined by the microbial enumeration technique. I. The persistance of drug susceptible tubercle bacilli in the tissues despite prolonged antimicrobial therapy. J Exp Med 104:737–762.

36. McCune RM Jr, Tompsett R, McDermott W. Fate of *Mycobacterium tuberculosis* in mouse tissues as determined by the microbial enumeration technique. II. The conversion of tuberculosis infection to the latent state by the administration of PZA and a companion drug. J Exp Med 104:763–802.

37. Grosset J, Grumbach F, Rist N (1978) Le rôle de la rifampicine dans la phase ultime du traitement de la tuberculose murine expérimentale. Rev Fr Mal Respir 6:515–520.

38. Medical Research Council Investigation (1948) Streptomycin treatment of pulmonary tuberculosis. Br Med J 2:769–782.

39. Medical Research Council Investigation (1950) Treatment of pulmonary tuberculosis with streptomycin and para-amino-salicylic acid. Br Med J 2:1073–1086.

40. Schatz A, Bugie E, Wasksman SA (1994) Streptomycin, a substance exhibiting antibiotic activity against Gram-positive and Gram-negative bacteria. Proc Soc Exp Biol Med 55:66–69.

41. Lehmann J (1946) Para-aminosalicylic acid in the treatment of tuberculosis. Lancet I:15–16.

42. Bernstein J, Lott WA, Steinberg BA, Yale HL (1952) Chemotherapy of experimental tuberculosis. V. Isonicotinic acid hydrazide (Nidrazid) and related compounds. Am Rev Tuberc 65:357–364.

43. Grünberg E, Schnitzer RJ (1952) Studies on the activity of hydrazine derivatives of isonicotinic acid in experimental tuberculosis of mice. Quart Bull Seaview Hosp 13:3–11.

44. Füresz S, Arioli V, Pallanza R (1965) Antimicrobial properties of new derivatives of rifamycin SV. Antimicrob Agents Chemother 4:770–774.

45. Grumbach F, Canetti G, Grosset J, Le Lirzin M (1967) Late results of long-term,

intermittent chemotherapy of advanced murine tuberculosis: Limits of the murine model. Tubercle 48:11–26.

46. Medical Research Council/Tuberculosis Chemotherapy Trials Committee (1962) Long-term chemotherapy in the treatment of chronic pulmonary tuberculosis with cavitation. Tubercle 43:201–267.

47. Canetti G, Grumbach F, Grosset J (1963) Long-term, two-stage chemotherapy of advanced experimental murine tuberculosis with intermittent regimes during the second stage. Tubercle 44:236–240.

48. Crofton J (1963) Etude comparative des résultats obtenus par l'emploi de la streptomycine quotidiennement, et par son emploi discontinu. Bull Union Int Tuberc 33:51–61.

49. Grumbach F (1962) Treatment of experimental murine tuberculosis with different combinations of isoniazid–streptomycin followed by isoniazid alone. Am Rev Respir Dis 86:211–215.

50. Grumbach F, Rist N (1967) Activité antituberculeuse expérimentale de la rifampicine, dérivé de la rifamycine SV. Rev Tuberc Pneum 31:749–762.

51. Batten J (1970) Rifampicin in the treatment of experimental tuberculosis in mice: sterilization of tubercle bacilli in the tissues. Tubercle 51:95–99.

52. Havel A, Adamek L, Simonova S (1970) Rifampicin in experimental investigations on mice. Antibiot Chemother 16:406–415.

53. Verbist L (1969) Rifampicin activity in vitro and in established tuberculosis in mice. Acta Tuberc Pneumol Belg 3–4:397–412.

54. Grumbach F, Canetti G, Le Lirzin M (1969) Rifampicin in daily and intermittent treatment of experimental murine tuberculosis, with emphasis on late results. Tubercle 50:380–393.

55. Grumbach F (1969) Experimental in vivo studies of new antituberculosis drugs: capreomycin, ethambutol, rifampicin. Tubercle 50(March Suppl): 12–21.

56. Grosset J, Truffot-Pernot Ch, Lecoeur H, Guelpa-Lauras CC (1983) Activité de la rifampicine administrée quotidiennement et d'une manière intermittente sur la tuberculose expérimentale de la souris. Pathol Biol 31:446–450.

57. Canetti G, Le Lirzin M, Porven G, Rist N, Grumbach F (1968) Some comparative aspects of rifampicin and isoniazid. Tubercle 49:367–376.

58. Mitchison DA (1985) The action of antituberculosis drugs in short-course chemotherapy. Tubercle 66:219–225.

59. Grumbach F (1975) La durée optimale de l'antibiothérapie par l'association isoniazide + rifampicine dans la tuberculose experimentale de la souris. Etude de la phase post-thérapeutique. Epreuve de la cortisone. Rev Fr Mal Respir 3:625–634.

60. Grosset J, Truffot Ch, Fermanian S, Lecoeur H (1982) Activité stérilisante des différents antibiotiques dans la tuberculose expérimentale de la souris. Pathol Biol 30:444–448.

61. World Health Organization Working Group (1991) Tuberculosis Research and Development. WHO/TB/91-162. Geneva: World Health Organization.

62. Culliton BJ (1993) Drug-resistant TB may bring epidemic. Nature (London) 356:473.

63. Frieden TR, Sterling T, Pablos-Mendez A, Kilburn JO, Cauthen GM, Dooley SW (1993) The emergence of drug-resistant tuberculosis in New York City. N Engl J Med 328:521–526.

64. Cohn ML, Kovitz C, Oda U, Middlebrook G (1954) Studies on isoniazid and tubercle bacilli. II. The growth requirements, catalase activities, and pathogenic properties of isoniazid resistant mutants. Am Rev Tuberc 70:641–664.

65. Selwyn PA, Hartel D, Lewis VA, Schoenbaum EE, Vermund SH, Klein RS, Walker AT, Friedland GH (1989) A prospective study of the risk of tuberculosis among intravenous drug users with human immunodeficiency virus infection. N Engl J Med 320:545–550.

66. Barnes PF, Bloch AB, Davidson PT, Snider De Jr (1991) Tuberculosis in patients with human immunodeficiency virus infection. N Engl J Med 24:1644–1650.

67. Narain JP, Raviglione MC, Kochi A (1992) HIV-associated tuberculosis in developing countries: epidemiology and strategies for prevention. Tuberc Lung Dis 73:311–321.

68. Yangco BG, Lackman-Smith C, Espinoza CG, Solomon DA, and Deresinski SC (1989) The hamster model of chronic *Mycobacterium avium* complex infection. J Infect Dis 159:556–561.

69. Collins FM, Stokes RW (1987) *Mycobacterium avium*-complex infections in normal and immunodeficient mice. Tubercle 68:127–136.

70. Kuze F (1984) Experimental chemotherapy in chronic *Mycobacterium avium–intracellulare* infection of mice. Am Rev Respir Dis 129:453–459.

71. Furney SK, Skinner PS, Farrer J, Orme IM (1995) Activities of rifabutin, clarithromycin, and ethambutol against two virulent strains of *Mycobacterium avium* in a mouse model. Actimicrob Agents Chemother 39:786–789.

72. Cohen Y, Perronne C, Lazard T, Truffot-Pernot C, Grosset J, Vilde JL, Pocidalo JJ (1995) Use of normal C57BL/6 mice with established *Mycobacterium avium* infection as an alternative model for evaluation of antibiotic activity. Antimicrob Agents Chemother 39:735–738.

73. Ji B, Lounis N, Truffot-Pernot C, Grosset J (1991) Susceptibility of the immunocompetent, beige and nude mice to *Mycobacterium avium* complex infection and response to clarithromycin. Program Abstracts 31st Interscience Conference on Antimicrobial Agent Chemotherapy.

74. Furney SK, Roberts AD, Orme IM (1990) Effect of rifabutin on disseminated *Mycobacterium avium* infection ib thymectomized CD4 T-cell-deficient mice. Antimicrob Agents Chemother 34:1629–1632.

75. Gangadharam PRJ, Edwards CK, Murthy PS, Pratt PF (1983) An acute infection model for *Mycobacterium intracellulare* disease using beige mice. Preliminary results. Am Rev Respir Dis 127:468–469.

76. Betram MA, Inderlied CB, Yadegar S, Kolanoski P, Yamada JP, Young LS (1986) Confirmation of the beige mouse model for study of disseminated infection with *Mycobacterium avium* complex. J Infect Dis 154:194–195.

77. Ji B, Lounis N, Truffot-Pernot C, Grosset J (1994) Effectiveness of various antimicrobial agents against *Mycobacterium avium* complex in the beige mouse model. Antimicrob Agents Chemother 38:2521–2529.

78. Gangadharam PRJ (1986) Animal models for nontuberculosis mycobacterial diseases. In: Zak O, Sande MA, eds. Experimental Models in Antimicrobial Chemotherapy. Volume 3, pp. 1–94. London: Academic Press.

79. Roder JC (1979) The beige mutation in the mouse. J Immunol 123:2168–2173.

80. Winn WA, Petroff SA (1933) Biological studies of the tubercle bacilli. II. A new conception of the pathology of experimental avium tuberculosis with special reference to the disease produced by dissociated variants. J Exp Med 57:239–270.

81. Moehring JM, Solotorowsky MR (1965) Relationship of colonial morphology to virulence to chickens of *Mycobacterium avium* and the nonphotochromogens. Am Rev Respir Dis 92:704–713.

82. Schaefer WB, Davis CL, Cohn ML (1970) Pathogenicity of transparent, opaque, and rough variants of *Mycobacterium avium* in chickens and mice. Am Rev Respir Dis 102:499–506.

83. Woodley CL, David HL (1976) Effect of temperature on the rate of the transparent to opaque colony type transition in *Mycobacterium avium*. Antimicrob Agents Chemother 9:113–119.

84. Saito H, Tomioka H (1988) Susceptibilities of transparent, opaque, and rough colonial variants of *Mycobacterium avium* complex to various fatty acids. Antimicrob Agents Chemother 32:400–402.

85. Gangadharam PRJ, Perumal VX, Jairam BT, Rao PN, Nguyen AK, Farhi DC, Iseman MD (1987) Activity of rifabutin alone or in combination with clofazimine or ethambutol or both against acute and chronic experimental *Mycobacterium intracellulare* infections. Am Rev Respir Dis 136:329–333.

86. Gangadharam PRJ, Perumal VX, Podapati NR, Kesavalu L, Iseman MD (1988) In vivo activity of amikacin alone or in combination with clofazimine or rifabutin or both against acute experimental *Mycobacterium avium* complex infection in beige mice. Antimicrob Agents Chemother 32:1400–1403.

87. Inderlied CB, Kolonoski PT, Wu M, Young LS (1989) Amikacin, ciprofloxacin, and imipenem treatment for disseminated *Mycobacterium avium* complex infection of beige mice. Antimicrob Agents Chemother 33:176–180.

88. Fernandes PR, Hardy DJ, McDaniel D, Hanson CW, Swanson RN (1989) In vitro and in vivo activities of clarithromycin against *Mycobacterium avium*. Antimicrob Agents Chemother 33:1531–1534.

89. Klemens SP, Cynamon MH (1991) In vivo activities of newer rifamycin analogs against *Mycobacterium avium* infection. Antimicrob Agents Chemother 35:2026–2030.

90. Cynamon MH, Klemens SP (1992) Activity of azithromycin against *Mycobacterium avium* infection in beige mice. Antimicrob Agents Chemother 36:1611–1613.

91. Klemens SP, Cynamon MH (1992) Activity of rifapentine against *Mycobacterium avium* infection in beige mice. J Antimicrob Chemother 29:555–561.

92. Kolonoski PT, Wu M, Inderlied CB, Young LS (1992) Azithromycin (AZM), ethambutol (EMB) and sparfloxacin (SP) for disseminated *M. avium* complex (MAC) infection in C57 beige mice. Program Abstracts 32nd Interscience Conference on Antimicrobial Agents Chemotherapy.

93. Ji B, Lounis N, Truffot-Pernot C, Grosset J (1992) Selection of resistant mutants of *Mycobacterium avium* in beige mice by clarithromycin monotherapy. Antimicrob Agents Chemother 36:2839–2840.

94. Lounis N, Ji B, Truffot-Pernot C, Grosset J (1995) Selection of clarithromycin-resistant *Mycobacterium avium* complex during combined therapy using the beige mouse model. Antimicrob Agents Chemother 39:608–612.

95. Ji B, Lounis N, Truffot-Pernot C, Grosset J. In vitro activities of clarithromycin-containing double- or triple-drug combinations against Mycobacterium avium complex. Program Abstracts First International Conference on Macrolides, Azalides and Streptogramins.

96. Klemens S, DeStefano MS, Cynamon MH (1992) Activity of clarithromycin against *Mycobacterium avium* complex infection in beige mice. Antimicrob Agents Chemother 36:2413–2417.

97. Dautzenberg B, Truffot C, Legris S, Meyohas M, Berlie HC, Mercat A, Chevret S, Grosset J (1991) Activity of clarithromycin against *Mycobacterium avium* infection in patients with the acquired immune deficiency syndrome. Am Rev Respir Dis 144:564–569.

98. Dautzenberg B, Saintmarc T, Meyohas MC, Eliaszewitch M, Haniez F, Rogues AM, Dewit S, Cotte L, Chauvin JP, Grosset J (1993) Clarithromycin and other antimicrobial agents in the treatment of disseminated *Mycobacterium avium* infections in patients with acquired immunodeficiency syndrome. Arch Intern Med 153:368–372.

4

Controlled Clinical Trials in Tuberculosis: Lessons to Be Drawn

S. Radhakrishna

Any drug combination has to be assessed ultimately in humans, irrespective of how successful it is in the test tube or in experimental animals. In other words, it must be subjected to a clinical trial of some sort. Only rarely does the investigator have in his possession a wonder drug such as a penicillin or an insulin, the efficacy of which is so obvious that it needs little rigorous investigation. More often than not, he is in the position of having to determine whether a new drug or regimen is indeed better (i.e., has greater efficacy or produces fewer side effects) than the standard, and the magnitude of the expected difference is not dramatically large. It is such situations that the randomized controlled trial (RCT) is designed to meet. One of the earliest, and still perhaps the most lucid, expositions of the controlled clinical trial was made by Bradford Hill (1). The ingredients of a good clinical trial include the following: a detailed description of the treatment regimen (drugs, dosage, rhythm, duration); a clear statement of the objectives (e.g., efficacy, toxicity, acceptability), together with the order of priorities; precise specification of the type of patients to be admitted (i.e., eligibility criteria, contraindications); the use of concurrent controls obtained by a process of random allocation; explicitly described procedures for ensuring similarity in the subsequent management of patients in the different treatment groups; and objective evaluation of the outcome measures, supported by tests of statistical significance.

Over the years, RCT has been widely used to investigate treatment regimens in both communicable and noncommunicable diseases. Also, the underlying principles and methodology have been employed to answer important questions pertaining to operational aspects of disease control programs and to evaluate the relative merits of different program policies. In the field of tuberculosis, the tool has been very extensively employed, leading not only to better treatment regimens in both developed and developing countries but also resulting in great advances in the principles of treatment of the disease (2). In the process, several valuable lessons have been drawn and these will now be highlighted,* mainly with examples from pulmonary tuberculosis with drug-susceptible organisms.

1. Tuberculosis Can Be Treated with Drugs

In the prechemotherapy era, tuberculosis used to be treated by providing patients with bed rest, nutritious food, and a salubrious environment, and by employing some procedures such as artificial pneumothorax or pneumoperitoneum. The discovery of streptomycin in 1944, para-aminosalicylic acid (PAS) in 1946, and isoniazid in 1952 changed the scenario and, through the medium of the controlled clinical trial, it was firmly established that tuberculosis can be successfully treated with drugs. The first trial was conducted in the United Kingdom by the British Medical Research Council (MRC) in 1947, and it evaluated the efficacy of streptomycin (2 g per day, given in four injections at 6-h intervals) for 4 months in patients under bed rest; an untreated group of patients (obtained by a process of random allocation of all eligible cases) on bed rest only constituted the control, a practice that was justified with the argument that streptomycin was the only drug available at the time and was in short supply. The study report (3) stated in the preamble that "the natural course of pulmonary tuberculosis is in fact so variable and unpredictable that evidence of improvement or cure following the use of a new drug in a few cases cannot be accepted as proof of the effect of that drug" This was a profoundly true statement, considering the fact that several unjustified and erroneous claims had been made on the basis of empirical (uncontrolled) evaluation of treatment modalities (e.g., gold treatment, blood-letting).

The main findings of the MRC Trial (3) are summarized in Table 4.1. By 6 months, only 7% of the streptomycin patients had died as compared with 27% of the control patients; these proportions increased to 22% and 46%, respectively, by the end of the year, but the gulf between the two series persisted and was statistically highly significant ($p = .01$). Moreover, among the survivors, 55% of the streptomycin patients as compared with only 11% of the control patients showed considerable x-ray progress ($p < .001$). Finally, culture negativity occurred significantly more often in the treated patients, the proportions being 19% and 2% in the two series at 3 months and 15% and 4% at 6 months. Thus, the trial, the first controlled trial of its kind to be reported, provided a very clear-cut

---

*The notations employed in the tables and text in this chapter are S for Streptomycin, P for para-aminosalicylic acid (PAS), H for isoniazid, M for matrix isoniazid (slow-release preparation), T for thioacetazone, R for rifampicin, Z for pyrazinamide, and E for ethambutol. The number entered before the drug(s) indicates the duration in months, and the subscript below the drug indicates the number of days in a week for which it is prescribed, if it is other than 7 (i.e., daily).

Table 4.1. First controlled trial of the efficacy of drug therapy in tuberculosis

	Streptomycin (4 months)	Control	Difference
Death by 6 months	7% of 55	27% of 52	20%
Among survivors			
Considerable X-ray progress	55% of 51	11% of 38	44%
Any x-ray deterioration	22% of 51	47% of 38	25%
Culture negative			
At 3 months	19% of 54	2% of 50	17%
At 6 months	15% of 54	4% of 50	11%
Death by 12 months	22% of 55	46% of 52	24%

Source: Ref. 3.

answer in the affirmative regarding the early efficacy of streptomycin in the treatment of tuberculosis and became a trendsetter for similar controlled trials in the future. A valuable lesson had been learned, not only regarding the efficacy of an antituberculosis drug but also about the scope and utility of the controlled clinical trial approach.

2. Addition of a Second Drug Enhances the Efficacy of Monotherapy

Although streptomycin prevented a large proportion of the deaths (over 70%) in the first 6 months and substantially enhanced the proportion of patients who showed considerable x-ray progress (Table 4.1), the results were not as satisfactory in terms of culture negativity, with only 15–20% of the patients having negative cultures by 3–6 months. The major reason for this was the early development of resistance to streptomycin; thus, 85% of the cultures tested showed streptomycin resistance in vitro (3). The addition of a second antituberculosis drug, PAS, improved the efficacy substantially, as was demonstrated in a controlled clinical trial in the United Kingdom (4), which also showed that the PAS was largely successful in preventing the emergence of resistance to streptomycin (Table 4.2). In a similar manner, controlled clinical trials in Madras (5) and East Africa (6) proved beyond doubt that the addition of PAS led to a substantial enhancement of the efficacy of monotherapy with isoniazid (Table 4.2) and that isoniazid resistance emerged much less frequently with combined chemotherapy. That two drugs are better than one was also confirmed by the outcome of controlled trials in the United Kingdom and the United States (7,8) that assessed the impact of administering streptomycin with isoniazid (Table 4.3). The main advantage of using concurrent controls and randomization was that not only was the conclusion

Table 4.2. Addition of second drug (PAS) enhances efficacy of monotherapy

Progress	PS (3 months)	S (3 months)	Difference	Ref.
Culture negativity	33% of 46	19% of 48	14%	4
X-ray improvement at 3 months	75% of 53	70% of 54	5%	
X-ray improvement at 6 months	87% of 53	74% of 54	13%	
Streptomycin resistance by 3 months	6% of 33	57% of 47	51%	
Progress	PH (1 year)	H (1 year)		
Moderate or more x-ray improvement	85% of 86	62% of 86	23%	5
Culture negativity	90% of 82	51% of 84	39%	
Favorable response	86% of 86	44% of 86	42%	
Isoniazid resistance at 3 months	9% of 22	75% of 48	66%	
Progress	PH (1 year)	H (1 year)		
Moderate or more x-ray improvement	82% of 34	63% of 38	19%	6
Culture negativity	84% of 31	57% of 30	27%	
Cavitation less or disappeared	100% of 29	89% of 28	11%	
Favorable response	81% of 31	36% of 36	45%	
Isoniazid resistance at 3 months	27% of 15	83% of 30	56%	

Table 4.3. Addition of second drug (streptomycin) enhances efficacy of monotherapy

Progress	SH daily (3 months)	H daily (3 months)	Difference	Ref.
Moderate or considerable x-ray improvement	38% of 142	22% of 120	16%	7
Culture negativity	67% of 117	37% of 101	30%	
H resistance	13% of 23	62% of 53	49%	
Progress	S_2H_7 (40 weeks)	H_7 (40 weeks)		
Moderate or marked x-ray improvement	59% of 195	51% of 205	8%	8
Culture negativity	79% of 195	65% of 205	14%	

S = streptomycin
H = Isoniazid
S_2H_7 = Streptomycin twice a week and isoniazid given daily, seven days a week
H_7 = Isoniazid given daily, seven days a week.

unbiased but also insight was gained regarding the mechanism by which two drugs proved to be more efficacious than single drugs in controlling the disease.

3.1 A Third Drug in the Initial Intensive Phase Chemotherapy Can Enhance Efficacy

The addition of a third drug (streptomycin) in the initial intensive phase (Table 4.4) enhanced the efficacy of two-drug regimens containing isoniazid and a weak second drug such as PAS or thioacetazone (9–11). However, in the case of a two-drug regimen containing two strong bactericidal drugs such as isoniazid and strep-

Table 4.4. A third drug in the initial intensive phase can enhance efficacy

Treatment	Duration of third drug	Favorable response	Difference	Ref.
PH(12)	—	84% of 81		9
SPH(12)	6 weeks	97% of 70	13%	
PH(8)	—	80% of 217		10
SPH(8)	6 weeks	88% of 241	8%	
TH(12)	—	79% of 181		11
STH(12)	2 months	90% of 162	11%	

Table 4.5. Duration of streptomycin in the initial intensive phase

	TH for 18 months with the following duration of streptomycin supplement			
	0 weeks	2 weeks	4 weeks	8 weeks
Moderate or more x-ray improvement at 12 months	89% (136)[a]	90% (153)	91% (152)	90% (153)
Cavitation disappeared by 12 months	55% (122)	62% (141)	59% (145)	55% (137)
Culture negative at 6 months	85% (144)	90% (156)	95% (151)	97% (154)
Culture negative at 12 months	88% (145)	89% (158)	93% (153)	96% (159)
Favorable response at 12 months[b]	88% (147)	90% (161)	94% (159)	96% (162)
Favorable response at 18 months[c]	87% (146)	87% (157)	93% (156)	96% (156)

Source: Ref. 12.
[a]Numbers in parentheses are the denominators on which the percentages are based.
[b]Based on sputum culture results at 10, 11, and 12 months.
[c]Based on sputum culture results at 15, 16, 17, and 18 months.

tomycin, the addition of a weak third drug such as PAS, even when given for the entire duration of treatment, did not prove to be beneficial (10); a corollary from this finding was that three drugs are not always better than two (which may be contrary to intuitive thinking), a conclusion that would not be readily accepted if it had not been ensured that all other things were equal, through the instrument of a randomized controlled trial.

Having demonstrated that a 2-month streptomycin supplement enhanced the efficacy of thioacetazone plus isoniazid (11), it was pertinent to inquire what the optimal duration should be for the streptomycin supplement. This was a question that could be answered only through a randomized controlled trial, such as the one conducted by the East African and British Medical Research Councils (12), the main findings of which are summarized in Table 4.5. There was an increasing trend in the response with an increasing duration of the streptomycin supplement from 2 weeks to 4 weeks to 8 weeks—for instance, the proportions of patients who were culture negative at 6 months were 85%, 90%, 95%, and 97% for streptomycin supplements of 0, 2, 4, and 8 weeks, respectively, a significant trend ($p < .001$). A similar trend was seen in the proportions classified as having a favorable response at the end of treatment, these being 88%, 90%, 94%, and 96%, respectively ($p < .01$). At 18 months, however, there appeared to be no impact from a 2-week supplement, and 8 weeks appeared to be optimal.

4. Isoniazid in a Single Dose Is Better Than in Two Divided Doses

In a concurrent comparison of a standard double-drug regimen of isoniazid plus PAS and three regimens of isoniazid alone (5), a special comparison of interest was between isoniazid 400 mg in a single daily dose and isoniazid in the same daily dosage but in two divided doses of 200 mg. The findings are summarized in Table 4.6 and clearly demonstrate that the single daily dose regimen was superior. This was a surprising finding at the time, especially because phar-

Table 4.6. Comparison of isoniazid in one or two daily doses

At one year	Isoniazid 400 mg daily for 12 months		
	Single dose (400 mg)	Two doses (200 mg × 2)	Difference
Considerable x-ray improvement	55% of 64	36% of 66	19%
Cavitation disappeared or less	87% of 63	72% of 61	15%
Culture negative	76% of 62	59% of 64	17%
Favorable response	73% of 64	58% of 66	15%

Source: Ref. 5.

macological studies showed that the duration for which a minimal inhibitory concentration of isoniazid (0.2 µg/ml) was maintained in rapid inactivators of isoniazid was markedly less for the single-dose regimen (12–18 h) than for the divided-dose regimen, namely 24–26 h (13). From this finding was born the idea that intermittent high doses of drugs would probably be as effective as conventional daily doses. This led to investigations of an intermittent regimen, which had the potential advantage that it could be administered on a fully supervised basis, thereby tiding over the knotty problem of noncompliance and the associated possibility of concealed irregularity in drug intake.

5. Supervised Intermittent Chemotherapy Is as Effective as Self-administered Daily Chemotherapy

The path-breaking classic investigation on this subject was reported from Madras as a preliminary communication in 1963, and a fuller report appeared in 1964 (14). In a randomized controlled trial of a fully supervised twice-weekly regimen of streptomycin plus high-dosage isoniazid and a conventional self-administered daily regimen of PAS plus isoniazid, a favorable response at 1 year was observed in 94% of the former and 85% of the latter (Table 4.7). Furthermore, over a 4-year period of follow-up, the proportions of patients who had a relapse requiring treatment were 5% and 10%, respectively (15), proving that the good results obtained initially with the twice-weekly regimen were maintained over a long period of follow-up.

Other randomized controlled trials followed (Table 4.7), which established the similarity of efficacies of twice-weekly streptomycin plus isoniazid and daily

Table 4.7. Supervised intermittent chemotherapy is equally effective

Treatment regimen (12 months)	No. of patients	Favorable response at one year		Ref.
PH	66	56	85%	14
S_2H_2	72	68	94%	
STH (1 month)/TH	136	127	93%	16
STH (1 month)/S_2H_2	99	87	88%	
SPH (2 weeks)/PH	83	72	87%	17
SPH (2 weeks)/P_2H_2	90	79	88%	
RH	39	36	92%	18
R_2H_2	41	36	88%	

thioacetazone plus isoniazid (16), twice-weekly and daily regimens of PAS plus isoniazid (17), and twice-weekly and daily regimens of rifampicin plus isoniazid (18). The net result of all these controlled trials was that the concept of fully supervised intermittent chemotherapy was established beyond all doubt. An important conclusion had been drawn and a major gain obtained in the management of tuberculosis patients.

The encouraging findings led to ambitious investigations of fully supervised once-weekly regimens, as they would be easier to organize under field conditions. Whereas once-weekly regimens from the very beginning proved to be unsatisfactory (Table 4.8), a regimen of streptomycin plus isoniazid daily for 4 weeks, followed by streptomycin plus high-dosage isoniazid once a week for the rest of the year yielded satisfactory results in general, and especially excellent results in slow inactivators of isoniazid (19). A fallout of this study was the recognition of the fact that the rate of inactivation of isoniazid is an important prognostic factor in once-weekly chemotherapy, unlike daily therapy. Subsequent controlled trials showed that the concomitant administration of PAS (20) or the use of matrix isoniazid [i.e., a slow-release preparation of isoniazid (21)] in the continuation phase (5–52 weeks) could not compensate for the inadequacy of the regimen in rapid inactivators of isoniazid, from which it followed that the universal application of once-weekly regimens to all patients was not a practicable proposition.

Table 4.8. Once-weekly regimens are not satisfactory in rapid inactivators of isoniazid

Regimen	Total patients	Favorable response at 1 year		Favorable response, by isoniazid inactivation status[a]			Ref
				Slow	Rapid	Difference	
S_2H_2	96	91	95%	97%	91%	6%	19
S_1H_1	77	51	66%	76%	56%	20%	
$S_1H_1Z_1$	101	76	75%	87%	53%	34%	
SH/S_1H_1	101	89	88%	95%	76%	19%	
SH/S_1H_1	176	149	85%	93%	72%	21%	20
$SPH/S_1P_1H_1$	170	148	87%	95%	76%	19%	
SH/S_1H_1	81	74	91%	98%	82%	16%	21
SH/S_1M_1	87	78	90%	98%	83%	15%	

Note: All regimens were of 12 months' duration. The initial intensive phase was 4 weeks and the continuation phase was 48 weeks for the fourth regimen in the first study (19) and for all regimens in the second and third studies (20,21).

[a]Determined by estimating the serum isoniazid concentration at $4\frac{1}{2}$ h after an intramuscular test dose of isoniazid 3 mg/kg body weight.

This was a valuable lesson, for it indicated that the interval between successive doses could not be stretched to more than 3–4 days in routine clinical practice.

6. Short-Course Chemotherapy of 6 Months' Duration Is as Effective as Standard-Duration Chemotherapy

A major practical problem with the application of standard-duration regimens is ensuring regularity in self-administration of drugs for long periods of time (12–18 months). Fully supervised intermittent chemotherapy, although effective, cannot always be organized, especially in rural settings. There was, therefore, a need for evolving shorter-duration regimens, which would not only lead to a favorable bacteriological status at the end of treatment but also maintain the satisfactory state after the discontinuation of treatment. Four short-course regimens of 6 months' duration were compared with a standard 18-month regimen in a large randomized controlled trial in East Africa (22), and the valuable lessons drawn from this classic study have altered management practices tremendously. The basic short-course regimen was 6 months of daily streptomycin plus isoniazid; the addition of rifampicin, pyrazinamide, or thioacetazone to this regimen constituted the other three short-course regimens. The most valuable lesson learned from this trial was that a short-course regimen of 6 months of streptomycin plus rifampicin plus isoniazid was as highly effective as the standard 18-month regimen (Table 4.9). Other important conclusions were that when pyrazinamide was substituted for rifampicin, the regimen was less effective ($p = .06$), although still satisfactory. Thioacetazone as the third drug was, however, ineffective, and 6 months of streptomycin plus isoniazid was very inadequate, the relapse rate over a 1-year follow-up period being as high as 27%. Such clear-cut conclusions could not have been drawn in the absence of a well-designed controlled trial. In a collaborative United States Public Health Service (USPHS) trial (23), a 6-month regimen of rifampicin plus isoniazid had a relapse rate of 9.2% during an observation phase of 21 months, which reduced to 1.6% in patients who had a 9-

Table 4.9. Short-course chemotherapy for 6 months is as effective as conventional standard chemotherapy

Regimen	Total	Success	Assessed for relapse	Relapse during 7–18 months	
6SHR	176	100%	156	4	3%
6SHZ	180	100%	166	13	8%
2STH/16TH	170	98%	139	3	2%
6SHT	154	100%	102	22	22%
6SH	180	99%	114	31	27%

Source: Ref. 2.

Table 4.10. Rifampicin or pyrazinamide is not necessary throughout the treatment period

Regimen	Total	Success	Assessed for relapse	Relapse during 7–30 months	
6SHR	181	100%	171	4	2%
6HR	183	99%	164	12	7%
2SHRZ/4TH	191	100%	179	13	7%
2SHRZ/4S$_2$H$_2$7$_{+2}$	179	99%	159	7	4%

Source: Ref. 24.

month continuation phase with isoniazid and ethambutol ($p < .001$). A valuable lesson from these two trials was that two-drug regimens for 6 months would be inadequate, and even a three-drug regimen would be satisfactory only if it included rifampicin.

A subsequent trial from East Africa (24) confirmed that the three-drug regimen of streptomycin, isoniazid, and rifampicin was highly effective (Table 4.10), and that a two-drug regimen (by omitting streptomycin) had a higher relapse rate ($p = .06$). It also suggested that it was not necessary that rifampicin or pyrazinamide be prescribed for the entire 6-month period, because 2 months of streptomycin, isoniazid, rifampicin, and pyrazinamide followed by 4 months of thioacetazone plus isoniazid had a relapse rate of 7%. Another valuable inference from this trial was that the initial phase of chemotherapy was particularly important, as both regimens with a four-drug initial intensive phase were reasonably satisfactory.

Further probes on the duration and number of drugs in the initial intensive phase, as well as the drugs, the rhythm, and duration of the continuation phase, were taken up in another well-designed controlled trial (25). This study was planned to investigate whether shortening the initial intensive phase (with four drugs) from 2 months to 1 month would affect the efficacy of the 6-month regimen assessed in an earlier trial (24), whether pyrazinamide was really necessary in the initial phase, whether continuation chemotherapy with twice-weekly streptomycin plus isoniazid plus pyrazinamide would be more effective than standard thioacetazone plus isoniazid, and whether extending the total duration of chemotherapy to 8 months (from 6 months) would appreciably lower the relapse rates with each of the regimens. That so many important questions could be investigated in one study is a tribute to the versatility of the controlled clinical trial approach, combined with an efficient experimental design. The findings (Table 4.11) showed that the 2SHRZ/6TH regimen was highly effective with a relapse rate of 0%, the 95% confidence limits being 0% to 4%. The factorial design of the trial also led to the conclusion that 8 months of treatment was better than 6 months to contain relapse ($p < .001$), a conclusion that was similar to that reached from a trial in Hong Kong (26). These findings suggested that the sterilizing role of short-course

Table 4.11. Short-course regimens with initial intensive phase of one or two months

Initial intensive phase	Continuation phase	6-Month regimen		8-Month regimen	
		Total	Relapse during 7–30 months	Total	Relapse during 9–30 months
2 SHRZ	4 (6) TH	75	10 13%	81	0 0%
1SHRZ	5 (7) TH	79	14 18%	58	4 7%
1SHRZ	5 (7^a) S$_2$H$_2$Z$_2$	75	7 9%	88	2 2%
2SHR	4 (6) TH	82	15 18%	77	5 6%

^{a} Five months of S$_2$H$_2$Z$_2$ followed by 2 months of TH.
Source: Ref. 25.
S = Streptomycin
H = Isoniazid
R = Rifampin
Z = Pyrazinamide
T = Thiacetazone
S$_2$H$_2$Z$_2$ = Streptomycin, Isoniazid, Pyrazinamide given twice a week.

chemotherapy was not only a function of the drugs in the regimen but also a function of the total duration of chemotherapy in regimens with a relatively weak continuation phase. Shortening the initial intensive phase to 1 month carried a therapeutic penalty, but this could be overcome by strengthening the regimen in the continuation phase (from TH to S$_2$H$_2$Z$_2$) [see footnote on page 99]; and finally, pyrazinamide did seem to have an important role in the initial intensive phase.

Most of the successful short-course regimens studied consisted of daily chemotherapy for at least a part of the total duration of treatment, and their success was probably dependent on regular drug intake, which was sometimes ensured by hospitalization. Fully supervised drug administration is an attractive alternative but is not easy to organize on a daily outpatient basis. However, if the short-course regimen is intermittent throughout, total supervision on an outpatient basis could be feasible. Several such 6-month regimens were investigated by the Tuberculosis Research Centre in Madras (27); the initial intensive phase consisted of 2 months of thrice-weekly or twice-weekly streptomycin, rifampicin, isoniazid, and pyrazinamide and this was followed by 4 months of twice-weekly chemotherapy (SRH, RH, SH) or once-weekly chemotherapy (SRH, RH). The relapse rates over a 5-year period were similar for the 10 regimens, the overall mean being 7.7% and the 95% confidence interval being 6.0% to 9.4% (Table 4.12). The controlled trial approach, supported by good experimental design, resulted in 10 effective regimens, the most attractive among them probably being 2S$_2$H$_2$R$_2$Z$_2$/4R$_1$H$_1$ that had a relapse rate of 8% over a 5-year period (i.e., in 54 months after stopping treatment). All the 10 regimens studied had a streptomycin component, which

Table 4.12. Fully supervised intermittent short-course regimens for 6 months are effective

Initial intensive phase (2 months)	Continuation phase (4 months)	Total patients assessed	Relapse during 7–60 months	
$2S_3R_3H_3Z_3$	$4R_2R_2H_2$	90	6	7%
	$4R_2H_2$	94	7	7%
	$4S_1R_1H_1$	99	6	6%
	$4R_1H_1$	96	5	5%
	$4S_2H_2$	89	7	8%
$2S_2R_2H_2Z_2$	$4S_2R_2H_2$	89	7	8%
	$4R_2H_2$	90	8	9%
	$4S_1R_1H_1$	103	7	7%
	$4R_1H_1$	92	7	8%
	$4S_2H_2$	95	12	13%
	Total	937	72	7.7%

Source: Ref. 27.

Table 4.13. Fully supervised oral intermittent chemotherapy for 6 months is effective

Rhythm	Drugs	Total		Relapse during 7–18 months	
Daily	6ZHRE	161	1	0.6%	
Thrice weekly	6ZHRE	164	4	2.4%	1.5%
Thrice weekly	6SZHRE	150	1	0.7%	of
Thrice weekly	6SZHR	150	2	1.3%	464
Thrice weekly	6SHRE	160	12	7.5%	

Source: Ref. 28.

meant that facilities for giving injections had to be available. In situations where this is not feasible, fully supervised oral intermittent regimens would be the obvious solution. A randomized controlled trial in Hong Kong (28) showed that fully supervised oral chemotherapy for 6 months with thrice-weekly regimens containing pyrazinamide was effective (regimens 2–4 in Table 4.13), the pooled relapse rate being 1.5% (95% confidence limits 0.4% to 2.6%) compared with 0.6% (95% confidence limits 0.0% to 1.8%) with a 6-month daily regimen of pyrazinamide, isoniazid, rifampicin, and ethambutol.

7. Duration of Short-Course Chemotherapy in Smear-Positive Patients Cannot Be Less Than 5 Months

The gratifying finding that short-course chemotherapy for 6 months was as effective as conventional duration chemotherapy raised hopes of developing regimens that would have even shorter duration, i.e., 3 or 4 months. A trial undertaken in South India by the Tuberculosis Research Centre in Madras and the National Tuberculosis Institute in Bangalore (29) showed that a 3-month daily regimen with streptomycin, rifampicin, isoniazid, and pyrazinamide had an unacceptably high relapse rate of 20% by 24 months (i.e., 21 months of follow-up), but this could be lowered to 4% if chemotherapy was given for 2 more months, on a twice-weekly basis, with streptomycin, isoniazid, and pyrazinamide (Table 4.14). This trial also showed that rifampicin in the initial intensive phase was absolutely essential, for the nonrifampicin regimen had a substantially higher relapse rate of 13%.

Table 4.14. Less than 5 Months of short-course chemotherapy is inadequate in smear-positive patients

Regimen	Total		Relapse		Ref.
3SRHZ	200	39	20%		29[a]
3SRHZ/2S$_2$H$_2$Z$_2$	187	8	4%		
3SHZ/2S$_2$H$_2$Z$_2$	199	25	13%		
					30[b]
SRHZ/RHZ	104	17	16%	13%	
SRHZ/RH	104	11	11%	of	
SRHZ/HZ	98	31	32%	208	
SRHZ/HZ	98	31	32%	34%	
SRHZ/H	105	32	30%	of	
RHZ/H	100	40	40%	303	
2SRHZ/2RHZ	79	9	11%	31[c]	
2SRHZ/4RHZ	78	0	0%		
2SRHZ/3S$_2$H$_2$Z$_2$	129	7	5%	32[d]	
2SRHZ/5S$_2$H$_2$Z$_2$	132	0	0%		
2SHZ/5S$_2$H$_2$Z$_2$	269	7	3%		

[a]Relapse was assessed up to 24 months.
[b]Initial intensive phase for 8 weeks followed by continuation phase for 9 weeks, making a total of 4 months; period of follow-up was 24 months after cessation of chemotherapy.
[c]Relapse was assessed up to 30 months.
[d]Relapse was assessed up to 24 months.

Five short-course regimens, each of 4 months' duration, were investigated in East Africa (30). Over a 2-year period of follow-up, all had disappointingly high relapse rates (Table 4.14), although the rate was substantially less in the regimens that contained rifampicin throughout (13% of 208) than in the others (34% of 303). The 4-month regimen with rifampicin had a high relapse rate in the study in Singapore also (11% of 79), which was reduced to 0% by extension of the continuation phase to 4 months (i.e., total duration of the regimen being 6 months) (31).

Finally, over a 2-year period of follow-up, a 5-month regimen consisting of daily streptomycin, rifampicin, isoniazid, and pyrazinamide for 2 months followed by 3 months of twice-weekly streptomycin plus isoniazid plus pyrazinamide had a relapse rate of 5% in Madras (Table 4.14), which was reduced to 0% by extending the continuation phase by 2 months (i.e., total duration of treatment being 7 months) (32). A 5-year report showed that the corresponding relapse rates were 7.1% and 4.0%, respectively (33); this trial also showed that the omission of rifampicin in the initial intensive phase increased the relapse rate over the 5-year period to 6.7%.

The net result of all these controlled trials was a clear-cut conclusion that the duration of chemotherapy for patients with smear-positive tuberculosis could not be reduced to less than 5 months.

8. Role of Maintenance Chemotherapy in the Prevention of Relapse

Although conventional regimens of 1-year duration produced very satisfactory results in patients with drug-susceptible organisms, it was important to know the extent to which the bacteriological quiescence attained was stable and whether further chemotherapy could reduce the likelihood of relapse. The role of maintenance chemotherapy with isoniazid in low dosage (200 mg) was therefore investigated, with concurrent placebo controls, in randomized trials (34,35). These established that isoniazid in a daily dosage of 200 mg in the second year was wholly adequate in patients with no residual cavitation at 1 year but was only 50% effective in those with residual cavitation (Table 4.15). A further trial (36) showed that increasing the daily dosage to 400 mg conferred no further benefit in the latter; also shown was that in patients with no residual cavitation, a shorter duration of isoniazid, even with a higher dose (300 mg), was not sufficient, the prophylactic efficacy being only 60%. (36). These were valuable lessons, but they had not completely answered the management problem, especially in patients with residual cavitation. Moreover, daily isoniazid would need to be self-administered by the patient, a practice fraught with the danger of noncompliance. Fully supervised once-weekly chemotherapy is an attractive alternative and was, therefore, investigated in a subsequent controlled trial (37). A 6-month regimen of once-weekly streptomycin plus high-dosage isoniazid was demonstrated to be as effective as daily isoniazid for 12 months in patients with no residual cavitation at 1 year, the relapse rates up to 5 years being 2% of 98 and 1% of 90, respectively;

Table 4.15 Role of maintenance chemotherapy with daily isoniazid in the second year

Cavitation status at 1 year	First study[a]				Second study[b]		
	Treatment in second year	Total	Relapse[a]		Treatment in second year	Total	Relapse requiring treatment[c]
Cavitated	Placebo	41	6	14.6%	Placebo	58	8	13.8%
	Isoniazid 200 mg for 12 months	55	4	7.3%	Isoniazid 400 mg for 12 months	63	4	6.3%
Noncavitated	Placebo	107	17	15.9%	Placebo	118	14	11.9%
	Isoniazid 200 mg for 12 months	103	1	1.0%	Isoniazid 300 mg for 6 months	127	6	4.75

[a]Data from Refs. 34 and 35.
[b]Data from Ref. 36.
[c]In second, third, fourth, and fifth years.

it was highly effective in patients with residual cavitation also if administered throughout the second year, the relapse rates being 3% of 87 as compared to 21% of 94 in patients who received placebo in the second year (37). The controlled clinical trial approach had thus provided invaluable information on the role of maintenance chemotherapy.

9. Domiciliary Treatment Is as Effective as Sanatorium Treatment

Although enough evidence was available about the high degree of effectiveness of chemotherapy in the mid-1950s, the general consensus at the time was that admission of the infectious patient to sanatorium was essential, because it would ensure that the patient had a balanced diet, adequate rest, and regular drug ingestion and the family contacts would not be further exposed to infection. However, the number of available sanatorium beds was woefully inadequate in most developing countries where tuberculosis was a major problem. Domiciliary chemotherapy seemed to be the obvious solution, but before it could be implemented as a policy, it was important to verify whether the immediate results of domiciliary outpatient treatment would be highly satisfactory, whether subsequent relapse rates would be acceptably low, and whether there would be any additional risk to the close family contacts from treating the infectious patient at home. A randomized controlled trial in South India (38), referred to as the "Madras Experiment," compared home treatment (i.e., outpatient treatment) with sanatorium treatment in about 200 patients, all of whom were prescribed a standard daily regimen of isoniazid plus PAS for 1 year. The radiographic and bacteriological progress of the patients in the two series are summarized in Table 4.16. In view of the unusually large differences between the two series (despite random allocation) in

Table 4.16. Comparison of Home Treatment and Sanatorium Treatment

Progress in 0–12 months	Observed percentage			Adjusted percentage[a]		
	Home (a)	Sanit. (b)	a/b	Home (c)	Sanit. (d)	c/d
Considerable or more x-ray improvement	41%	52%	79%	44%	49%	90%
Cavitation disappeared	37%	54%	68%	43%	48%	89%
Culture negative (at 12 months)	87%	94%	93%	90%	94%	96%
Favorable response at the end of treatment	84%	92%	91%	86%	92%	94%
Total patients	82	81		82	81	

[a]Percentage adjusted statistically for unusually large differences between the home and the sanatorium series in the extent of cavitation and lung disease involvement at the start of treatment.
Source: Ref. 38.

the initial condition of the patients (the home patients were worse off in terms of the extent of cavitation and the number of lung zones involved in disease), the observed results were adjusted statistically and the results are also tabulated in the right-hand section of Table 4.16. These indicate that the response to home treatment was only slightly less good than the results of sanatorium treatment. Further, among males in the two series who were in very similar pretreatment condition, the results were virtually identical (38).

Over a 4-year period of follow-up, the relapse rate was 7% in 57 home patients and 10% in 69 sanatorium patients (34). Over the 5-year period, 5% of the 91 allocated home patients and 6% of the 94 allocated sanatorium patients had died of tuberculosis; 4% in each series had bacteriologically active disease at 5 years and 90% and 89%, respectively, had bacteriologically quiescent disease (34).

In initially uninfected (tuberculin negative) contacts, the attack rate of tuberculosis over a 5-year period was 10.5% of 86 if the infectious patient was treated at home and 11.5% of 87 if he/she was segregated (and treated) in a sanatorium, a finding which indicated that the contacts of home patients were at no greater risk of acquiring tuberculosis than the contacts of patients isolated in a sanatorium (39).

Other noteworthy findings of this policy study were that diet was not important, either for the attainment of bacteriological quiescence (40) or its maintenance over a 4-year period of follow-up (41), and that strenuous physical activity and overcrowded living conditions for home patients and the stress of returning to an unfavorable home environment in the case of sanatorium patients had no bearing on the likelihood of bacteriological relapse (41).

The net effect of the innumerable valuable lessons learned from this controlled investigation was that domiciliary outpatient chemotherapy became the sheet anchor for the tuberculosis program in several developing countries. This probably is the most beneficial outcome from the controlled trial approach, from the point of view of global containment of tuberculosis.

10. Policy of Home Visits to Outpatients to Promote Compliance

Failure of outpatients to collect drugs and consume them, especially after the disappearance of symptoms, is a chronic problem in domiciliary chemotherapy, and different techniques are empirically employed to monitor and promote compliance. One such method, employed in East Africa, was to make two surprise home visits per month for 18 months to check on drug regularity, motivate the patient to be compliant, and remind him of the date of his next clinic attendance. The findings, as quoted from Ref. 2, are that the proportion of defaulters was 17% at 12 months, 20% at 18 months, and 35% at 36 months. These figures were not appreciably lower than the corresponding proportions of 22%, 24%, and 38%, respectively, in patients who received no such home visits, indicating that there was no great merit in organizing a prophylactic home visiting program. The proportion of patients with quiescent disease at the different time points also confirmed this conclusion (2). Two reasons can be advanced for these findings which go against the grain of intuition. One is that it was possible to organize the surprise home visits on only 71% of the occasions, and of these, the patient was actually seen in 69% of the visits (i.e., in an overall proportion of only 49% of the total intended number of occasions) (42). The other is that some patients who were visited may have been compliant in any case and not in need of any motivating home visits. The lesson from this policy study was that the limited resources available should be targeted to retrieving defaulters.

11. Policy for Managing Default in Outpatients

In many countries, a recommended procedure for retrieving defaulters is to post a reminder letter on the fourth day after the missed appointment and follow this up, if necessary, with a home visit by a health visitor or social worker 7 days later. The effectiveness of the policy was investigated in South India, and as there were reasons to believe that this policy may not be highly effective on account of high illiteracy rates, it was compared with a more intensive policy that consisted of a personalized home visit (on the fourth day), followed, if necessary, by three more home visits (43). A randomized controlled clinical trial approach was em-

Table 4.17. Two policies of default management in outpatients

	Routine policy[a]	Intensive policy
Retrieved after first defaulter action (all episodes)	56% of 193	65% of 132
Mean number of defaults	3.7	2.9
All 6 collections in first 6 months	41% of 75	68% of 75
Mean number of collections in 12 months	8.6	9.8
12 collections in 15 months	52% of 67	69% of 71

[a]For details of routine and intensive policies, see text.
Source: Ref. 43.

ployed, and the findings are summarized in Table 4.17. It can be seen that retrieval after the first defaulter action was about 10% better with the intensive policy (home visit) than the routine policy (letter-posting), also that the mean number of defaults was reduced by over 20%, from 3.7 to 2.9. Further, in terms of drug collections, 68% of the intensive-policy patients as compared with only 41% of the routine-policy patients collected all drug supplies in the first 6 months ($p <$ .01), and the mean numbers of drug collections in the year were 9.8 and 8.6, respectively ($p = .03$). Finally, 12 collections in a period of 15 months (satisfactory completion of the regimen under program conditions) were made by 69% of the intensive-policy patients and 52% of the routine-policy patients ($p = .07$). The important lesson from this policy study was that speedy personalized contact with the defaulting patient would be more effective than impersonal letters in the post.

12. Policy of Prescribing Vitamins or Antihistamines as Prophylactics

Thioacetazone, an inexpensive drug, was found to be a useful companion drug to isoniazid in several studies, but there was some apprehension regarding the incidence of side effects, which led to the routine prescription of prophylactics in the form of vitamins and antihistamines in some parts of Asia. The efficacy of such supplements is difficult to determine in routine clinical practice because some of the side effects may resolve on their own, whereas others may well be coincidental. The use of randomized concurrent controls and the double-blind technique is the most satisfactory method of evaluating the supplements, and this was exactly what Miller et al. did in an international cooperative investigation (44), the findings of which are summarized in Table 4.18. Over a 12-week period,

Table 4.18. No benefit from vitamin and antihistamine supplements

Regimen	Group	Total patients	Patients with side effects[b]		Major departures from prescribed chemotherapy	
TH	Additives[a]	584	217	37.2%	55	9.4%
	No additives	581	223	38.4%	55	9.5%
STH	Additives	707	357	50.5%	146	20.7%
	No additives	689	345	50.1%	131	19.0%
SH	Additives	696	221	31.8%	43	6.2%
	No additives	711	216	30.4%	43	6.0%

[a]Vitamins plus antihistamine for 12 weeks.
[b]Gastric, cutaneous, vestibular, neurological, hematological, and miscellaneous.
Source: Ref. 44.

the additives failed to reduce the overall incidence of side effects, as did the frequency of major departures from the prescribed regimen. Nor was there any evidence that the more serious side effects, i.e. rashes, jaundice, and agranulocytosis, were reduced by the additives. Consequently, and in view of the high cost of introducing the supplements and other operational problems, the authors concluded that the use of prophylactic supplements could not be recommended.

Studies in Madras had shown that a vitamin B complex preparation, which provided a daily dosage of 60 mg of aneurine hydrochloride, 30 mg of riboflavine, 300 mg of nicotinamide, 6 mg of pyridoxine, 18 mg of panthenol, and 6 μg of cyanocobalamin, was effective in the treatment of peripheral neuropathy due to isoniazid (45). Because the incidence of peripheral neuropathy was over 50% in slow inactivators of isoniazid who received isoniazid 15 mg/kg body weight (45), a natural question was whether it would be beneficial to have a policy of routinely prescribing B complex (as a prophylactic) to all patients. However, such a policy would have greatly added to the cost of treatment and would not have been affordable in several developing countries. A controlled clinical trial was therefore undertaken to verify whether it was the pyridoxine in the B complex that was responsible for the protective effect (46). This study showed that peripheral neuropathy developed in 6 of 16 slow inactivators of isoniazid and 1 of 8 rapid inactivators who received B complex without the pyridoxine 6-mg component, as compared to none of 24 patients (14 slow, 10 rapid) who received B complex with the pyridoxine and none of 26 patients (18 slow, 8 rapid) who received only 6 mg of pyridoxine. Thus, by means of a randomized controlled trial and good experimental design, pyridoxine 6 mg was identified as the effective agent in the B-complex preparation that prevented isoniazid-induced peripheral neuropathy and was routinely incorporated in the isoniazid tablets in all subsequent trials.

13. Policy Study of the Role of Pretreatment Drug Susceptibility Tests in the Choice of Treatment Regimens

To commence treatment with a triple-drug regimen of streptomycin plus PAS plus isoniazid in all bacteriologically confirmed cases and review the chemotherapeutic regimen when the results of drug susceptibility tests become available was standard clinical practice in the days of conventional daily regimens. Although this policy might be very beneficial at the level of the individual patient, it is not easy to administer on a community basis, for the facilities and expertise required and work load involved in collecting and testing the specimens can be stupendous. Whether the benefit at the *community level* would be *sufficiently large* to warrant the extra infrastructure and work load could be an important question in several *developing* countries. It was so in Hong Kong in the early 1970s, for the levels of initial drug resistance were high, i.e., about 30%. A novel study, using the controlled clinical trial approach, was therefore undertaken in which three management policies were compared (47). For patients allocated to Policy A, streptomycin plus PAS plus isoniazid was given daily for 3 or 6 months at random, followed by isoniazid plus PAS for the rest of the year, without reference to the outcome of the pretreatment drug susceptibility results. Policy B consisted of treating with daily streptomycin, PAS, and isoniazid until the results of indirect susceptibility tests became available (by 6–7 weeks usually, but not exceeding 13 weeks); ethionamide was substituted for the period of triple chemotherapy if pretreatment resistance to any one of the three standard drugs was found, and three reserve drugs followed by two reserve drugs were given in cases where resistance to two or all three standard drugs was noted. If sensitive, treatment was the same as in Policy A. Patients allocated to Policy C had a rapid slide culture susceptibility test *before* the start of treatment; those with sensitive organisms were prescribed standard chemotherapy (SPH/PH) and others were managed as in Policy B. The outcome of these management policies is summarized in Table 4.19. A favorable status was noted in 89% of Policy A patients, 92% of Policy B patients, and 94% of Policy C patients; the differences are small and nonsignificant. Further, when allowance was made statistically for differences between the three series in the proportions of initially drug-resistant patients (33% for A, 25% for B, 33% for C), the difference became even smaller, the adjusted (standardized) proportions being 90% for Policy A, 91% for Policy B, and 93% for Policy C patients. Thus, this study based on the principles of controlled clinical trial approach came to the rather unexpected conclusion that the benefit from organizing *mass* chemotherapy programs on the basis of pretreatment sensitivity tests is, at most, very small.

14. Policy Study of Treatment of Smear-Negative Tuberculosis

In most countries, the numbers of patients who are treated for smear-negative (but radiologically active) tuberculosis far exceed the numbers that are treated for

Table 4.19. Policy study of the role of pretreatment drug susceptibility tests

Pretreatment susceptibility	Policy A[a]		Policy B		Policy C	
	Total patients	Favorable response at 1 year	Total patients	Favorable response at 1 year	Total patients	Favorable response at 1 year
Susceptible to all drugs	125	122 98%	144	133 92%	125	119 95%
Resistant to 1 or more drugs	62	45 73%	48	43 90%	62	56 90%
Total	187	167 89%	192	176 92%	187	175 94%
Favourable response (standardized by direct method)		90%		91%		93%

[a]Policy A: Treatment not dependent on pretreatment drug-susceptibility tests; policy B: Treatment based on results of indirect drug-susceptibility tests; policy C: Treatment based on rapid slide culture tests.
Source: Ref. 47.

smear-positive tuberculosis, and this is often a great drain on the limited resources available. Study of policies of treatment for the former are of considerable value, both to public health administrators and program managers. The controlled trial approach has been employed with benefit to deal with this problem in Hong Kong (48) where Chinese patients with radiologically active pulmonary tuberculosis but at least five sputum smears negative were randomly allocated to selective chemotherapy (i.e., chemotherapy with $3SPH/9S_2H_2$ started only if culture became positive or radiographic or clinical deterioration occurred during follow-up), or to streptomycin plus isoniazid plus rifampicin plus pyrazinamide daily for 2 months or for 3 months, or to a standard 12-month control regimen of daily streptomycin plus PAS plus isoniazid for 3 months followed by twice-weekly isoniazid plus streptomycin for 9 months. Among patients with at least one positive culture on admission, neither the 2-month nor the 3-month short-course regimen was satisfactory, as the failure rates were 32% and 13%, respectively, over a 5-year period of follow-up, as compared to only 5% with the conventional daily regimen (Table 4.20). In patients with all negative cultures on admission, the corresponding rates were 11%, 7%, and 2%, respectively. Finally, in the selective chemotherapy series, 57% of the patients had to have chemotherapy started during the 5-year period because their disease was confirmed to be active. Such a high proportion justified the routine policy in Hong Kong of treating all patients in whom active pulmonary tuberculosis was diagnosed clinically and radiographically by experienced physicians, even if as many as five sputum smears were negative. The findings in the other three series clearly demonstrated the inadequacy of short-course regimens of 2 or 3 months' duration, even if four potent drugs are employed. In a subsequent controlled trial (49), all patients were prescribed streptomycin, isoniazid, rifampicin, and pyrazinamide, either on a daily

Table 4.20. First policy study of treatment of smear-negative tuberculosis: findings at 5 years

(a) All cultures negative on admission

Treatment policy	Total patients assessed	Confirmed active disease or relapse		Bacteriologically confirmed	
Selective chemotherapy[a]	173	99	57%	71	41%
2SHRZ	161	17	11%	10	6%
3SHRZ	161	11	7%	5	3%
$3SPH/9S_2H_2$	160	3	2%	1	1%

(b) At least one positive culture on admission

	Total patients assessed	Failure (on C) or relapse (after A or B)		Bacteriologically confirmed	
(A) 2SHRZ	71	23	32%	16	23%
(B) 3SHRZ	68	9	13%	7	10%
(C) $3SPH/9S_2H_2$	75	4	5%	1	1%

[a]Antituberculosis chemotherapy was started only if culture became positive or radiographic or clinic deterioration occurred during follow-up.
Source: Ref. 48.

Table 4.21. Second policy study of treatment of smear-negative tuberculosis; findings at 5 years

Treatment	Total patients assessed	Relapse after end of treatment		Bacteriologically confirmed	
(a) All cultures negative on admission					
3SHRZ	364	21	5.8%	10	2.7%
$3S_3H_3R_3Z_3$	345	27	7.8%	10	2.9%
$4S_3H_3R_3Z_3$	325	12	3.7%	4	1.2%
(b) At least one positive culture on admission					
4SHRZ	157	4	2.5%	4	2.5%
$4S_3H_3R_3Z_3$	136	3	2.2%	1	0.7%
$6S_3H_3R_3Z_3$	166	8	4.8%	3	1.8%

Source: Ref. 49.

basis or a thrice-weekly basis; the duration of the daily regimen was 3 months for culture-negative patients and 4 months for culture-positive patients, and the duration of the intermittent regimen was 3 or 4 months for the former and 4 or 6 months for the latter (Table 4.21). This study showed that whereas 3-month regimens were inadequate even in culture-negative patients, a 4-month regimen consisting of streptomycin, isoniazid, rifampicin, and pyrazinamide (daily or thrice a week) was satisfactory in both culture-negative and culture-positive patients; consequently, it was recommended that the country could embark on a routine policy of 4 months of treatment for all smear-negative patients, irrespective of their initial culture status.

15. Management Policy for Patients with Tuberculosis of Doubtful Activity

The management of patients with tuberculosis of doubtful activity (e.g., assessed on mass miniature radiography) had posed a problem—whether patients should be kept on observation and the activity reassessed by serial radiographs at periodic intervals or whether they should be treated with antituberculosis drugs to prevent possible breakdown. A controlled clinical trial was undertaken by the Tuberculosis Society of Scotland (50) in which such patients were randomly allocated to at least 6 months of daily isoniazid plus PAS or to no treatment. Although there was some evidence that those prescribed isoniazid plus PAS responded better (Table 4.22), the differences were relatively small. Moreover, the advice tendered to patients to take prophylactic drugs was not always accepted; for instance, six of the seven patients in the treatment group who deteriorated failed to take their chemotherapy as requested. In view of these findings, a policy decision was taken not to routinely recommend chemotherapy for all doubtfully active cases.

Table 4.22. Pulmonary tuberculosis of doubtful activity

	Cumulative probability at 2 years[a]	
	Control group	PH group
Radiographic deterioration	12.%	8.1%
Radiographic or bacteriological deterioration	17.8%	8.5%
Any radiographic improvement	18.3%	30.6%
Moderate or more radiographic improvement	8.2%	18.0%
Patients in analysis	95	94

[a]Based on modified life-table calculations
Source: Ref. 50.

16. General Discussion

The studies described in the preceding pages have clearly portrayed the scope of the controlled clinical trial and documented some of the valuable lessons that have been learned by scientists and program managers in the field of tuberculosis. Nevertheless, some may wonder what the dangers are in not strictly adhering to the principles of controlled clinical trial methodology. These will now be discussed.

16.1. Importance of Controls and Scope for Erroneous Conclusions in Their Absence

EFFICACY OF GOLD TREATMENT

Numerous clinical reports had been published and exaggerated claims were made for gold treatment in tuberculosis. However, most of the trials lacked methodological rigor and were, therefore, not really conclusive. To provide a clear verdict, a controlled clinical trial (perhaps the first of its kind) was undertaken in 1931 by Amberson and his colleagues (51). On the basis of clinical, x-ray, and laboratory findings, the authors identified 24 cases, matched them into 12 pairs, and allocated one member of each pair "by the flip of a coin" to Sanocrysin (containing 37.4% gold) injections at intervals of 4 days, for a total dosage of 3.1–6.1 g; the other member of the pair received distilled-water injections at the same intervals. The treatment administered was known only to the nurse-in-charge and two physicians, not to the patient. At the end of treatment, 6 Sanocrysin and 7 control patients showed improvement, 0 and 3 showed no change, 1 and 1 became slightly worse, and 4 and 0 became much worse; one patient (Sanocrysin) died of parenchymatous degeneration of the liver which the authors interpreted as gold poisoning, and one (control) had to be transferred to another hospital for surgical treatment. Thus, there was no benefit from Sanocrysin, and in view of possible harm, the authors came to a firm conclusion that its use was not justified.

UTILITY OF STEROIDS

Steroids are prescribed as adjuvants to chemotherapy in the expectation that their immunosuppressive effect on the host might lead to multiplication of dormant bacilli, thereby rendering them susceptible to the bactericidal effect of antituberculosis drugs. Although the results of studies with conventional daily regimens were equivocal, interest in steroids resurfaced with the advent of short-course chemotherapy, consisting of three or four powerful drugs in the initial intensive phase. A randomized controlled trial in Madras (32) examined the value of the concomitant administration of steroids in the first 2 months, with three short-course regimens that included streptomycin, pyrazinamide, and isoniazid (and, in about half the patients, rifampicin). Although all patients responded well and sputum conversion was speedy and this could easily have been attributed to the steroids, the presence of a randomized concurrent control group of patients who did not receive any steroids showed that there was no real benefit, leading the authors to conclude that steroids have little place in short-course chemotherapy, especially under program conditions in developing countries.

DURATION OF TREATMENT

Suppose a group of patients has been treated for, say, 3 years, and very few or no relapses are observed over the next 2 years. A tempting inference could be that all patients need to be treated for 3 years. (Recommendations for very prolonged treatment have been made in the literature, on the basis of such evidence). However, whether the third year of treatment is really necessary can be really decided only if the relapse rate is known in patients who had only 2 years (or less) of treatment. In trials undertaken by the Tuberculosis Chemotherapy Centre in Madras, 1 of 77 bacteriologically quiescent patients who received maintenance therapy with isoniazid in the third year had a bacteriological relapse by 5 years (34,35); however, in the same trial, no relapse was observed in a parallel group of 74 patients who had only 2 years of treatment, demonstrating that the third year of treatment was unnecessary.

INCIDENCE OF SIDE EFFECTS TO DRUGS

Giddiness, a known side effect of streptomycin, was reported by 35% of 78 patients who received a regimen of isoniazid plus streptomycin (14). A likely inference, albeit hasty, could be that one in three patients exhibit streptomycin toxicity. In a concurrent control group of 70 patients on isoniazid plus PAS in the same trial, 11% complained of giddiness. The difference between the two proportions (24%, i.e., one in four) is a better estimate of the toxicity due to streptomycin if one notes that giddiness is not usually associated with PAS. Similarly, in an international study of side effects due to thioacetazone (52), 21.4% of the patients on a regimen of streptomycin, thioacetazone, and isoniazid had at least one side effect, 6.5% had an interruption of treatment, and 3.4% had treatment stopped. In a concurrent control group of patients treated with streptomycin plus isoniazid, the corresponding proportions were 7.8%, 1.6%, and 0.9%, respectively. Better estimates for incidence of side effects due to thioacetazone, treatment interruption, and treatment termination are obtained by subtracting the latter proportions from the former, yielding figures of 13.6% for side effects, 4.9% for treatment interruption, and 2.5% for treatment termination.

EFFECTIVENESS OF DOMICILIARY CHEMOTHERAPY AT THE COMMUNITY
LEVEL

To study the effectiveness of a domiciliary chemotherapy program in the control
of tuberculosis, a study was undertaken in 12 towns in South India (53), which
were randomly divided into 2 comparable groups of 6 towns, after taking into
account the findings of a baseline random sample survey. In one group, designated
as "treatment towns," intensive case-finding was undertaken by two x-ray surveys
followed by bacteriological investigations where appropriate, and all bacillary
cases detected were offered treatment with isoniazid alone or with isoniazid plus
PAS. A population survey was undertaken after 2½ years and yet another after 4
more years. The mean prevalence of bacillary tuberculosis (age-standardized) was
6.81 per 1000 initially (survey I) and decreased to 5.01 in survey II and 4.83 in
survey III; the decreases were highly significant statistically ($p < .01$). These
findings would normally be interpreted as evidence of effectiveness of the treatment control program. However, this interpretation was radically altered when
the findings in the six control towns (where no efforts were made by the research
team to diagnose or treat tuberculosis) were also taken into account; in these
towns, the baseline random sample survey and only the second population survey
had been undertaken. Table 4.23 shows that the prevalence decreased in four
treatment towns and five control towns, and increased in two and one, respectively.
Also, the mean decrease of 1.1 per 1000 in the 6 treatment towns was slightly
less than the corresponding decrease of 1.5 in the 6 control towns. It was rightly
concluded that the treatment program per se had made no impact on the prevalence
of bacillary tuberculosis in the community, and the authors then proceeded to
discuss the reasons for this rather surprising finding.

16.2. Dangers of Employing Retrospective Controls

The need for the controls being a concurrent group is not always appreciated.
Thus, we sometimes have situations (especially in hospitals) where the efficacy
of a new regimen is determined in patients admitted in one year, and a comparison
made with the efficacy of the control regimen in patients treated in previous years.
Such comparisons are unwise, as there are many factors that could vary from one
point in time to another. For instance, in hospital trials, the disease condition of
the patients might be different in different years because of changes in hospital
policy, bed strength, or diagnostic techniques. In outpatient trials, the cooperation
shown by the patients in attending the clinic to collect drug supplies or self-
administering them subsequently may vary from one year to another because of
various socioeconomic factors. It is also likely that the overall levels of clinic
supervision and laboratory standards for different tests may vary with time, es-

Table 4.23. Effectiveness of domiciliary chemotherapy at community level

	Base-line survey[a] (a)	Resurvey[b] (b)	Difference (a-b)
	Prevalence of smear-positive tuberculosis (per 1000 adults)		
Treatment towns			
A	4.3	4.6	−0.3
B	3.2	2.9	0.3
C	10.0	6.4	3.6
D	10.3	6.9	3.4
E	3.2	2.5	0.7
F	4.5	5.4	−0.9
Mean	5.9	4.8	1.1
Control towns			
G	11.2	5.5	5.7
H	6.0	2.1	3.9
I	1.8	1.5	0.3
J	5.6	5.5	0.1
K	7.9	6.7	1.2
L	1.8	3.9	−2.1
Mean	5.7	4.2	1.5

[a]20–25% random sample survey in 1959.
[b]Population survey during 1965–1968.
Source: Ref. 53.

pecially with changes in personnel. The most satisfactory way out of these problems is the use of concurrent controls.

16.3. *Importance of Random Allocation*

In any controlled clinical trial, the mode of deciding the treatment regimen for any individual patient must not only be free of bias but also appear to be free of bias. If the choice is left to the clinician, there is a possibility that more ill patients may not be admitted to the test treatment (e.g., domiciliary chemotherapy in the 1950s) and this could result in two groups of patients (e.g., home and sanatorium patients) who are not comparable initially; on the other hand, if the clinician believes that the test treatment is better, the bias could work in the opposite direction. To investigators who believe they are absolutely objective, it would be relevant to recall Sir John Crofton's warning, "Let no investigator say that he is unbiased; he is only making it clear that he has no insight."

Alternation is sometimes employed, but it is, again, a procedure that could lend itself to bias. Random allocation (i.e. allocation based on numbers generated by a random process) is the ideal method and is now invariably employed in all controlled trials. The great advantage of random allocation is that it is highly likely to result in the construction of groups which are similar in all characteristics: (a) known and measured, (b) known but immeasurable or not measured, and (c) the unknown. In trials that are not based on random allocation, similarity by characteristics in category (a) can be verified but not those in category (b) or (c); thus, doubt will always remain about the interpretation of any differences in response between the groups. (In the case of known and measured characteristics that have prognostic importance, a further precaution could be to stratify the patients into two or more groups—noncavitated and cavitated—and undertake the allocation from separate series of sealed envelopes, one for each group.) Apart from preventing personal bias, random allocation fulfills another important role; that is, it provides validity for the tests of statistical significance undertaken on the outcome measures.

16.4. Dangers of Bias in the Management of the Patients After Admission to Trial

Despite random allocation, the findings of a controlled clinical trial can become unacceptable if there has been a failure to ensure similarity in the subsequent management of the patients in the two series. It requires a great deal of planning and effort to ensure similarity during treatment. Thus, it is necessary to set out in advance (1) the frequency of examinations during treatment—clinical, radiographic, and bacteriological, (2) the nature and frequency of checks on drug regularity, (3) procedures for dealing with defaulters, and (4) procedures for the recording of symptoms of toxicity. Finally, and most important, the circumstances under which a patient may be withdrawn from the study must be specified in advance. For instance, the criteria could be serious radiographic or clinical deterioration in the presence of a positive sputum or major drug toxicity. It must be emphasized that all these procedures must be implemented alike for all patients, regardless of the treatment regimen.

In situations where the determination of therapeutic efficacy is the most important aim of a controlled trial and can be done objectively, the physician invariably knows the details of the treatment regimen (drugs, dosages, and duration) prescribed to any individual patient. As stated above, bias in the management of the patients receiving different regimens is usually prevented by careful planning and standardization of various procedures. However, if the assessment of toxicity is the main aim, knowledge of the treatment prescribed may be undesirable as such assessments are often subjective; for instance, giddiness due to streptomycin or neurologic symptoms due to isoniazid. In these circumstances, the only way to prevent bias is to withhold from the physician and the individual patient the identity of the treatment regimen prescribed to that patient. Such trials are called "double-blind" trials and require a great deal of planning and ingenious procedures to be really satisfactory.

16.5. Care to Be Taken at the Evaluation Stage

Although random allocation can be expected to yield two series which are very similar in the initial condition, analyses should nevertheless be undertaken to check whether the two series are, in fact, similar. If, by chance, there are any unusually large differences, these must be allowed for by statistical procedures such as standardization or analysis of covariance. In some trials, appreciable losses in follow-up could occur for various reasons, and such patients are then excluded from the analyses. In such situations, it should be verified that the reasons for exclusion are unrelated to the treatment regimen, that exclusions have occurred to similar extent in the two series, and that the baseline characteristics of the exclusions are not very different from those of the inclusions. Similarity in subsequent management also needs to be verified by direct analyses. Finally, when assessing radiographic progress, the assessor should be a person unconnected with the day-to-day management of the patients, and the radiographs must be put up to him in a random order. Finally, all differences of interest should be assessed for their statistical significance.

17. Concluding Remarks

The controlled clinical trial approach has made tremendous contributions to our understanding of the principles of the chemotherapy of tuberculosis, to the evolution of effective, nontoxic, and acceptable treatment regimens, and to the formulation of rational policy decisions in tuberculosis programs in several countries. With increasing awareness of its potential and strict conformity to its rigorous methodology, it is bound to have substantial impact on policy decisions in the future.

References

1. Hill AB (1961) Principles of Medical Statistics, 7th Ed. London: The Lancet Limited.
2. Fox W (1971) The scope of the controlled clinical trial illustrated by studies of pulmonary tuberculosis. Bull WHO 45:559.

3. Medical Research Council (1948) Streptomycin treatment of pulmonary tuberculosis. Br Med J 2:769.

4. Medical Research Council (1950) Treatment of pulmonary tuberculosis with streptomycin and para-aminosalicylic acid. Br Med J 2:1073.

5. Tuberculosis Chemotherapy Centre, Madras (1960) A concurrent comparison of isoniazid plus PAS with three regimens of isoniazid alone in the domiciliary treatment of pulmonary tuberculosis in South India. Bull WHO 23:535.

6. East African/British Medical Research Council Isoniazid Investigation (1960) Comparative trial of isoniazid alone in low and high dosage and isoniazid plus PAS in the treatment of acute pulmonary tuberculosis in East Africans. Tubercle 41:83.

7. Medical Research Council (1953) Isoniazid in the treatment of pulmonary tuberculosis—Second report. Br Med J 1:521.

8. Mount FW, Jenkins BE, Ferebee SH (1953) Control study of comparative efficacy of isoniazid, streptomycin–isoniazid, and streptomycin–para-aminosalicylic acid in pulmonary tuberculosis therapy. Am Rev Tuberc 68:264.

9. British Medical Research Council (1962) Long-term chemotherapy in the treatment of chronic pulmonary tuberculosis with cavitation. Tubercle 43:201.

10. MacDonald FW (1968) Study of triple versus double drug therapy of cavitary tuberculosis. As quoted by Fox W (1968) The John Barnwell lecture—Changing concepts in the chemotherapy of pulmonary tuberculosis. Am Rev Respir Dis 97:767.

11. East African/British Medical Research Council Third Thiacetazone Investigation (1966) Isoniazid with thiacetazone (thioacetazone) in the treatment of pulmonary tuberculosis in East Africa—Third investigation: The effect of an initial streptomycin supplement. Tubercle 47:1.

12. East African/British Medical Research Council Fifth Thiacetazone Investigation (1970) Isoniazid with thiacetazone (thioacetazone) in the treatment of pulmonary tuberculosis in East Africa—First and second reports. Tubercle 51:123, 353.

13. Gangadharam PRJ, Devadatta S, Fox W, Narayanan Nair C, Selkon JB (1961) Rate of inactivation of isoniazid in South Indian patients with pulmonary tuberculosis. 3. Serum concentrations of isoniazid produced by three regimens of isoniazid alone and one of isoniazid plus PAS. Bull WHO 25:793.

14. Tuberculosis Chemotherapy Centre, Madras (1964) A concurrent comparison of intermittent (twice-weekly) isoniazid plus streptomycin and daily isoniazid plus PAS in the domiciliary treatment of pulmonary tuberculosis. Bull WHO 31:247.

15. Ramakrishnan GV, Devadatta S, Evans C, Fox W, Menon NK, Nazareth O, Radhakrishna S, Sambamoorthy S, Stott H, Tripathy SP, Velu S (1969) A four-year follow-up of patients with quiescent pulmonary tuberculosis at the end of a year of chemotherapy with twice-weekly isoniazid plus streptomycin or daily isoniazid plus PAS. Tubercle 50:115.

16. International Union Against Tuberculosis Investigation (1970) A controlled trial of three regimens of self-administered and supervised chemotherapy for pulmonary tuberculosis. Bull Int. Union Against Tuberc 44:8.

17. Tuberculosis Chemotherapy Centre, Madras (1973) Controlled comparison of oral twice-weekly and oral daily isoniazid plus PAS in newly diagnosed pulmonary tuberculosis. Br Med J 2:7.

18. Decroix G, Kreis B, Sors C, Birenbaum, Le Lirzin M, Canetti G (1971) Etude comparative du traitement de la tuberculose pulmonaire par l' association rifampicine–isoniazide administree quotidiennement et deux fois par semaine pendant une annee. Rev Tuberc Pneunol 35:39.

19. Tuberculosis Chemotherapy Centre, Madras (1970) A controlled comparison of a twice-weekly and three once-weekly regimens in the initial treatment of pulmonary tuberculosis. Bull WHO 43:143.

20. Tuberculosis Chemotherapy Centre, Madras (1973) A controlled comparison of two fully supervised once-weekly regimens in the treatment of newly diagnosed pulmonary tuberculosis. Tubercle 54:23.

21. Tuberculosis Chemotherapy Centre, Madras (1974) Comparison of streptomycin plus slow-release isoniazid (matrix isoniazid) with streptomycin plus ordinary isoniazid, given once-weekly, in the initial treatment of pulmonary tuberculosis. Report on Research Activities during 1974, p. 4.

22. East African/British Medical Research Councils (1973) Controlled clinical trial of four short course (6-month) regimens of chemotherapy for treatment of pulmonary tuberculosis. Second Report. Lancet 1:1331.

23. Snider DE, Jr, Long MW, Cross FS, Farer LS (1984) Six months isoniazid–rifampicin therapy for pulmonary tuberculosis. Report of a United States Public Health Service cooperative trial. Am Rev Respir Dis 129:573.

24. Second East African/British Medical Research Council Study (1976) Controlled clinical trial of four 6-month regimens of chemotherapy for pulmonary tuberculosis. Second report. Am Rev Respir Dis 114:471.

25. Third East African/British Medical Research Council Study (1980) Controlled clinical trial of four short-course regimens of chemotherapy for two durations in the treatment of pulmonary tuberculosis—Second report. Tubercle 61:59.

26. Hong Kong Chest Service/British Medical Research Council (1979) Controlled trial of 6-month and 8-month regimens in the treatment of pulmonary tuberculosis: The results up to 24 months. Tubercle 60:201.

27. Balasubramanian R (1991) Fully intermittent six month regimens for pulmonary tuberculosis in South India. Indian J Tuberc 38:51.

28. Hong Kong Chest Service/British Medical Research Council (1981) Controlled trial of four thrice-weekly regimens and a daily regimen, all given for 6 months, for pulmonary tuberculosis. Lancet 1:171.

29. Tuberculosis Research Centre, Madras/National Tuberculosis Institute, Bangalore (1986) A controlled clinical trial of 3- and 5-month regimens in the treatment of sputum-positive pulmonary tuberculosis in South India. Am Rev Respir Dis 134:27.

30. East African/British Medical Research Councils Study (1981) Controlled clinical trial of five short-course (4-month) chemotherapy regimens in pulmonary tuberculosis—Second report of the 4th study. Am Rev Respir Dis 123:165.

31. Singapore Tuberculosis Service/British Medical Research Council (1981) Clinical trial of six-month and four-month regimens of chemotherapy in the treatment of pulmonary tuberculosis: the results up to 30 months. Tubercle 62:95.

32. Tuberculosis Research Centre, Madras (1983) Study of chemotherapy regimens of 5 and 7 months' duration and the role of corticosteroids in the treatment of sputum-positive patients with pulmonary tuberculosis in South India. Tubercle 64:73.

33. Santha T, Nazareth O, Krishnamurthy MS, Balasubramanian R, Vijayan VK, Janardhanam B, Venkataraman P, Tripathy SP, Prabhakar R (1989) Treatment of pulmonary tuberculosis with short course chemotherapy in South India—5-year follow-up. Tubercle 70:229.

34. Dawson JJY, Devadatta S, Fox W, Radhakrishna S, Ramakrishnan CV, Somasundaram PR, Stott H, Tripathy SP, Velu S (1966) A 5-year study of patients with pulmonary tuberculosis in a concurrent comparison of home and sanatorium treatment for one year with isoniazid plus PAS. Bull WHO 34:533.

35. Evans C, Devadatta S, Fox W, Gangadharam PRJ, Menon NK, Ramakrishnan CV, Sivasubramanian S, Somasundaram PR, Stott H, Velu S (1969) A 5-year study of patients with pulmonary tuberculosis treated at home in a controlled comparison of isoniazid plus PAS with 3 regimens of isoniazid alone. Bull WHO 41:1.

36. Nazareth O, Devadatta S, Fox W, Menon NK, Radhakrishna S, Rajappa D, Ramakrishnan CV, Somasundaram PR, Stott H, Subbammal S, Velu S (1971) Two controlled studies of the efficacy of isoniazid alone in preventing relapse in patients with bacteriologically quiescent pulmonary tuberculosis at the end of one year of chemotherapy. Bull WHO 45:603.

37. Nazareth O, Parthasarathy R, Ramakrishnan CV, Santha T, Sivasubramanian S, Somasundaram PR, Subbammal S, Tripathy SP (1977) Efficacy of once-weekly isoniazid–streptomycin in preventing relapse of pulmonary tuberculosis. Indian J Med Res 65:35.

38. Tuberculosis Chemotherapy Centre, Madras (1959) A concurrent comparison of home and sanatorium treatment of pulmonary tuberculosis in South India. Bull WHO 21:51.

39. Kamat SR, Dawson JJY, Devadatta S, Fox W, Janardhanam B, Radhakrishna S, Ramakrishnan CV, Somasundaram PR, Stott H, Velu S (1966) A controlled study of the influence of segregation of tuberculous patients for one year on the attack rate of tuberculosis in a 5-year period in close family contacts in South India. Bull WHO 34:517.

40. Ramakrishnan CV, Rajendran K, Jacob PG, Fox W, Radhakrishna S (1961) The role of diet in the treatment of pulmonary tuberculosis. An evaluation in a controlled chemotherapy study in home and sanatorium patients in South India. Bull WHO 25:339.

41. Ramakrishnan CV, Rajendran K, Mohan K, Fox W, Radhakrishna S (1966) The diet, physical activity and accommodation of patients with quiescent pulmonary tuberculosis in a poor South Indian community. Bull WHO 34:553.

42. East African/British Medical Research Council's Investigation (1969) The results from

twelve to thirty-six months in patients submitted to two studies of primary chemotherapy for pulmonary tuberculosis in East Africa. Tubercle 50:233.

43. Krishnaswami KV, Somasundaram PR, Tripathy SP, Vaidyanathan B, Radhakrishna S, Fox W (1981) A randomised study of two policies for managing default in outpatients collecting supplies of drugs for pulmonary tuberculosis in a large city in South India. Tubercle 61:103.

44. Miller AB, Nunn AJ, Robinson DK, Ferguson GC, Fox W, Tall R (1970) A second international co-operative investigation into thioacetazone side-effects. 1. The influence of a vitamin and antihistamine supplement. Bull WHO 43:107.

45. Tuberculosis Chemotherapy Centre, Madras (1963) The prevention and treatment of isoniazid toxicity in the therapy of pulmonary tuberculosis 1: An assessment of two vitamin B preparations and glutamic acid. Bull WHO 28:455.

46. Tuberculosis Chemotherapy Centre, Madras (1963) The prevention and treatment of isoniazid toxicity in the therapy of pulmonary tuberculosis 2: An assessment of the prophylactic effect of pyridoxine in low dosage. Bull WHO 29:457.

47. Hong Kong Tuberculosis Treatment Services/British Medical Research Council Investigation (1972) A study in Hong Kong to evaluate the role of pretreatment susceptibility tests in the selection of regimens of chemotherapy for pulmonary tuberculosis. Am Rev Respir Dis 106:1.

48. Hong Kong Chest Service/Tuberculosis Research Centre, Madras/British Medical Research Council (1984) A controlled trial of 2-month, 3-month, and 12-month regimens of chemotherapy for sputum-smear-negative pulmonary tuberculosis. Am Rev Respir Dis 130:23.

49. Hong Kong Chest Service/Tuberculosis Research Centre, Madras/British Medical Research Council (1989) A controlled trial of 3-month, 4-month, and 6-month regimens of chemotherapy for sputum-smear-negative pulmonary tuberculosis. Am Rev Respir Dis 139:871.

50. Tuberculosis Society of Scotland (1958) A controlled trial of chemotherapy in pulmonary tuberculosis of doubtful activity. Tubercle 39:129.

51. Amberson JB, Jr, McMahon BT, Pinner M (1931) A clinical trial of sanocrysin in pulmonary tuberculosis. Am Rev Tuberc 24:401.

52. Miller AB, Fox W, Tall R (1966) An international cooperative investigation into thiacetazone (thioacetazone) side-effects. Tubercle 47:33.

53. Frimodl-Möller J, Acharyulu GS, Kesava Pillai K (1981) A controlled study of the effect of a domiciliary tuberculosis chemotherapy programme in a rural community in South India. Indian J Med Res 73(Suppl):1.

5

Chemotherapy of Tuberculosis in Developed Countries

Asim K. Dutt and Jay B. Mehta

1. Introduction

The chemotherapeutic era of tuberculosis began with the discovery of streptomycin (SM) in 1944 and its trial in 1945 (1,2). Soon after the introduction of isoniazid (INH) in 1952 (3), drug therapy was adopted as an important weapon in the treatment of this ancient disease. In the early stages of chemotherapy, it became apparent that drug resistance and treatment failure resulted when monotherapy with SM was used to treat active disease harboring large bacillary populations (4). The addition of a second drug to the regimen, such as para-aminosalicylic acid (PAS) or INH, prevented the emergence of drug resistance and treatment failure (5,6). Soon, the efficacy of chemotherapy was firmly established. Provided that the appropriate drug combinations were utilized against susceptible pretreatment organisms, tuberculosis could be cured within 18–24 months.

It very rapidly became clear that treatment of tuberculosis could be summarized in one word—chemotherapy. Mitchison's statement that host factors, such as type of disease and patient's immune response, decrease in importance as the potency of chemotherapeutic agents increase has proven to be true with the passage of time (7). The bacilli are killed by the drugs and not by the patient. Thus, several recognized aspects of therapy, such as prolonged bed rest, surgery, and nutritional supplements, soon became irrelevant in curing the disease.

During the past three decades, the treatment of tuberculosis has undergone a revolution with the availability of more powerful bactericidal drugs. The disease now can be cured with only 6–9 months of therapy (8). In certain situations, treatment might be completed in even less than 6 months (9).

Although several effective antituberculosis drugs are currently available, selection of appropriate therapeutic regimens and duration of therapy are revised frequently due to changing epidemiology. In developed countries, human immunodeficiency virus (HIV) infection, deteriorating social conditions, such as homelessness, poverty, and drug abuse, and immigration from countries with a high prevalence of drug-resistant disease have adversely impacted the outcome of disease. Such factors have led to problems with compliance of therapy, resulting in the resurgence of drug-resistant disease. Because potent new antituberculosis drugs have not been added to the drug armamentarium, frequent reshuffling of drug regimens and their durations are made according to existing situations. Additional tuberculosis control measures are recommended for prevention of drug-resistant disease and its transmission. Directly observed therapy (DOT) is being advocated more frequently. Thus, a thorough understanding of the proper use of chemotherapeutic agents, including the use of newer drugs and other treatment modalities, is essential for achieving favorable outcome and eventual eradication of tuberculosis.

2. Basics of Chemotherapy

Three aspects of disease must be clearly recognized in planning chemotherapy for tuberculosis: (1) tubercle bacilli, (2) tuberculosis lesions, and (3) action of antituberculosis drugs.

2.1. Tubercle Bacilli

Mycobacterium tuberculosis is an obligate aerobe, which thrives best in physiologic pH of 7.4 and a pO_2 of 100–140 mm of Hg. For this reason, large populations of organisms are found inside cavities where oxygen tension is high and the medium is slightly alkaline. Organisms are in active metabolic state and divide frequently in this environment (10). In contrast, the populations of bacilli in closed caseous lesions or within phagocytes are small and metabolize less actively due to reduced oxygen tension and a less favorable pH.

Unlike other common bacteria, tubercle bacilli divide only every 16–20 h, making the infection chronic and indolent. For this reason, initial daily intensive drug therapy is necessary in order to kill large bacterial populations. After control of the disease is achieved, fewer drugs and less frequent doses for an extended period are required to fully eradicate the bacilli from the tissues. The concept of two-phase chemotherapy, (1) initial intensive therapy to eliminate large bacillary populations followed by (2) less intense therapy for a longer period after the population has been substantially reduced (continuation phase), was developed to deal with this characteristic of the bacilli (11).

Another trait of the bacilli involves the natural occurrence of resistant mutants, irrespective of exposure to drugs (12). Even in virgin populations of tubercle bacilli, resistant mutants to any single effective drug occur with a frequency of 10^{-5}–10^{-6} (13). The frequency of resistant mutants for rifampin (RIF) is 1 in 10^8, whereas it is 1 in 10^6 for INH and SM. In cavitary disease with positive sputum smears, the bacillary population may be on the order of 10^7–10^{10}; there may be 10^3–10^4 organisms which are resistant to any single effective drug. Fortunately, the chance of an organism being resistant to two drugs simultaneously (10^{-10}–10^{-12}) is virtually negligible. Monotherapy or improperly administered two-drug therapy allows selective survival of drug-resistant mutants with emergence of drug-resistant disease leading to therapy failure.

2.2. Tuberculosis Lesions

Mycobacterium tuberculosis exists in four separate populations in tuberculous lesions (Fig. 5.1), and a knowledge of this fact helps when selecting appropriate drugs for treatment. Large bacterial populations are present extracellularly in the walls of cavitary lesions which actively multiply in a neutral or slightly alkaline medium (14). As the cavities and caseous lesions communicate with the bronchi,

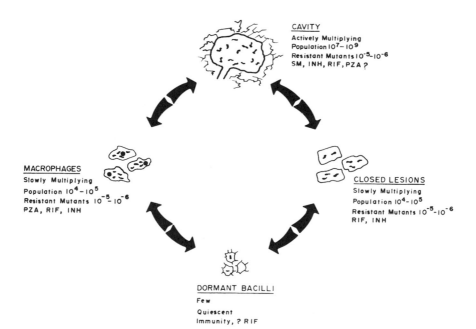

Figure 5.1. Bacterial populations in tuberculosis lesion.

the large actively multiplying bacillary population can be recognized by a positive smear in the bronchial secretions. This population contains a significant number of resistant mutants to a single drug. Effective chemotherapy is vital in eliminating this large population in order to achieve control of the infection without permitting the survival of drug-resistant mutants (bactericidal action). Such action is achieved by combining potent bactericidal drugs. Second, a smaller population (10^2–10^5) of bacilli is located in closed caseous lesions where the multiplication is slow or intermittent. Third, a similar bacillary population is present inside the acid milieu of macrophages. These slowly or intermittently multiplying bacilli are captive and are not reflected in the sputum examination. Prolonged chemotherapy (continuation phase) is required to ensure elimination (sterilizing action) of these latter two bacillary populations. If not eradicated completely, these "persisters" may be the cause of relapse by means of late growth. Finally, a few truly dormant bacilli occasionally remain even after effective chemotherapy. These are generally eliminated by the body's immune defenses.

2.3. Antituberculosis Drugs

Present chemotherapy for tuberculosis must be based on a combination of effective bactericidal drugs (14–16). The most common of these and their respective dosages and side effects are outlined in Table 5.1. Bactericidal drugs (first-line drugs) of today include INH, RIF, pyrazinamide (PZA), SM, and ethambutol (EMB). Due to their potency, minimal side effects, and ease of administration, the duration of therapy is shortened. Each bactericidal drug has its own mechanism and individual site of action on the bacterial population (Table 5.1; Fig. 5.1) (10,14). Although RIF has better bioavailability on an empty stomach, the time of the medication can be modified to improve compliance. Drugs such as cycloserine are better tolerated when taken on a full stomach. Bacteriostatic drugs (second-line drugs) have no place in the initial treatment of tuberculosis. They prevent selection of drug-resistant mutants but have very little effect on the bactericidal activity of a drug regimen—an essential element in reducing the duration of therapy. With their increased side effects and low potency, bacteriostatic agents generally are reserved for the treatment of drug-resistant disease or when first-line drugs are prohibited due to toxicity. The duration of therapy is prolonged with the use of second-line drugs.

Streptomycin is bactericidal for the large actively multiplying extracellular bacillary population in the slightly alkaline medium of the liquid caseous material and in the walls of cavities (17,18). Both INH and RIF are effective against actively multiplying large extracellular populations, as well as against the smaller slowly metabolizing population located in closed caseous lesions and within macrophages (14,15,19). Rifampin is particularly successful in eliminating bacilli exhibiting minimal intermittent metabolic activity and, thus, has a powerful ster-

Table 5.1. Antituberculosis drugs: Dosage, side effects, and mode of action

Drug	Daily dosage	Twice weekly dosage	Side effects	Mode of action
First-Line Drugs				
Streptomycin and other aminoglycosides	10–15 mg/kg (usually 0.5–1.0 g), 5 days/week	20–25 mg/kg (usually 1.0–1.5 g) IM	Cranial nerve VIII damage (vestibular and auditory) nephrotoxicity, allergic fever, rash	Active against rapidly multiplying bacilli in neutral or slightly alkaline extracellular medium
Capreomycin	Same as above	Same as above	Same as above	Same as above
Isoniazid (INH)	5 mg/kg (usually 300 mg), PO or IM	15 mg/kg (usually 900 mg) PO	Peripheral neuritis, hepatotoxicity, allergic fever and rash, lupus erythematosus phenomenon	Acts strongly on rapidly dividing extracellular bacilli: acts weakly on slowly multiplying intracellular bacilli
Rifampin (RIF)	10 mg/kg (usually 450–600 mg), PO	10 mg/kg (usually 450–600 mg) PO	Hepatotoxicity, nausea and vomiting, allergic fever and rash, flulike syndrome, petechiae with thrombocytopenia or acute renal failure during intermittent therapy	Acts on both rapidly and slowly multiplying bacilli either extracellular or intracellular, particularly on slowly multiplying "persisters"
Rifabutin (Ansamycin)	150–300 mg, PO			Same as above
Pyrazinamide (PZA)	25–30 mg/kg (usually 2.5 g), PO	45–50 mg/kg (usually 3.0–3.5 mg) PO	Hyperuricemia, hepatotoxicity, allergic fever and rash	Active in acid pH medium on intracellular bacilli
Ethambutol (EMB)	15–25 mg/kg (usually 800–1600 mg), PO	50 mg/kg PO	Optic neuritis, skin rash, hyperuricemia	Weakly active against both extracellular and intracellular bacilli to inhibit the development of resistant bacilli

Table 5.1. (continued) Antituberculosis drugs: Dosage, side effects, and mode of action

Drug	Daily dosage	Twice weekly dosage	Side effects	Mode of action
Second-Line Drugs				
Ethionamide	10–15 mg/kg (usually 500–750 mg) in divided doses, PO	Not used	Nausea, vomiting, anorexia, allergic fever and rash, hepatotoxic, neurotoxic	Same as above
Cycloserine	15–20 mg/kg (usually 0.75–1.0 g) in divided doses with 200 mg of pyridoxine, PO	Not used	Personality changes, psychosis, convulsions, rash	Same as above
Para-aminosalicylic acid (PAS)	150 mg/kg (usually 12 g) in divided doses, PO	Not used	Nausea, vomiting, diarrhea, hepatotoxicity, allergic rash and fever	Weak action on extracellular bacilli: inhibits development of drug-resistant organisms
Thiocetazone (not available in the United States)	150 mg daily, PO	Not used	Allergic rash and fever, Stevens–Johnson syndrome, blood disorders, nausea and vomiting	Same as above
Clofazimine (antileprosy)	100 mg TID, PO	Not used	Pigmentation of skin, abdominal pain	Against *Mycobacterium intracellulare*
Newer agents				
Ofloxacin	400 mg, q12	Not used	Gastrointestinal: diarrhea, nausea, abdominal pain, anorexia, CNS effects: dizziness, restlessness, nightmares, ataxia, seizures	Rapidly multiplying bacilli at neutral or alkaline pH
Ciprofloxacin	750 mg, q12	Not used	Same as above	Same as above
Azithromycin	500 mg/day for up to 30 days	Not used	Gastrointestinal: diarrhea, nausea, abdominal pain, elevation of liver enzymes	Rapidly multiplying bacilli in macrophages, against *M. intracellulare*
Clarithromycin	1 g, q12	Not used	Same as above	Same as above

ilizing effect (20). Isoniazid is inferior in this aspect. Pyrazinamide has very little effect against extracellular populations but is bactericidal for the intracellular bacilli within macrophages in an acidic milieu (21). In clinical application, PZA appears to be more effective in the initial phase of chemotherapy, possibly by eliminating intracellular bacilli. Isoniazid and RIF are also quite effective against intracellular bacilli (17,19). However, SM does not show much activity against organisms in the acid medium (17). Ethambutol, when given in doses of 25 mg/ kg body weight daily, contributes marginally to the bactericidal drug regimen. A combination of INH and RIF is, therefore, bactericidal for all three bactericidal populations in a tuberculous lesion (Fig. 5.1). Additionally, the body's immune system may play a role in eliminating the few remaining truly dormant persisters which might have escaped during therapy.

The majority of patients with tuberculosis are successfully treated with some combination of INH, RIF, PZA, SM, and EMB—standard antituberculosis therapy.

3. Principles of Chemotherapy

As mentioned earlier, chemotherapy of tuberculosis today is based on the use of bactericidal drug combinations. Due to the rapid destruction of bacilli and sterilization of lesions, the duration of therapy has been drastically reduced. The use of second-line drugs is generally limited to the treatment of drug-resistant disease when the duration of therapy is prolonged to 1½–2 years.

The principles of modern chemotherapy are shown in Figure 5.2. Isoniazid, RIF, SM, EMB, and PZA are individually capable of eliminating actively multiplying bacilli. A combination of INH and RIF is a powerful bactericidal combination, provided the organisms are susceptible (22). The addition of PZA and EMB or SM enhances the efficacy of the regimen and achieves sputum conversion to negativity in a shorter period of time (23). When treated adequately, the extracellular bacilli are eliminated rapidly through the bactericidal action of these drugs. Inadequate therapy would result in treatment failure with persistence of positive sputum smears beyond 3 or 4 months of therapy, generally due to development of drug-resistant organisms. After the initial phase of 1–2 months of intensive therapy, INH and RIF are administered daily or twice weekly with or without PZA. They kill the slowly multiplying bacilli within macrophages and in closed caseous lesions (sterilization). With adequate therapy, the lesions are sterilized and any lingering bacilli are eliminated, facilitating a lasting cure of the disease. If adequate therapy for appropriate duration is not given, relapse can occur after therapy has ended due to late growth of persisters which are generally drug-sensitive organisms.

Thus, a combination of INH and RIF might cure drug-sensitive disease in 9

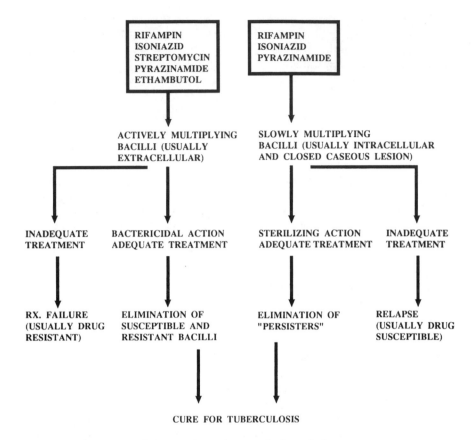

Figure 5.2. Principles of modern chemotherapy of tuberculosis.

months (22), whereas initial intensive therapy with a combination of three or four bactericidal drugs followed by INH and RIF further reduces the duration of therapy to 6 months (23). The two-drug combination of INH and RIF provides insufficient protection against the presence of drug-resistant bacilli to one of them; however, a combination of three or four bactericidal drugs guards against the presence of drug-resistant organisms with a favorable outcome.

4.1 Assessment of Drug Resistance Before Initiation of Therapy

Although effective antituberculosis drugs are available to treat drug-resistant disease, they carry the added problems of increased side effects, lengthy treatment, and high cost. The strategy should therefore be to choose initial drug combinations which cure the disease on the first attempt, thereby precluding any opportunity for drug-resistant disease to develop. Physicians should be aware of the prevalence of drug-resistant organisms in their particular locales of practice when selecting a drug regimen. In the United States, primary resistance rose to 9% within 4 years (1982–1986)—mostly to INH (5.3%) and SM (4.9%) (24). Isolation of drug-resistant organisms to other drugs was rare. However, there are specific geographic areas where the primary resistance is more common, e.g., southern Texas, where primary resistance to one drug was found in 20% of patients and resistance to RIF in approximately 10% (25). Similarly, drug resistance, primary or secondary, was found to be increasing in a New York institution to at least 31% to one drug and 15% to RIF (26).

Prior chemotherapy with INH or SM is strong evidence for the presence of drug-resistant bacilli (24,27). In one study, the risk of relapse with INH-resistant organisms increased by about 4% per each month of prior therapy; even 1 month of treatment with INH may cause INH-resistant disease of up to 23% (28,29). In one southern California hospital, drug-resistant organisms were found in 71% of patients with cavitary disease and who had received prior treatment (30). The prevalence of resistance to INH and/or SM is very high in developing countries— as much as 50% in some regions (31,32). In a Centers for Disease Control (CDC) survey (1982–1986), the rate of both primary and acquired drug resistance were two times higher among foreign-born persons than persons born in the United States (24). Foreign-born persons and those with infection acquired in developing countries, (e.g., Asia, Latin American, and Africa) must be suspected of harboring drug-resistant organisms. In the United States, foreign-born persons contributed to 30% of the total caseload in 1993 (33). Overall, non-U.S.-born individuals accounted for 60% of the increase in cases from 1986 through 1992 (33). History of previous contact with drug-resistant disease should also increase suspicion of drug-resistant bacilli (33–35).

In developed countries, drug-resistant disease often develops in or is acquired by the homeless, drug abusers, and HIV-infected persons. Nardell et al. reported over 60% drug-resistant disease among homeless persons diagnosed with tuberculosis (TB) in a Boston hostel (36). A study of social conditions of TB patients in New York (1982–1987) revealed resistant bacilli in 8% of patients living in homes, 21% without homes, and 42% of African Americans without homes. In New York City, overall drug-resistant disease has been detected in a high proportion of cases—33% to one drug and 19% to INH and RIF. Several microepidemics of multidrug-resistant tuberculosis (MDR-TB) with mortality rates of 80% or more have been reported, mostly affecting HIV-infected persons and members of other high-risk groups in hospitals, prisons, and residential facilities (27).

In patients who harbor drug-resistant bacilli, use of a drug regimen containing only two bactericidal agents can result in development of drug-resistance to the second drug as well. Recognized clinical situations with risk factors for drug-resistant bacilli, particularly to INH, are profiled in Table 5.2. In such situations, chemotherapy must be initiated with a combination of at least four bactericidal drugs (INH, RIF, PZA, SM, or EMB) for 6–8 weeks until the susceptibility results are known (22). Sometimes, regimens containing five or more drugs may be recommended, depending on the prevalence of MDR-TB in a particular geographic region (37).

With a combination of four bactericidal drugs, at least two will be active against all bacterial populations in the lesions, even when they harbor INH-resistant bacilli. Rifampin and SM or EMB act against such bacilli in the large extracellular bacterial populations, whereas RIF and PZA are effective against the occasional

Table 5.2. Conditions with increased risk of drug-resistant TB

1. History of previous treatment with anti-TB drugs including preventive therapy
2. Patients from areas with high prevalence of initial or primary drug resistance (>4%), e.g., urban population in the northeast United States, Florida, California, U.S.–Mexican border, etc.
3. Foreign-born persons from areas with high prevalence of drug resistant TB, e.g., Southeast Asia, Mexico, South America, Africa, etc.
4. Contacts of persons with drug-resistant disease
5. Disease in persons who are homeless, drug abusers, and HIV infected
6. Persons with positive sputum smears and cultures after 3 months of chemotherapy

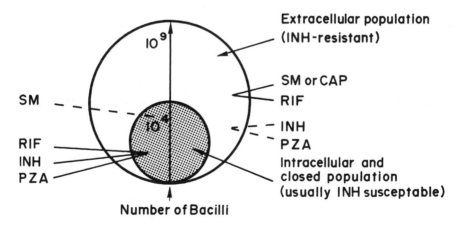

Figure 5.3. Drug actions in the retreatment regimen used in cases of acquired INH resistance: A solid line indicates that a drug can reach a population and is effective. A broken line indicates that a drug either does not reach a population or is ineffective at the site. Thus, two or three drugs will be effective against resistant bacilli in both extracellular and intracellular/closed lesion populations.

small populations of resistant bacilli found intracellularly or that exist in closed caseous lesions (Fig. 5.3) (38). After susceptibility results are known, the therapy can be changed accordingly. In a study by Mitchison and Nunn, initial resistance to INH or SM did not compromise the results with 6-month, four-drug intensive treatment. However, cases with initial RIF resistance responded less successfully than did those with resistance to INH (39).

When resistance is found to two or more drugs, including INH and RIF (MDR-TB), two or more agents to which the organisms are susceptible must replace the regimen. The drugs should be administered in full tolerable doses and for a prolonged period. If EMB is used, the dose should be 25 mg/kg body weight daily, or 50 mg/kg body weight twice weekly. For daily therapy, the dose of EMB can be reduced to 15 mg/kg after the conversion of sputum bacteriology to negativity in a few months. Multidrug-resistant tuberculosis cases often call for second-line drugs, which are more toxic and less effective. Some of the more difficult instances require newer antituberculosis agents and/or surgery in selected circumstances.

5. Chemotherapeutic Regimens and Their Selection

Tuberculosis is a treatable and curable disease with the same principles of treatment for persons of all ages (22). Complete recovery is expected with appropriate and timely drug therapy. Although principles of drug therapy remain the same, several countries prefer to have a national policy on drug regimens. In contrast, U.S. physicians individualize treatment regimens according to the circumstances and needs of each patient. Various drug regimens and durations of treatment are shown in Table 5.3.

5.1. Nine-Month Regimen

Several studies in the United States, England, and other countries have established that combinations of two bactericidal drugs (INH and RIF) are highly successful in curing tuberculosis (8,40,41). In England and France, a 9-month regimen was adopted, consisting of INH and RIF with an initial supplement of SM or EMB for 3 months; it was proven highly effective with almost no relapse (42). The addition of a third drug (EMB or SM) did not add significantly to the bactericidal action of INH and RIF. However, the supplement ensures against failure in the event of resistance to INH. The combination of INH 300 mg and RIF 600 mg daily by mouth on an empty stomach for 9 months' duration has been recommended for all forms of drug-sensitive tuberculosis (8). Based on information that appeared in 1975, the authors were the first in the United States to adopt a regimen for short-course therapy. This formula, consisting of INH 300

Table 5.3. Chemotherapy regimens for treatment of tuberculosis

9-Month regimen
   INH[a] 300 mg + RIF 600 mg daily for 9 months
or
   INH 300 mg + RIF 600 mg daily for 1 month
   INH 900 mg + RIF 600 mg twice weekly for 8 months

6-Month regimen
   INH 300 mg + RIF 600 mg + PZA 25–30 mg/kg/body weight daily for 2 months
      followed by INH 300 mg + RIF 600 daily for 4 months
or
   INH 900 mg + RIF 600 mg twice weekly for 4 months
For *suspected drug resistance:* ethambutol 25 mg/kg daily or streptomycin 0.5–0.75 g
      (five times/week) initially for 2 months to above regimen or until drug-
      susceptibility results are known

[a]INH, isoniazid; RIF, rifampin; PZA, pyrazinamide.

mg and RIF 600 mg given daily for 1 month followed by INH 900 mg and RIF 600 mg twice weekly for another 8 months (43), was successful in more than 95% of the cases in which we conducted long-term follow-up (44). Emergence of drug resistance has not been a problem and toxicity is minimal. Our experience has shown the regimen to be effective in patients with tuberculosis and concomitant conditions such as diabetes, alcoholism, corticosteroid therapy, malignancy, and cytotoxic drugs (45). The regimen is also effective against all forms of extrapulmonary tuberculosis (46).

This two-drug, twice weekly, largely intermittent 9-month drug regimen has the advantages of very low cost and ease of supervision, when indicated. The regimen has few side effects and can be easily supervised by nursing home staff or by friends/family of patients living at home. The same applies to pediatric tuberculosis (47). Likewise, this regimen is most suitable for elderly patients, where drug resistance is rare, because drug resistance is rare in the recrudescent disease arising from old infection acquired many years ago when drug resistance was uncommon. On the other hand, this regimen is not recommended in areas where initial drug resistance to INH is high (>3%) or expected to be so. (Table 5.2). In such situations, it is mandatory to add additional drugs (PZA and/or EMB) until the drug-susceptibility results are known.

5.2. Six-Month Regimen

Several studies conducted by the British Medical Research Council in the early 1960s revealed that intensive daily therapy with INH, RIF, PZA, and SM initially for 2 months followed by INH and RIF either daily or twice weekly for 4 months

Table 5.4. Regimen options for the initial treatment of TB among children and adults

TB without HIV infection			
Option 1	Option 2	Option 3	TB with HIV infection
INH, RIF, and PZA (if initial INH resistance is < 4%) daily for 8 weeks, followed by INH and RIF daily or twice weekly for 16 weeks	INH, RIF, PZA, EMB or SM daily for 2 weeks, then 2 times/week for 6 weeks (by DOT) and subsequently INH and RIF twice weekly for 16 weeks (by DOT)	INH, RIF, PZA, EMB or SM 3 times/week for 6 months (by DOT)	Options 1, 2, or 3 can be used for a total of 9 months and at least 6 months beyond culture conversion
Add EMB or SM if resistance is > 4%			
Total treatment: 6 months (at least 3 months past culture conversion)	Total treatment: 6 months	Total treatment: 6 months	Total treatment: 6–9 months

Source: Ref. 52.

was highly successful in curing tuberculosis (16). Conversion of the sputum bacteriology to negativity was faster and relapses were rare after discontinuation of therapy. Later, it was demonstrated that results with INH, RIF, and PZA for 2 months followed by INH and RIF daily or twice weekly for 4 months were not improved by adding either SM or EMB in drug-susceptible disease (48). The American Thoracic Society and the Centers for Disease Control (CDC) recommend an initial 2 months' therapy with INH, RIF, and PZA, followed by INH and RIF for 4 months either daily or twice weekly for bacteriologically positive disease (8). Supplementation of PZA to INH and RIF guards against the development of INH-resistant organisms. The impact of INH resistance on the success is not documented but appears to be satisfactory. However, initial RIF resistance can result in a high rate of failure (39,49). The regimen is successful in reducing the duration of therapy and has acceptable side effects (50). Pyrazinamide is generally beneficial in the first 2 or 3 months of therapy, after which it is discontinued (51). The regimen is a suitable alternative to 9 months of INH and RIF; however, the addition of PZA increases the cost of therapy. The Advisory Council for the Elimination of Tuberculosis recommends the regimen when the drug resistance is less than 4% (Table 5.4) (52).

5.3. Six-Month Therapy in Suspected Initial Resistance

When drug resistance is suspected or likely (Table 5.2), therapy must be initiated with four bactericidal drugs consisting of INH 300 mg, RIF 600 mg, PZA 25–30 mg/kg body weight, and EMB 25 mg/kg body weight or SM 0.5–1 g intramuscularly 5 days a week (Table 5.4) (22,53). The therapy is modified later, depending on the results of drug-susceptibility testing, generally available in 6–8 weeks. If the organisms are found to be susceptible, therapy is completed with INH and RIF daily or twice weekly for another 4 months. This regimen is generally successful in cases of INH resistance (16,39). However, therapy may be continued with RIF, EMB, and PZA for another 6–7 months. Because persisters are normally sensitive to INH, this drug can be continued in the regimen. In RIF-resistant cases, another two or three drugs should be administered for an additional 12–18 months to avoid relapse.

6. Treatment of Multidrug-Resistant Tuberculosis

In geographic areas with a high prevalence of multidrug resistance and HIV infection, it may be necessary to initiate therapy with five to seven drugs, including second-line agents, until drug-susceptibility results become available (37). This is particularly applicable for large urban areas of the United States (e.g., New York, New Jersey, Miami, San Francisco, etc.).

Treatment of MDR-TB, including resistance to INH and RIF, is most often unsuccessful even with the best available therapy (53–55). The drugs available are second-line agents, which are less potent, more toxic, and require administration for a prolonged period of time. The patient's ability to complete the full course of therapy is frequently complicated by side effects and lack of adherence to the regimen for the prescribed period.

There are important principles which must be followed while treating MDR-TB: (1) A single drug must not be added to a failing regimen; (2) two or three new drugs to which the patient has not been exposed should replace the existing drug regimen until the susceptibility results are known; (3) therapy should be prolonged to 24 months or more; (4) the drug regimen should consist of an injectable drug which improves compliance and needs to be continued for at least 4 months after bacteriological conversion; (5) compliance should be ensured by DOT; (6) repeat susceptibility testing should be performed if the culture remains positive after 3 months of therapy.

Generally, second-line drugs (ethionamide, cycloserine, PAS, kanamycin, and capreomycin) are required to treat MDR-TB; major side effects and dosages of these drugs are shown in Table 5.1. Such medications all have considerable side effects which need close monitoring. Early minor adjustments in time of administration, dosage, symptomatic treatment of minor symptoms, and reassurance from physicians may help in adherence to compliance. New drugs such as fluoroquinolones (ciprofloxacin, ofloxacin, Levofloxacin) and amikacin have shown good in vitro activity against M. tuberculosis. Fluoroquinolones, noted for their limited side effects, are used increasingly in treatment, although their effectiveness is not proven by controlled studies. In vitro susceptibility studies indicate that levofloxacin has perhaps greater inhibitory and bactericidal activities than other quinolone compounds against either extracellular or intracellular tubercle bacilli (56). Clofazimine, an antileprosy drug whose efficacy is unproven in the treatment of tuberculosis, is sometimes added to the drug regimen. Not available in the United States, thiacetazone is a toxic agent sometimes added to the drug armamentarium in very difficult situations. In HIV-infected patients, it should not be used due to increased toxicity. This drug may be obtained from overseas with permission from the U.S. Food and Drug Administration as an investigational drug.

As mentioned previously, the response to therapy is usually slow, sometimes weeks before any clinical improvement is observed. Bacteriological monitoring of sputum should be regular, at least one specimen per month. Isolation of patients should be strictly followed to avoid transmission of drug-resistant infection to others.

In selected patients with MDR-TB, surgery may be indicated occasionally for removal of destroyed lung tissue after the administration of drug therapy for a few months (37). In one study, a combination of therapy and surgery yielded better results than did medical therapy alone (57).

Preventive therapy for contacts of drug-resistant tuberculosis is important but controversial. Although INH prophylaxis is very effective in preventing disease among contacts infected with drug-susceptible bacilli (58,59), its role in INH-resistant infection is unknown. Rifampin is suggested for INH-resistant but RIF-susceptible infection (60). Two CDC recommended regimens for MDR-TB contacts include (1) PZA and EMB and (2) PZA and ciprofloxacin or ofloxacin for 12–24 months. Due to the uncertain efficacy of these regimens, strict medical and radiological checkups should be performed throughout the duration of therapy (61).

Although the BCG vaccine has been used in developing countries for decades, most tuberculosis experts in the United States remain skeptical of its efficacy and reliability in preventing pulmonary tuberculosis among the general population. Current CDC recommendations for BCG vaccination (62) include newborns in developing countries or geographic areas in which risk for tuberculosis is very high ($\geq 1\%$ per year), and health care personnel working in certain locales where MDR-TB is prevalent.

7. Treatment in HIV-infected Persons

Therapy of drug-susceptible tuberculosis in HIV-infected individuals is uniformly successful; even those with AIDS respond well to treatment. Relapse rates are almost the same in HIV-infected patients as in seronegative patients (63).

Currently recommended is the administration of INH, RIF, PZA, and EMB or SM for 2 months, followed by INH and RIF daily or twice weekly for a total of 9 months and for at least 6 months following conversion of sputum culture to negative (Table 5.4) (52,64). However, in clinical practice, treatment for 6 months appears to be adequate (63). Adverse reactions to antituberculosis drugs occur more frequently and mortality is higher in HIV-infected persons than seronegative patients. Intermittent, two or three times weekly, therapy may be better tolerated and remain effective in HIV-infected patients (Table 5.4) (52).

Another significant problem is the growing prevalence of MDR-TB among HIV-infected persons (65–67). Treatment involves use of many second-line drugs, DOT, and intermittent therapy (67,68). The World Health Organization (WHO) has reported the success of intermittent DOT in an urban setting (68a) as a practical and effective approach to this problem. This topic is discussed in detail in Chapter 8.

8. Treatment of Extrapulmonary Tuberculosis

Generally, the bacterial load in extrapulmonary tuberculosis lesions is much less than in cavitary pulmonary disease. Treatment regimens for 6–9 months, as suggested for pulmonary tuberculosis, are also adequate for extrapulmonary disease (8,46). We treated tuberculosis of various extrapulmonary sites effectively with INH and RIF for 9 months (46). A 6-month regimen should be effective in extrapulmonary tuberculosis as in the pulmonary disease (8). Tuberculosis of the vertebrae (Pott's disease) may require longer therapy, along with debridement and removal of necrotic caseous material from the site. Diagnostic approaches and treatment of various sites of extrapulmonary tuberculosis are summarized in Table 5.5.

9. Treatment of Pediatric Tuberculosis

More often than not, tuberculosis in children is a manifestation of primary infection. The bacterial population is low in primary tuberculosis unless associated with extensive infiltration in the lungs. Preventive therapy with a single drug may be sufficient in cases of a positive Mantoux test but with no radiological evidence of disease. Many studies have shown that 6–9 months of therapy is successful for most forms of tuberculosis in children, including pulmonary disease. Abernathy et al. found that INH (10 mg/kg) and RIF (10–20 mg/kg) daily for 1 month followed by INH (20–40 mg/kg) and RIF (10–20 mg/kg) twice weekly for another 8 months (9 months total) was very effective (47). The duration of therapy can be reduced to 6 months with initial use of INH, RIF, and PZA (69). The American Academy of Pediatrics recommends 6 months' therapy with

Table 5.5. Diagnosis of suspected tuberculosis. Abbreviations: S = smear, C = culture for mycobacteria, H = histology, Br = bronchial, Bronch = bronchoscopy, CSF = cerebrospinal fluid, Bx = biopsy, neg = negative, Gu = genitourinary; * therapy started in suspected cases, awaiting cultural results and/or clinical response.

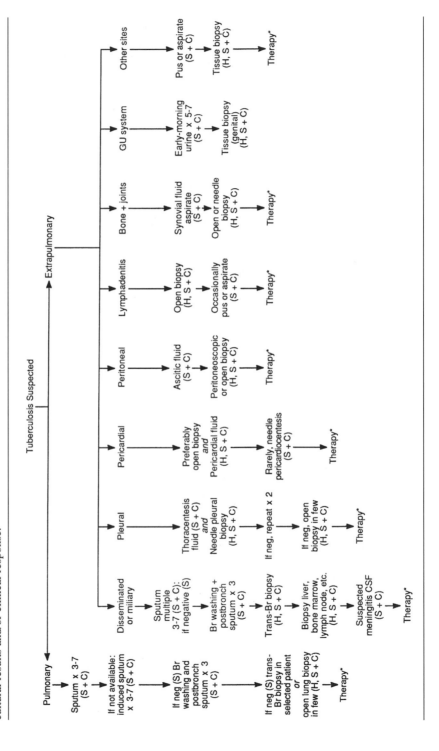

INH and RIF supplemented by PZA for 2 months initially (70). Direct observation of therapy may be necessary in cases involving difficult social issues where reliable self-administration of drugs is doubtful.

10. Smear-Negative Tuberculosis

Positive sputum smears signify a large bacterial population in the lung lesions, usually indicating cavitary disease (12). In noncavitary fibrotic lesions and in disease from recently acquired infection (primary), however, the smears are often negative (71). Narain et al. found that if not treated, 35% of patients with positive smears died within 18 months, whereas 15% of smear-negative/culture-negative patients expired within the same period, indicating a correlation between severity of disease and positive microscopy (72). Thus, smear-negative cases do not require the same intensity and duration of treatment as do smear-positive cases (73).

The British Medical Research Council in a 5-year study found that intensive treatment with SM, INH, RIF, and PZA for 4 months was highly successful, with a 2% relapse rate in smear-negative culture-positive cases (74). Likewise, a Hong Kong study with such patients, found that 4 months of three times weekly treatment with SM, INH, RIF, and PZA showed relapse in only 4% of the cases (75) and is now routinely used in that country. In smear-positive cases, the addition of SM to the regimen of INH, RIF, and PZA offers no benefit (47), and use of PZA beyond 2 months has no advantage (74). In the United States, smear-positive tuberculosis with susceptible bacilli is currently treated with INH, RIF, and PZA for 2 months followed by INH and RIF for another 4 months (76). However, for the treatment of smear-negative, culture-positive disease, the 4-month regimen is adequate.

In Arkansas, we have treated smear-negative, culture-positive disease with INH and RIF for 6 months, achieving success in over 97% of cases (77). Tuberculous pleural effusion is also adequately treated with INH and RIF for 6 months, even when associated with smear-negative, culture-positive tuberculosis (78). Because the initial resistance to drug(s) is low in Arkansas, additional drugs are not needed for therapy. For areas with a high prevalence of drug-resistant organisms, a four-drug bactericidal regimen should be administered until the drug-susceptibility results are known.

Six months' therapy with INH and RIF with initial supplement of PZA for 2 months has been reported to be effective in smear-negative, culture-negative tuberculosis and pleural effusion (79). For smear-negative culture-negative disease, we found that 4 months' therapy with INH and RIF is sufficient (80). The British Medical Research Council recommends that 4 months of daily or three times weekly therapy with a four-drug regimen is adequate for smear-negative, culture-negative disease (75).

There are many advantages in treating smear-negative cases early. The duration of therapy is shorter than that for smear-positive disease. Therapy prevents transmission of infection during future bacteriological relapse. In developed countries, early treatment achieves better control of tuberculosis.

11. Chemotherapy in Special Conditions

11.1. Pregnancy

Tuberculosis during pregnancy is not an indication for abortion; INH and RIF therapy for 9 months is safe. If INH resistance is suspected, EMB should be added to the regimen (81). Pyrazinamide is not recommended at present due to inadequate data on teratogenicity, but should not be withheld if needed. Streptomycin is to be avoided because it may cause high-frequency hearing loss in the fetus. This is also the case with kanamycin, capreomycin, and amikacin. Ethionamide is not recommended because of its teratogenic effects. Due to the negligible amount of drug secreted in the milk (82), there is no contraindication for breast-feeding while the mother is receiving antituberculosis treatment, regardless of which regimen is followed. Isoniazid prophylaxis in pregnancy may be associated with a slightly increased risk of hepatotoxicity. Monitoring for drug side effects must be done carefully and on a regular basis (83).

11.2. Liver Disease

Alcoholic liver disease does not preclude use of antituberculosis drugs. However, as RIF, INH, PZA, PAS, and ethionamide are potentially hepatotoxic agents, regular monitoring for side effects is essential (84). In frank liver failure, initial therapy should consist of INH and EMB until the liver function returns to normal. At that time, RIF and/or PZA may be added to the drug regimen, if required.

11.3. Renal Failure

Isoniazid and RIF are the drugs of choice in patients with renal failure because they are excreted by the liver. In renal dialysis patients, the drugs should be administered after dialysis (85). The dosage of EMB must be reduced to 8–10 mg/kg body weight in advanced renal failure. All nephrotoxic drugs, including SM, kanamycin, amikacin, and capreomycin, should be avoided unless absolutely necessary and levels should be monitored if used in very unusual circumstances. Pyrazinamide may be used in reduced dosages of 15–20 mg/kg body weight.

11.4. Immunodeficiency States

Patients with immunosuppression due to malnutrition, chronic renal failure, hematologic and reticuloendothelial malignancies, diabetes, or immunosuppressive drugs respond well to 6 or 9 months of therapy. A 6-month regimen should consist of INH and RIF, with PZA and EMB for the first 2 months. The latter two drugs can be discontinued after 2 months if the susceptibility tests support this. Nine months' therapy with INH and RIF in immunosuppressed patients is also effective (45). Immunosuppressive therapy should be continued when indicated for underlying disease (86,87).

12. Corticosteroid Therapy

Steroids are not routinely used in the treatment of tuberculosis. In patients who are markedly toxic and severely malnourished, a moderate dose of prednisone, 20–30 mg/day, may improve general sense of well-being, increase appetite, reduce fever, and improve nutrition. The drug is tapered off gradually in 4–8 weeks. Severe hypoxemia and respiratory failure in miliary or disseminated tuberculosis may be improved by prednisone 40–60 mg/day. Disseminated tuberculosis in AIDS patients has been treated successfully with corticosteroids (88). Most authorities believe that complicated tuberculous meningitis should be treated with prednisone 60–80 mg/day, which is slowly tapered off in 8–12 weeks. Although some advocate the use of steroid therapy for all cases of tuberculous pericarditis to prevent constrictive pericarditis (89), we do not recommend it routinely.

12.1. Directly Observed Therapy in Noncompliant Patients

Persons whose lives are overwhelmed by poverty, homelessness, alcoholism, drug abuse, or HIV infection may not take responsibility for completing therapy without supervision. Such cases usually result in treatment failure, development of drug-resistant disease, and transmission of drug-resistant infection (90,91). Directly observed therapy (DOT) is mandatory in such situations. The advantages of DOT are clear in one study comparing the successful control of tuberculosis in Nicaragua, where resources are limited and DOT is used routinely, to treatment failure in New York City, where DOT is less commonly practiced (92). Directly observed therapy for tuberculosis leads to significant reduction in the frequency of primary drug resistance, acquired drug resistance, and relapse (93). The fact that most 6-month regimens may be given intermittently has led to development of regimens suitable for DOT. One method, the Denver regimen, consists of daily INH, RIF, PZA, EMB, or SM for 2 weeks followed by twice-weekly doses for 6 weeks, then twice-weekly administration of INH and RIF for another 16 weeks (94). A second regimen suitable for DOT involves INH, RIF, PZA, and EMB or SM three times a week for 6 months (52,76). These intermittent regimens are far less costly in comparison to the astronomical expense incurred in treating a single case of MDR-TB and the accompanying transmission of infection to the community. Thus, it is very difficult to argue against DOT when compliance is in doubt (95).

Each dose of medication should be observed by a trained health care worker a reliable family member or friend. Although some patients may find this implication of mistrust and insulting, DOT is the best method to assure compliance. In denying the patient the opportunity to take medications selectively, the risk of developing resistant disease is avoided. Should the patient prematurely discontinue treatment or develop toxicity to the drug regimen, the physician is promptly notified and can then take appropriate medical or legal steps to deal with the situation.

As the duration of therapy necessary for effective treatment steadily diminishes, compliance is becoming an ever more important issue. Intermittent regimens involving as few as 62 doses over a 6-month period claim the cure rates equal to the daily regimens (180 doses). However, when the duration is shortened or the total number of doses reduced so drastically, it is essential that compliance be as close to perfect as possible. This can be achieved by DOT. Although DOT increases the cost of health care delivery, it is well worth the price in the long run.

To establish DOT in any tuberculosis control program, a core public health staff is required. These members commonly include a clerk, a nurse, and a community worker, all trained and supported by an experienced physician. Nongovernment organizations (NGOs) can assist such efforts. Volunteer workers for NGOs are of particular importance in large cities where patients of various ethnic diversity require added socioeconomic support.

Although routine hospitalization for tuberculosis patients is no longer popular, supervised therapy can, nevertheless, improve treatment success in some sections of developed countries. This is especially important for patients residing in rural areas far removed from treatment facilities, the homeless, alcoholics, and drug addicts. Severely ill patients or those with MDR-TB may benefit from hospitalization during the initial phase. However, routine hospitalization to provide DOT may not be justified, particularly in view of the high cost of inpatient care.

13. Strategy to Improve Compliance

One of the leading causes of treatment failure and development of drug-resistant tuberculosis is lack of compliance. Various methods to improve compliance are outlined in Table 5.6. Newer strategy is required to assure compliance. In developed nations, the personal cost of noncompliance is high when tuberculosis control programs have to compete with other public health programs for state funding. To accomplish DOT in a cost-effective manner, some states have established mobile home communities to house active tuberculosis cases in geograph-

Table 5.6. Methods to improve patient compliance

• Convenient time and location of clinic
• Free transportation to appointments
• Other incentives (coffee, snacks, babysitting service, etc.)
• Appointment reminders, follow-up
• Housing facilities for homeless or difficult cases (mobile home communities, motels, hospitals, jails, etc.)
• Directly observed therapy in the workplace, home, and clinic
• Assistance from family and friends
• Patient education
• Improved communication (foreign-language pamphlets)

---

ically contained areas. The amount of staff time needed to administer DOT is substantially reduced when noncompliant patients are provided lodging and boarding in a convenient location.

14. Use of Combined Preparations of Antituberculosis Drugs

Administration of combination preparations of antituberculosis drugs prevents the development of drug resistance by avoiding ingestion of a single bactericidal drug (96). In the United States, two commercial combinations are available: Rifamate® and Rifater® from Marion Merrel Dow. Rifamate is a combination capsule of INH 150 mg and RIF 300 mg; the recommended dosage is two capsules daily. Rifater contains INH 50 mg, RIF 120 mg, and PZA 300 mg in each tablet; the recommended dosage is five to six tablets daily. The combination preparations provide adequate plasma concentrations. We strongly recommend the use of combination preparations as a safeguard against the development of drug resistance, particularly for patients who are not on DOT.

15. Monitoring and Follow-up of Patients

Bacteriological monitoring is an important aspect in the management of tuberculosis. We recommend initial smear and culture examination of three to five specimens of bronchial secretions (sputum) in addition to drug-susceptibility testing. During therapy, at least one specimen of sputum should be submitted every 2 weeks until conversion to negative occurs. Thereafter, one specimen a month is adequate for the duration of therapy. This permits early detection of noncompliance and impending failure. After completion of therapy, one specimen every three months ($\times$ 3) should be cultured, after which the patient is discharged from supervision with the advice to return if symptoms recur. Such an approach enables the detection of early relapse. Gradual decline of the bacterial population in the sputum indicates a satisfactory response to treatment and suggests compliance of therapy. With intensive chemotherapy, bacteriological negativity is expected in 2–3 months. Persistence of bacilli on microscopy and culture beyond 3 months or reversion to positive bacteriology after conversion to negative should arouse suspicion of failure. Reevaluation of the patient's compliance and repeat drug-susceptibility testing are then indicated for determination of the best retreatment regimen with DOT. Frequent chest roentgenograms are not required if the bacteriology is monitored as suggested. Compliance in taking the medication must be evaluated constantly in all patients. This may be accomplished through frequent interviews with the patient, by keeping records of attendance for clinic appointments, whether drugs are picked up in a timely manner, surprise pill counting, and examination of urine for color and excretion of drug (e.g., Rifampin). When noncompliance is suspected, DOT must be utilized. Intermittent, twice or three times weekly, administration of drugs facilitates direct supervision.

Drug side effects should be monitored clinically on a monthly basis after initial explanation to the patient about the symptoms to watch for (i.e., nausea, vomiting, anorexia, yellow skin or urine, etc.). Collection of blood for baseline studies, including complete blood count and renal and hepatic functions, should be done initially. Beyond routine baseline biochemical tests, regular biochemical monitoring leads to more confusion than enlightenment (96a). Transient asymptomatic elevation of hepatic enzymes is common during therapy (97,98). Patients are advised to discontinue medication when symptomatic and to report for repeat blood studies at that time. Elevation of liver enzymes of more than five times the base level accompanied by symptoms is very likely due to drug toxicity, and the therapy must be discontinued. After the symptoms have subsided and hepatic enzymes have returned to baseline, drugs should be reintroduced one at a time, starting with half-dosages and monitoring of hepatic enzymes (99). The offending drug may be identified in this manner and the therapy changed accordingly. Most drugs can be reintroduced without further adverse effects (100).

During therapy, the health care team must remain in close contact with patients so that any suspicious symptoms can be reported quickly. Patients should be told to discontinue therapy if symptoms of adverse side effects occur and to report to the clinic promptly for assessment. To monitor potential side effects of EMB, vision and color studies are performed monthly. Monthly examination of balance and possible hearing loss is necessary with use of SM.

16. Support of Health Department

Most health departments in developed countries provide facilities for the collection of sputum specimens for bacteriological examination. Expert advice is offered in difficult cases and public health nurses are available for monitoring patient compliance and side effects. When indicated, nurses also conduct contact evaluation including tuberculin testing and chest radiography. In the last few years, the "infrastructure" that supported the TB program has become weak, either because of the reduced resources or diverting them to the AIDS program. Reestablishing the "infrastructure" is crucial to the success of public health programs.

It is required by law that confirmed cases of tuberculosis be reported to the public health department. Delay in such notification can result in tragedy, particularly for children, in whom the disease can progress rapidly from infection to death. Close cooperation between clinicians and health department personnel is, therefore, vital in the treatment of tuberculosis and prevention of new disease.

References

1. Waksman SA, Bugie E, Schatz A (1944) Isolation of antibiotic substance from soil microorganisms with special reference to streptothricin and streptomycin. Mayo Clin Proc 19:537–548.
2. Schatz A, Waksman SA (1944) Effect of streptomycin and other substances upon mycobacterium tuberculosis and related organisms. Proc Soc Br Med 57:244–248.
3. Steenken W Jr, Wolinsky E (1952) Antituberculous properties of hydrazines of isonicotinic acid (Rimifon, Marzilid). Am Rev Tuberc 65:365–375.
4. British Medical Research Council (1948) Streptomycin treatment of pulmonary tuberculosis: a Medical Research Council investigation. Br Med J 2:769–782.
5. British Medical Research Council (1950) Treatment of pulmonary tuberculosis with streptomycin and para-amino-salicylic acid: a Medical Research Council investigation. Br Med J 2:1073–1085.
6. British Medical Research Council (1952) Treatment of pulmonary tuberculosis with isoniazid: an interim report to Medical Research Council by their Tuberculosis Chemotherapy Trials Committee. Br Med J 2:735–746.
7. Mitchison DA (1965) Chemotherapy of tuberculosis: a bacteriologist's viewpoint. Br Med J 1:1333–1340.
8. American Thoracic Society/Centers for Disease Control (1986) Treatment of tuberculosis and tuberculosis infection in adults and children. Am Rev Respir Dis 134:355–363.
9. Cowie RL, Langton ME, Escreet BC (1985) Ultrashort course chemotherapy for culture-negative pulmonary tuberculosis: a qualified success. S African Med J 68:879–880.
10. Grosset J (1980) Bacteriologic basis of short course chemotherapy for tuberculosis. Clin Chest Med 1:231–241.
11. Fox W (1968) The John Barnwell Lecture: changing concepts in the chemotherapy of pulmonary tuberculosis. Am Rev Respir Dis 97:767–790.

12. Canneti G (1965) The J. Burn's Amberson Lecture: present aspects of bacterial resistance in tuberculosis. Am Rev Respir Dis 92:687–703.

13. David HL (1980) Drug resistance to *M. tuberculosis* and other mycobacteria. Clin Chest Med 11:227–230.

14. Mitchison DA, Dickinson JM (1978) Bactericidal mechanisms in short course chemotherapy of tuberculosis. Bull Int Union Against Tuberc 53:270–274.

15. Grosset J (1978) The sterilizing value of rifampicin and pyrazinamide in experimental short course chemotherapy. Tubercle 59:287–297.

16. Fox W (1981) Whither short course chemotherapy? Br J Dis Chest 75:331–357.

17. Mackaness GB, Smith N (1953) The bactericidal action of isoniazid, streptomycin and terramycin in extracellular and intracellular tubercle bacilli. Am Rev Respir Dis 67:322–340.

18. Dickinson JM, Aber VR, Mitchison DA (1977) Bactericidal activity of streptomycin, isoniazid, rifampin, ethambutol and pyrazinamide alone and in combination against *Mycobacterium tuberculosis*. Am Rev Respir Dis 116:627–635.

19. Clini V, Grassi C (1970) The action of new antituberculosis drugs on intracellular tubercle bacilli. Antibiot Chemother 16:20–26.

20. Dickinson JM, Mitchison DA (1981) Experimental models to explain the high sterilizing activity of rifampin in the chemotherapy of tuberculosis. Am Rev Respir Dis 123:367–371.

21. Mackaness GB (1950) The intracellular activation of pyrazinamide and nicotinamide. Am Rev Tuberc 74:718–721.

22. Dutt AK, Stead WW (1982) Medical perspective: present chemotherapy for tuberculosis. J Infect Dis 146:698–704.

23. Fox W (1979) The current status of short course chemotherapy. Tubercle 60:177–190.

24. Snider DE Jr, Cauthen GM, Farer LS, Kelly GD, Kilburn JO, Grood RC, Dooley SW (1991) Drug resistant tuberculosis (letter). Am Rev Respir Dis 144:732.

25. Carpenter JL, Obnibene AJ, Gorby EW, Neimes RE, Koch JR, Perkins WL (1983) Antituberculosis drug resistance in South Texas. Am Rev Respir Dis 128:1055–1058.

26. Chawla PK, Klapper PJ, Kamholz SL, Pollock AH, Heurich AE (1992) Drug-resistant tuberculosis in an urban population including patients at risk for human immunodeficiency virus infection. Am Rev Respir Dis 146:280–284.

27. Frieden TR, Sterling T, Pablos-Mendez A, Kilburn JO, Canthen GH, Dooley SW (1993) The emergence of drug-resistant tuberculosis in New York City. N Engl J Med 328:521–526.

28. Costello HD, Caras GJ, Snider DE Jr (1980) Drug resistance among previously treated tuberculosis patients: a brief report. Am Rev Respir Dis 121:313–316.

29. Centers for Disease Control and Prevention (1994) Expanded tuberculosis surveillance and morbidity—United States, 1993. Morbid Mortal Wkly Rep 43:361–366.

30. Ben-Dov I, Mason GR (1987) Drug-resistant tuberculosis in a Southern California hospital: trends from 1969 to 1984. Am Rev Respir Dis 135:1307–1310.

31. Barnes PF (1987) The influence of epidemiologic factors on drug resistance rates in tuberculosis. Am Rev Respir Dis 136:325–328.

32. Riley LW, Arathoon E, Loverde VD (1989) The epidemiologic patterns of drug-resistant *Mycobacterium tuberculosis:* a community-based study. Am Rev Respir Dis 139:1282–1285.

33. Cantwell MF, Snider DE Jr, Cauthen GM, Onovato IM (1994) Epidemiology of tuberculosis in the United States, 1985 through 1992. J Am Med Assoc 272:532–539.

34. Snider DE Jr, Kelly GD, Cauthen GM, Thompson NJ, Kilburn JO (1985) Infection and disease among contacts of tuberculosis cases with drug-resistant and drug-susceptible bacilli. Am Rev Respir Dis 132:125–132.

35. Steiner M, Zimmerman R, Park BH, Shinali SR, Schmidt H (1968) Primary tuberculosis in children: correlation of susceptibility patterns from *M. tuberculosis* isolated from children with those isolated from source case as an index of drug resistant infection in a community. Am Rev Respir Dis 8:201–209.

36. Nardell E, McInnis B, Thomas B, Weidhass S (1986) Exogenous reinfection with tuberculosis in a shelter for the homeless. N Engl J Med 315:1570–1575.

37. Iseman MD (1993) Treatment of multidrug-resistant tuberculosis. N Engl J Med 329:784–791.

38. Stead WW, Dutt AK (1982) Chemotherapy for tuberculosis today. Am Rev Respir Dis 125:94–101.

39. Mitchison DA, Nunn AJ (1986) Influence of initial drug resistance in the response to short course chemotherapy of pulmonary tuberculosis. Am Rev Respir Dis 133:423–430.

40. Mitchison DA (1992) Understanding the chemotherapy of tuberculosis: current problem. J Antimicrob Chemother 29:477–493.

41. Dutt AK, Stead WW (1980) Chemotherapy of tuberculosis for the 1980s. Clin Chest Med 1:243–252.

42. British Thoracic and Tuberculosis Association (1976) Short-course chemotherapy in pulmonary tuberculosis: a controlled trial. Lancet 2:1102–1104.

43. Dutt AK, Jones L, Stead WW (1979) Short-course chemotherapy for tuberculosis with largely twice-weekly isoniazid-rifampin. Chest 75:441–447.

44. Dutt AK, Moers D, Stead WW (1984) Short course chemotherapy for tuberculosis with mainly twice weekly isoniazid and rifampin: community physician's seven years experience with mainly outpatients. Am J Med 77:233–242.

45. Dutt AK, Moers D, Stead WW (1980) Results of short course chemotherapy for tuberculosis in patients with complicating medical disorders. Chest 78:514 (abstract).

46. Dutt AK, Moers D. Stead WW (1986) Short course chemotherapy for extrapulmonary tuberculosis. Ann Intern Med 104:7–12.

47. Abernathy RS, Dutt AK, Stead WW, Moers DJ (1983) Short course chemotherapy for tuberculosis in children. Pediatrics 12:801–806.

48. Snider DE Jr, Zierski M, Graczyk J, et al. (1986) Short-course tuberculosis chemotherapy studies conducted in Poland during the past decade. Eur J Respir Dis 68:12–18.

49. Manalo F, Tan F, Sbarbaro JA, Iseman MD (1990) Community-based short-course treatment of pulmonary tuberculosis in a developing nation. Am Rev Respir Dis 142:1301–1305.

50. Steele MA, Des Prez RM (1988) The role of pyrazinamide in tuberculosis chemotherapy. Chest 94:845–850.

51. Girling DJ (1989) The chemotherapy of tuberculosis. Biol Mycobact 3:285–294.

52. Centers for Disease Control (1993) Initial therapy for tuberculosis in the era of multidrug resistance: recommendations of the Advisory Council for the Elimination of Tuberculosis. Morbid Mortal Wkly Rep 42(RR-7):1–8.

53. Goble M (1986) Drug resistant tuberculosis. Semin Respir Infect 1:220–229.

54. Iseman MD, Madsen LA (1989) Drug-resistant tuberculosis. Clin Chest Med 10:341–353.

55. Goble M, Iseman MD, Madsen LA, Waite D, Ackerson L, Horsburgh CRIV, (1993) Treatment of 171 patients with pulmonary tuberculosis resistant to isoniazid and rifampin. N Engl J Med 328:527–532.

56. Imamura N, Vanderkolk J, Heifets L (1994) Inhibitory and bactericidal activities of levofloxacin against Mycobacterium tuberculosis in vitro and in human macrophages. Antimicrob. Agents Chemotherap. 38(5):1161–1164.

57. Iseman MD, Madsen L, Goble M, Pomerantz M (1990) Surgical intervention in the treatment of pulmonary disease caused by drug-resistant *Mycobacterium tuberculosis.* Am Rev Respir Dis 141:623–625.

58. Stead WW, To T, Harrison RW, Abraham JH III (1987) Benefit–risk considerations in preventive treatment for tuberculosis in elderly persons. Ann Intern Med 107:843–845.

59. International Union Against Tuberculosis Committee on Prophylaxis (1982) Efficacy of various duration of isoniazid preventive therapy for tuberculosis: five year follow-up in the IVAT trial. Bull WHO 60:555–564.

60. Koplan JP, Farer LS (1980) Choice of preventive treatment for isoniazid-resistant tuberculosis infection. J Am Med Assoc 244:2736–2740.

61. Centers for Disease Control (1992) Management of persons exposed to multidrug-resistant tuberculosis. Morbid Mortal Wkly Rep 41(RR-11):61–74.

62. Centers for Disease Control. The role of BCG vaccine in the prevention and control of tuberculosis in the United States: a joint statement by the advisory council for the elimination of tuberculosis. MMWR 1996;45(RR-4):1–18.

63. Small PM, Schecter GF, Goodman PC, Sande MA, Chaisson RE, Hopewell PC (1992) Treatment of tuberculosis in patients with advanced human immunodeficiency virus infection. N Engl J Med 324:289–294.

64. Jones BE, Otaya M, Antoniskis D, Sian S, Wang F, Mercado A, Davidson PT, Barnes PF (1994) A prospective evaluation of antituberculosis therapy in patients with human immunodeficiency virus infection. Am J Respir Crit Care Med 150:1499–1502.

65. Centers for Disease Control (1991) Nosocomial transmission of multidrug-resistant tuberculosis among HIV-infected persons: Florida and New York 1988–1991. Morbid Mortal Wkly Rep 40(34):585–591.

66. Edlin BR, Tokars JI, Grieco MH, Crawford JT, William J, Sordillo EM, Ong KR, Kilburn J Dooley SW, Castro KG, Jarvis WR, Holmberg SD (1992) An outbreak of multidrug-resistant tuberculosis among hospitalized patients with the acquired immunodeficiency syndrome. N Engl J Med 326:1514–1521.

67. Alwood K, Kevuly J, Moore-Rice K, Stanton DL, Chaulk CD, Chaisson RE (1994) Effectiveness of supervised intermittent therapy for tuberculosis in HIV infection. AIDS 8:1103–1108.

68. Fischl MA, Daikos GL, Uttamchandani RD, Poblete RB, Moreno JN, Reyes RR, Boota AM, Thompson LM, Cleary TJ, Oldham SA (1992) Clinical presentation and outcome of patients with HIV infection and tuberculosis caused by multiple drug resistant bacilli. Ann Intern Med 117:184–190.

68a. A comparative study of daily and twice-weekly continuation regimens of tuberculosis chemotherapy, including a comparison of two durations of sanatorium treatment. I. First report: the results at 12 months. Bull World Health Organ 1971;45(5):573–593.

69. Starke JR (1990) Multidrug therapy for tuberculosis in children. Pediatr Infect Dis J 9:785–793.

70. Committee on Infectious Diseases (1991) Report of the Committee on Infectious Diseases, 22nd Ed., Elk Grove, IL: American Academy of Pediatrics.

71. Anonymous: Smear negative pulmonary tuberculosis. Tubercle 1980;61:113–115.

72. Narain R, Nair SS, Naganna K (1968) Problem of defining a case of pulmonary tuberculosis in prevalence survey. Bull WHO 39:701–729.

73. Dutt AK, Stead WW (1994) Smear-negative pulmonary tuberculosis. Semin Respir Infect 9(2):113–119.

74. Singapore Tuberculosis Service/British Medical Research Council (1986) Long-term follow-up of a clinical trial of six-month and four-month regimens of chemotherapy in the treatment of pulmonary tuberculosis. Am Rev Respir Dis 133:779–783.

75. Hong Kong Chest Service/Tuberculosis Research Center Madras/British Medical Research Council (1989) A controlled trial of 3-month, 4-month and 6-month regimens of chemotherapy for sputum smear-negative pulmonary tuberculosis: results at 5 years. Am Rev Respir Dis 139:871–876.

76. American Thoracic Society/Centers for Disease Control (1994) Recommendations for the treatment and preventive treatment of tuberculosis. Am J Respir Crit Car Med 151:1359–1374.

77. Dutt AK, Moers D, Stead WW (1990) Smear-negative culture-positive pulmonary

tuberculosis: six months chemotherapy with isoniazid and rifampin. Am Rev Respir Dis 144:1232–1235.

78. Dutt AK, Moers D, Stead WW (1992) Tuberculous pleural effusion: six month therapy with isoniazid and rifampin. Am Rev Respir Dis 145:1429–1432.

79. Ormerod LP, McCarthy OR, Rudd RM, Horsfield N, (1995) Short course chemotherapy for tuberculous pleural effusion and culture-negative pulmonary tuberculosis. Tuber Lung Dis 76:25–27.

80. Dutt AK, Moers D, Stead WW (1989) Smear- and culture-negative pulmonary tuberculosis: four months short course chemotherapy. Am Rev Respir Dis 139:867–870.

81. Snider DE Jr, Layde RN, Johnson MW, Lyle MA (1980) Treatment of tuberculosis during pregnancy. Am Rev Respir Dis 122:65–79.

82. Snider DE Jr, Powell KE (1984) Should women taking antituberculosis drugs breast feed? Arch Intern Med 144:589–590.

83. Hamadeh MA, Glassroth J (1992) Tuberculosis and pregnancy. Chest 101:1114–1120.

84. Cross FS, Long MW, Banner AS, Snider DE, Jr (1980) Rifampin-isoniazid therapy of alcoholic and non-alcoholic tuberculosis patients in a U.S. Public Health Service Cooperative Therapy trial. Am Rev Respir Dis 122:349–353.

85. Andrew OT, Schoenfeld PY, Hopewell PC, Humphrey's MH (1980) Tuberculosis in patients with end-stage renal disease. Am J Med 68:59–65.

86. Dautzenberg B, Grosset J, Fechner J, Luccianni, Debra P, Hersons S, Truffot C, Sors C (1984) The management of thirty immunocompromised patients with tuberculosis. Am Rev Respir Dis 129:494–496.

87. Davidson PT, Le HQ: Drug treatment of tuberculosis—1992. Drugs 43:651–673.

88. Masud T, Kemp E (1988) Corticosteroids in treatment of disseminated tuberculosis in patients with HIV infection. Br Med J 296:464–465.

89. Strang JIG, Kakaza MMS, Gibson DG, Allen BW, Mitchison DA, Evans DJ, Girling DJ, Nunn AJ, Fox, W (1988) Controlled clinical trial of complete open surgical drainage and of prednisolone in treatment of tuberculous pericardial effusion in Transkei. Lancet 2:759–764.

90. Brudney K, Dobkin J (1991) Resurgent tuberculosis in New York City: human immunodeficiency virus, homelessness and the decline of tuberculosis program. Am Rev Respir Dis 144:745–749.

91. Mahmoudi A, Iseman MD (1993) Pitfalls in the care of patients with tuberculosis: common errors and their association with the acquisition of drug resistance. J Am Med Assoc 270:65–68.

92. Brudney K, Dobkin J (1991) A tale of two cities: tuberculosis in Nicaragua and New York City. Semin Respir Infect 6:261–272.

93. Weis SE, Slocum PC, Blais FX, King B, Matney GB, Gomez E, Foresman BH (1994) The effect of directly observed therapy on the rate of drug resistance and relapse in tuberculosis. N Engl J Med 330:1179–1184.

94. Cohn DL, Catlin BJ, Peterson KL, Judson FN, Sbarbaro JA, (1990) A 62-dose, six-month therapy for pulmonary and extrapulmonary tuberculosis: a twice weekly, directly observed and cost-effective regimen. Ann Intern Med 112:407–415.

95. Iseman MD, Cohn DL, Sbarbaro JA (1993) Directly observed treatment of tuberculosis: we can't afford not to try it. N Engl J Med 328:576–578.

96. Moulding T, Dutt AK, Reichman LB (1998) Fixed dose combinations of antituberculosis medications to prevent drug resistance. Ann Intern Med 122:951–954.

96a. Van-den-Brande P, Van-Steenbergen W, Vervoort G, Demedts M (1995) Aging and hepatotoxicity of isoniazid and rifampin in pulmonary tuberculosis Am J Respir Crit Care Med 152, 1705–1708.

97. Byrd RB, Horn BR, Soloman DA, Griggs GA, (1979) Toxic effects of isoniazid in tuberculosis chemoprophylaxis: role of biochemical monitoring in 1000 patients. J Am Med Assoc 241:1239–1241.

98. Bailey WC, Taylor SC, Dascomb WH, Greenberg HB, Ziskind MM (1973) Disturbed hepatic function during isoniazid chemoprophylaxis. Am Rev Respir Dis 107:523–529.

99. Dutt AK, Moers D, Stead WW (1983) Undesirable side effects of isoniazid and rifampin in largely twice-weekly short-course chemotherapy for tuberculosis. Am Rev Respir Dis 128:419–424.

100. Girling DJ (1978) The hepatic toxicity of antituberculosis regimens containing isoniazid, rifampicin and pyrazinamide. Tubercle 59:13–32.

6

Treatment of Tuberculosis in Low-Income Countries

Donald A. Enarson and Philip C. Hopewell

1. Introduction

Lung diseases, including tuberculosis, account for an estimated 13% of all disability-adjusted life years lost as a result of disease; tuberculosis is second only to childhood pneumonia as a cause of death due to lung disease and the most frequent cause of death from a single infectious agent in the productive years of life (ages 15–49 years) (1,2). Nearly all deaths from tuberculosis occur in low-income countries (3). In addition to these effects on mortality, a high proportion of all patients with tuberculosis in many industrialized countries (such as Canada, Switzerland, the United States, and the Scandinavian countries) were born in low-income countries. Based on the above observations, for both humanitarian and strategic reasons, the control of tuberculosis in low-income countries should have a high priority for the world as a whole. Control of tuberculosis in industrialized, high-income countries cannot be accomplished without effective control programs in low-income countries. For this reason, it is essential that tuberculosis control be viewed in a global context.

The control of tuberculosis is accomplished mainly through the early identification and treatment of patients who are infectious. Effective treatment decreases the probability and duration of exposure to the causative organism, *Mycobacterium tuberculosis*. To understand the impact of treatment in reducing the amount of tuberculosis in a community requires an understanding of the clinical course of the disease in the absence of treatment, of the effectiveness of treatment (through standard methods such as the blinded, randomized controlled trial) and, finally, of the evaluation of the impact of treatment at the community level. Because the evaluation of the effectiveness of treatment is covered elsewhere in this book, we will not review them again here. In this chapter, we will discuss the treatment of tuberculosis in low-income countries, emphasizing the impact of treatment in changing the clinical course of the disease and reducing its impact within the community.

2. The Clinical Course of Tuberculosis Without Treatment

The mode of transmission of *M. tuberculosis* has been studied extensively in the context of investigations of contacts of patients with active tuberculosis (4–6). From these studies, we know that the most efficient source of transmission is the person with tuberculosis whose lungs contain a sufficient number of organisms to be seen by microscopic examination of sputum smears. Virtually all such patients have symptoms of disease and a high proportion of them seek medical attention, even under the most stringent socioeconomic conditions (7). It is thus possible to identify a large proportion of all highly infectious patients by systematic examination of persons presenting themselves to the routine health services. Patients with tuberculosis who do not have acid-fast organisms detected by microscopic examinations of sputum but who have *M. tuberculosis* isolated in culture are less infectious than sputum-smear-positive patients. Nevertheless, such patients, because they may go undetected for relatively longer periods of time, may also be the source of a substantial number of new tuberculous infections. When infectious patients are not detected and treated, it has been estimated that, on average, 10 individuals who have been in contact with the patient will become infected with *M. tuberculosis.*

The likelihood of a person who is in contact with an infectious case becoming infected varies with the amount of time the infectious patient remains infectious and the closeness and duration of the exposure. Prevention of exposure was one of the reasons for isolating infectious patients in sanatoria prior to the time when specific chemotherapy became available. In addition to the likelihood of transmission, another determinant of the epidemiologic impact of an infectious case is the number of uninfected individuals who are exposed. The number of exposed persons varies with the size of the family and the age of the source case. Younger adults with infectious tuberculosis are more likely to have a number of small children who are uninfected and, therefore, highly susceptible living in the same household. For this reason, mother-to-child transmission is likely to play an important role in communities where tuberculosis is common. The probability of transmission also varies with the characteristics of the dwelling in which a person with infectious tuberculosis lives; small dwellings with little ventilation were often the rule in North America and northern Europe in the past century, when tuberculosis rates reached enormous levels (8). At the same time that case rates were very high in Europe and North America, tuberculosis was relatively unknown in central Africa (9) and even today, the rates of tuberculosis in Africa, although some of the highest in the world, are much lower than those in Europe and America 100 years ago. Variations in factors related to the probability and extent of exposure to *M. tuberculosis* may explain variations in the rates of tuberculosis in different geographic locations and during different periods.

Information concerning the clinical course of untreated tuberculosis is available from a variety of sources, including evaluations of patients prior to the development of specific chemotherapy (10–12), and, in one instance, since the introduction of chemotherapy, a report in which patients were followed without treatment (13). On average, by 18 months following diagnosis, 30% of patients had died; by 36 months, a further 10% had died, and by 60 months, a further 9%, making a cumulative case fatality rate of 49%. In a certain proportion of cases, the disease became quiescent and bacteriologically negative: 28% at 18 months after diagnosis, 35% at 36 months, and 33% at 60 months. The remaining patients continued to have positive sputum smears and to be infectious: 42% at 18 months after diagnosis, 25% at 36 months, and 18% at 60 months. Thus, even without treatment by specific chemotherapy, an important proportion of cases were "cured," although, overall, a high proportion of patients died of their disease. The trend in proportion of cases remaining infectious progressively declined with time when no treatment was given.

3. The Impact of Chemotherapy

Chemotherapy rapidly reduces the potential for transmission of *M. tuberculosis* by rendering the patient noninfectious (14). However, a permanent cure requires an appropriate combination of medications taken for a sufficient period of time (15). The discovery and utilization of specific chemotherapy resulted in dramatic improvement in the clinical condition of patients with tuberculosis (16). The case fatality ratio fell from nearly 50% to around 10%. This impact is detectable in the trends in mortality from tuberculosis in most industrialized countries beginning in the mid-1940s (17).

In many countries of Europe immediately after the introduction of chemotherapy with streptomycin alone in 1947, large numbers of patients survived only to become chronically infectious and the total number of patients was actually increased rather than reduced. The pioneering work of the British Medical Research Council and individuals such as Sir John Crofton and his colleagues in Edinburgh (18) demonstrated that multiple-drug chemotherapy was necessary to bring about a permanent cure and to prevent the development of resistance to the antimicrobial agents used. Those countries that did not promptly adopt multiple-drug therapy as standard policy [such as Poland (19)] continued to create large numbers of patients with chronic tuberculosis. It was only the introduction of a new powerful medication, isoniazid, that enabled the rapid reduction in the number of these chronic patients, with a concomitant reduction in the spread of the disease in the community.

How important are patients who have chronically positive sputum in the transmission of tuberculous infection? It has been suggested by a number of investigators that patients with chronic tuberculosis are of little importance in the transmission of *M. tuberculosis*. Evidence for this point of view is the observation that tubercle bacilli that are drug resistant are less vigorous than those which are fully susceptible and, in consequence, it was proposed that they might be less virulent (20). Moreover, studies of contacts of patients with chronic tuberculosis indicated that, in comparison with contacts of new cases, no additional infections could be documented to have taken place, probably because the source patient had already infected all contacts likely to become infected by the time of the initial diagnosis and that no further infections were therefore likely to occur (21). Others postulated that patients with chronic tuberculosis were unlikely to live long, they were aware of their disease and were likely to take precautions against spreading it, and, finally, they were likely to be confined due to the severity of their condition. Although little systematic evidence concerning these points is available, it is clear, for example, by reviewing a series of such patients in Poland (19), that they are often very active, continue to work and may live for a very long time.

A modeling exercise to address the question of the contribution of patients with chronic tuberculosis to the transmission of *M. tuberculosis* was undertaken using the results of the successive prevalence surveys reported from Taiwan and Korea (22). Within the modeling exercise, an estimation of the dynamics of transmission of new and of chronic cases was attempted, using certain transmission parameters and the "marker" of drug resistance for indicating the transmission from a source case who was "chronic" (had previously been treated, but unsuccessfully). From this information, a set of equations was developed which modeled transmission dynamics and to which the question of what type of case was the more effective "transmitter" of infection, the new or the chronic case, was addressed. From the equations, it was determined that the likelihood of transmission was greater for chronic cases than for new cases. The model was developed on the data from Taiwan and then applied to data from Korea; in both analyses, the chronic cases appeared to be more important sources of subsequent new cases. The results of this modeling exercise, emphasizing the importance of chronic (resistant) cases as sources of infection, have been supported by studies of nosocomial transmission of tuberculosis in settings where both tuberculosis and human immunodeficiency virus (HIV) are present (making transmission of infection and development of consequent disease much more efficient and therefore much easier to study) (23).

4. Treatment Programs for Tuberculosis in Low-Income Countries

A standardized approach to the management of tuberculosis was developed by the World Health Organization in 1974 (24). This approach emphasized diagnosis based on bacteriological examination, standardized multidrug chemotherapy regimens, ambulatory care, and integration of diagnostic and treatment services into the general health services. These recommendations were derived from a broadbased experience of program implementation in a large number of industrialized countries, as well as numerous experiments undertaken with collaborating scientific institutions, most notably the National Tuberculosis Institute in Bangalore, India, where extensive investigations of the structure of health services and the sociological components of tuberculosis had been undertaken (25).

Utilizing the WHO recommendations, national tuberculosis programs were introduced into many low-income countries. Subsequently, evaluations were undertaken to determine the impact of these programs on the tuberculosis situation in the countries. At the very outset of application of the regimens used so successfully in industrialized countries to low-income countries, the results of treatment were unsatisfactory. Frimodt-Möller (26) reported on the results of prevalence surveys in India undertaken 7 years apart, beginning with the introduction of chemotherapy in 1948. The prevalence survey included radiographic and bacteriologic examination and tuberculin skin testing. A total of 32,000 persons were examined, representing 64% of the total population of the study area over the age of 4 years. During this period, the prevalence of tuberculosis actually *increased*. The increase was greatest in women aged 20–30 years of age (in whom the prevalence increased by 80%), with a slight decline in children under 10 years of age. Frimodt-Moller felt that the explanation of these results was that chemotherapy had saved the lives of patients who would otherwise have died and that, as expected, tuberculosis in adults is usually the result of infection in childhood; therefore, the impact of treatment could only be seen in the children who had not already been infected prior to the implementation of the treatment program.

In Peru, an evaluation was undertaken of the treatment results in a cohort of 2863 consecutive cases of pulmonary tuberculosis whose records were available, whose sputum was positive on direct microscopic examination for acid-fast bacilli, and who were enrolled in 1980 on treatment with a "standard" regimen of 12 months of chemotherapy not containing rifampin (27). This evaluation excluded 24% of the cases whose records were not available for evaluation and 5% who were treated with "nonstandard" regimens. Of the total whose records were available, 9% were transferred to another facility and their eventual results were unknown; 4% were known to have died, 7% remained, or became again, smear positive 6 months or more after the commencement of treatment, 36% failed to return to collect their medications for 2 months or more while on treatment (and judged as having absconded from their treatment), and 36% completed their treatment and were, apparently, cured.

A similar evaluation was undertaken in 1990 in a region of Haiti where the same chemotherapy regimens were utilized and the tuberculosis control program recommended by the WHO had been implemented throughout the country for a number of years (28). Virtually the same results were observed: More patients abandoned their treatment before completion (45%) than were cured (35%), even after many years of operation of the program.

In Korea (29) and Taiwan (30), periodic national prevalence surveys of tuberculosis have been undertaken during the past 30 years, and in China (31) during the past 10 years. These surveys have been carefully designed to identify a representative sample of the entire population. The examinations consisted of chest radiographs, bacteriologic examination of sputum samples, and tuberculin skin testing. These surveys showed that although case fatality ratios were less than would have been expected with no treatment, the rate of cure was not commensurately increased. Those patients who did not die but were not cured remained alive and infectious. It was discovered that more than one-half of all the smear-positive cases had previously been treated for tuberculosis.

The most immediate effect of chemotherapy for tuberculosis in the programs surveyed was to reduce death rates without increasing rates of cure. It is relatively easy to save lives with therapy, but much more difficult to ensure that patients become noninfectious. In poorly organized programs, treatment has saved lives, thereby enabling chronically infectious patients to continue to spread *M. tuberculosis*. The impact of various approaches to tuberculosis treatment on the subsequent tendency in prevalence of infectious tuberculosis is illustrated in Figure 6.1.

The national tuberculosis programs in Korea, China, Taiwan, Peru, and Haiti at the time of these evaluations had utilized treatment regimens of at least 12 months' duration that did not include rifampin. Is it possible that these poor results are simply due to the absence of rifampin in treatment? A recent study from south India has provided information relative to this question. In an evaluation of the results of treatment within a defined population in India, the authors evaluated the treatment results of an enrollment cohort of 3357 consecutive patients with smear-positive pulmonary tuberculosis enrolled from 1986 to 1988 and evaluated 36 months later (32). The majority (69%) had been treated with a rifampin-containing regimen; 24% were unavailable for evaluation. Of the remaining cases available for evaluation, 28% had died and 31% remained bacteriologically positive. It is clear from this example that a total management package, not just powerful treatment regimens, is required (33).

A more sinister accompaniment to chronic tuberculosis is the increased likelihood that such patients will carry tubercle bacilli that are resistant to antimycobacterial agents. In the surveys from Korea and China, approximately three out

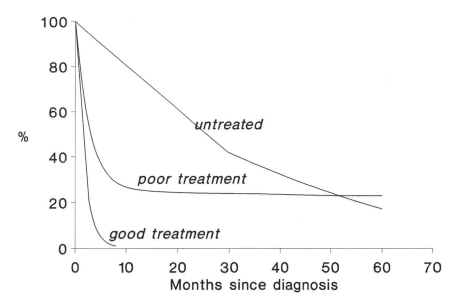

Figure 6.1. Impact of various types of treatment on the prevalence of infectious (smear positive) cases of tuberculosis.

of four patients with chronic tuberculosis had resistant organisms. Persons who are infected with these resistant organisms, if they subsequently develop tuberculosis, will also have disease resistant to treatment. Thus, not only will the number of cases be increased by poor treatment but the cases will be dangerous, drug-resistant sources of new infection in the community and will, additionally, be very difficult to cure. This eventuality is serious enough in those programs utilizing regimens not containing rifampin, resulting in the development of resistance to isoniazid or isoniazid and streptomycin. Although this is a serious development, such patients might still be cured using a straightforward retreatment regimen. Where regimens containing rifampin are used and the treatment results are poor, the grave danger of development of resistance to both isoniazid and rifampin (multidrug resistance) becomes a reality and the likelihood of cure of such patients is markedly decreased (30).

The cost of treatment of each patient with tuberculosis caused by organisms resistant to both isoniazid and rifampin is so high [estimated at approximately US$ 100,000 per case (34)], that they essentially become untreatable in most low-income countries. Drug resistance should be viewed as a "toxic side effect" of a poor treatment program; patients who never receive treatment are extremely unlikely to have disease due to resistant bacilli and all clinically important drug resistance is caused by improper treatment.

5. Model Programs for Tuberculosis Control in Low-Income Countries

The record of tuberculosis control programs in low-income countries has obviously not been as positive as those in many industrialized countries. Much of the difficulty has been related to the inability to achieve satisfactory treatment results in patients with active tuberculosis. This difficulty has recently been overcome by an approach to tuberculosis services developed by the International Union Against Tuberculosis and Lung Disease (IUATLD) in collaboration with the health authorities of some developing countries (especially Tanzania, Malawi, Mozambique, Benin, and Nicaragua) and with donor partners (notably the government of the Kingdom of Norway, the government of Switzerland, the Royal Netherlands Anti-Tuberculosis Association, the Norwegian National Health Association, the Norwegian Heart and Lung Fund, the Japan Antituberculosis Association, and the Swiss Antituberculosis Association).

Within programs in Tanzania, Malawi, Mozambique, Benin, and Nicaragua, more than 70,000 cases of tuberculosis are diagnosed and treated each year (more than 1 out of every 10 new cases estimated to occur in sub-Saharan Africa), and the majority (more than 75%) are documented to be cured. The introduction of such a "model" program is illustrated by the example of Tanzania. The technical cooperation began in 1979. At that time, a standardized 12-month treatment regimen utilizing thioacetazone and isoniazid in a combined tablet supplemented by streptomycin in the initial 2 months of treatment was adopted. A separate and more powerful treatment regimen was used for patients who had been treated previously for as little as 4 weeks. This regimen used isoniazid, rifampin, pyrazinamide, and ethambutol for 3 months, given daily and supplemented in the first 2 months by streptomycin, followed by 5 months of isoniazid, rifampin, and ethambutol given three times weekly.

When the program was first introduced in Tanzania, a cohort evaluation of treatment results of untreated smear-positive cases demonstrated the sorts of results already described from Peru and Haiti: 5% died while on treatment; 18% remained chronically sputum positive; 33% either abandoned treatment or were transferred to another area; and only 44% completed treatment and were presumed to be cured. Expansion of the strengthened program throughout the country improved the results somewhat by 1982: 52% completed treatment and were documented or presumed cured. Only after an initial treatment regimen containing rifampin was introduced did the program achieve satisfactory results: by 1987, 79% were presumed cured. The trend in treatment results is illustrated in Figure 6.2. The importance of the use of rifampin in achieving satisfactory results is shown in Figure 6.3. The introduction and expansion of a rifampin-containing regimen was associated with a progressive increase in the rates of favorable outcome of treatment with results that reached the levels required to have a positive epidemiological impact. Similar observations have been made in other programs

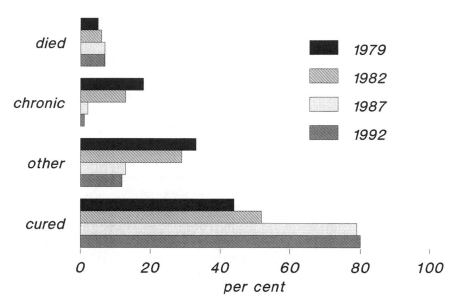

Figure 6.2. Trend from 1979 to 1992 in results of treatment of smear-positive cases of tuberculosis not previously treated, in Tanzania.

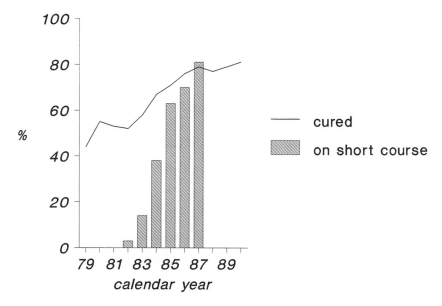

Figure 6.3. The association of introduction and expansion of the use of rifampin-containing regimens and the results of tuberculosis treatment in smear-positive cases never previously treated in Tanzania.

assisted by the IUATLD: by 1989, of 109,691 cases of smear-positive tuberculosis never previously treated (52,840 of whom were treated with rifampin-containing regimens), 79% of those on rifampin-containing regimens were cured compared with 56% of those on regimens not containing rifampin (35).

Approximately 60% of all the costs of these successful programs have been met by the budgets of the national governments, whereas 40% represented hard currency requirements which were difficult to obtain and which were provided by donors. The technical implementation was supervised by the IUATLD with regular (usually twice yearly) visits to the programs, followed by evaluation and written recommendations. In an evaluation conducted by the World Bank, these programs were judged to be among the most cost-effective of any health intervention in developing countries (36). The cost is equivalent to less than US$ 50 per case. Since the development and favorable assessment of this "model" program, at least 29 other developing countries have secured donor partners and are implementing programs similar to those developed by the IUATLD and now adopted by the World Health Organization in its new Global Program (37). Of the donor support given to assist the national tuberculosis programs in 1991, 60% was spent on medications and 20% on materials for diagnosis. A significant amount, however, was invested in supervision (14%), training (4%), evaluation (2%), and transport (1%). Without these latter investments, it would not have been possible to achieve the results described.

6. Components of Model Programs

The IUATLD-assisted national tuberculosis programs are based on a rather tight set of the basic requirements as follows (33).

6.1. Political Commitment on the Part of the Government

As noted above, the major part of the expenditure for national tuberculosis programs (NTPs), even with donor support, is borne by the government. The World Bank assesses these programs as being so cost-effective that governments cannot afford not to invest in them. However, it is impossible, even with large donor investments, to implement an effective program without the support of the government authorities.

The most important aspect of governmental support is the development of a structure for delivery of the services. The care of the patients must be provided by general health personnel within the general health services to which all patients come for care. These personnel must be properly trained and supervised in order to ensure that the patients are cared for correctly. The training and supervision of health personnel in the specific aspects of tuberculosis control must be effected by a supervisory staff at the intermediate level, among whose special skills is a clear understanding of the basis and operation of tuberculosis control. In addition, a unit specialized in tuberculosis (sometimes also responsible for other specialized functions) at the central (national) level of the country must be established to deal with the planning, coordination, supply, and training within the program. The central unit must consist of a full-time director with support personnel, transport, and supplies. It is absolutely essential that the central unit constantly represent the interests of tuberculosis patients in central level planning, determining budgets, and ensuring that tuberculosis patients are not discriminated against (e.g., refused admission to hospitals where necessary). Diagnostic and treatment services for tuberculosis should be horizontally integrated into general primary health care services and carried out by multifunctional providers. However, a vertically oriented technical support and supervision program that includes training is essential for effective program operation.

6.2. A Secure System of Supplies

Because of the relatively long treatment period, tuberculosis programs are more vulnerable than many other programs to lapses in supply of medications and materials. A reliable system often requires donor assistance because of the need for hard currency to purchase medications on world markets. In addition, the projected amounts of medication needed should be calculated and a reserve stock sufficient to prevent the interruption of the treatment for individual patients at all levels should be provided. One of the most important factors responsible for improving adherence of patients to treatment in many countries is the assurance that the health service will be able to consistently provide diagnostic examinations and treatment. The long-term impact of patient confidence in the tuberculosis program has been demonstrated in Tanzania (Fig. 6.4). With no change in the program but simply a better awareness on the part of the patients of the reliability of the service, the proportion of patients absconding from their treatment steadily declined over a number of years. Moreover, the unavailability of some medications while others remain available is an important and avoidable cause of resistance to medications. A system for regular planning and provision of supplies was shown to be possible even under war conditions in Nicaragua and Mozambique in the 1980s.

6.3. Diagnosis Based on Bacteriologic Examination with a System of Quality Control

Because the focus of the NTP is the cessation of transmission of *M. tuberculosis,* accurate sputum microscopy to identify the infectious (smear positive) cases and to monitor the results of treatment is vital to the program. The network of

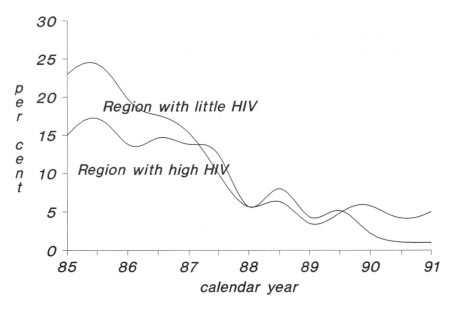

Figure 6.4. Trend in patient adherence to treatment over time in two regions of Tanzania.

diagnostic centers, located within the general-purpose laboratory of routine health institutions, must serve no less than 50,000 persons in order to maintain sufficient experience to ensure expertise of the laboratory technician conducting the examinations. Accuracy in microscopy must be assured through a program of training and routine evaluation of a sample of smears in a reference center. Based on this evaluation, close attention can be paid to the training of technicians who do not perform at a sufficient level of accuracy. To maintain this system of quality assurance, regular supervision visits to the laboratories are essential.

6.4. *Proper Recording and Reporting of Patients*

The proper recording and reporting of patients and the results of their treatment are indispensable to the achievement of the targets established for NTPs. Standardized assessments of treatment results using cohort analyses, as established by the IUATLD, are vital to ensure that the targets are being met (14). In addition to the use of cohort analyses to monitor program performance, accurate reporting and recording are also necessary for the purposes of planning, budgeting, and determining the amounts of supplies needed. Without reliable information, it is not possible to obtain necessary funds for operation of programs, to plan efficient distribution of supplies, or to make regular plans for supervision and training.

6.5. *Adequate Control and Monitoring of the Use of Rifampin*

The two drugs isoniazid and rifampin are the most potent medications currently available for the treatment of tuberculosis. In addition to these two medications, there is only a limited number of effective agents that are available currently. Because isoniazid has been widely used, often in a chaotic fashion in the past, resistance to isoniazid is present in most countries. Rifampin, on the other hand, has not been widely used (or abused) in many low-income countries; therefore, resistance to it is quite uncommon (although in many middle-income countries and some areas of high-income countries, this is not the case [38]). Rifampin is highly valued in many low-income countries; failing to control it often leads to its appearance in the general market (legally or illegally) and the danger of its misuse. The availability of rifampin on the general market, often without prescription could lead to erratic and uncontrolled use in patients with tuberculosis and, to the development of rifampin resistance.

It is absolutely essential that the health staff are well acquainted with the principles of treatment for tuberculosis and are well supervised. Even in a tuberculosis control program, rifampin may be misused or poorly supervised, resulting in resistance to the drug. The creation of resistance to rifampin (and, especially, resistance to both isoniazid and rifampin) must be avoided at all costs if control of tuberculosis is to be an achievable goal. Before rifampin is introduced into a program, it is necessary to ensure a high level of compliance with the treatment regimen. Compliance is best achieved by directly observed therapy. It is best to provide the rifampin only in combined tablets together with isoniazid in order to ensure that it is never consumed as an individual medication. These preparations, however, must be purchased from suppliers who can provide clear evidence that the bioavailability of the components is satisfactory, as a number of preparations have been shown to be substandard (39,40).

7. Regimens Used in the Model Programs

Because *Mycobacterium tuberculosis* has a great tendency to develop resistance to medications and the number of active compounds effective against tuberculosis is limited, the treatment of tuberculosis should be standardized; that is, there should be guidelines prepared by a group of experts to indicate how (with few exceptions) tuberculosis patients should be treated in a particular setting. To treat patients in a manner other than that recommended by the guidelines developed for the particular area is potentially harmful to the patient and to the community.

Numerous effective regimens have been precisely defined for the treatment of tuberculosis (41) and recommendations on effective treatment regimens have been provided by the IUATLD (42), the World Health Organization (WHO) (43), and the American Thoracic Society/Centers for Disease Control (44). Nevertheless, specific recommendations of standardized regimens must be prepared for each locality, based on the international recommendations. If specific recommendations are not made and followed, confusion remains in the minds of practitioners as to the correct treatment to give (most persons caring for tuberculosis patients in low-income countries are nonspecialists and are often paramedical personnel), and it becomes virtually impossible to plan the procurement and distribution of drugs.

The factors of importance in the selection of a standard regimen include cost of the drugs and their administration, effectiveness of the regimens, and appropriateness of the treatment program to the local circumstances. Priority must always be given to treatment of the smear-positive patient and careful attention must always be paid to determining which patients have received previous treatment for tuberculosis (45).

The recommendations published by the Treatment Committee of the IUATLD as well as those from WHO and the American Thoracic Society/Centers for Disease Control indicate that it is desirable for rifampin to be given throughout the course of a 6-month regimen. The regimen should consist of an initial daily phase of isoniazid, rifampin, pyrazinamide, and either streptomycin or ethambutol for 2 months, followed by 4 months of isoniazid and rifampin given twice weekly. In developing countries, however, because of cost and the inability to directly supervise a regimen that contains rifampin throughout, such a treatment program may not be feasible. Under such circumstances, the regimen used for patients with positive sputum smears in the IUATLD model programs is both effective and inexpensive. This regimen consists of the same initial daily phase for 2 months or until the sputum is smear negative, followed by 6 months of isoniazid and thioacetazone given daily. For smear-positive patients who were previously treated for as much as 1 month in the past, daily isoniazid, rifampin, pyrazinamide, and ethambutol should be given for 3 months, supplemented with streptomycin in the first 2 months, followed by thrice-weekly isoniazid, rifampin, and ethambutol for a further 5 months.

In countries or areas where resources are limited, cases diagnosed on clinical or radiographic criteria and who are smear negative can be treated with daily isoniazid and thiacetazone given for 12 months, supplemented, where feasible, with streptomycin in the first 2 months.

This group of regimens has been proven to be efficacious on a countrywide basis in Tanzania. Regimens containing thiacetazone may not be acceptable, however, in populations having a high frequency of adverse reactions to the drug. In particular, Asians may have difficulty with thiacetazone and severe adverse reactions (Stevens–Johnson syndrome) occur frequently in persons with HIV infection. Thus, depending on the patient population, modifications of the treatment regimen may be necessary (as is currently the case in China, for example).

These regimens have been utilized by the IUATLD based on the following rationale. The first priority of the treatment is to ensure the conversion from positive to negative sputum smears, without creating resistance to any of the medications used. In low-income countries, it is not feasible to base the choice of a treatment regimen on the results of cultures and of drug-susceptibility tests. For this reason, the regimens that are chosen must have the least potential for inducing drug resistance while yielding a high cure rate. In choosing regimens with the lowest potential for generating drug resistance, it must be ensured that there is no possibility of "effective monotherapy" by the administration of rifampin and isoniazid without other medications when a patient may have tuberculosis caused by organisms that are resistant to isoniazid. At least 1 in 15 of all smear-positive patients never previously treated in many low-income countries has disease caused by isoniazid-resistant organisms (45,46). Moreover, in selecting a retreatment regimen, it is assumed that individuals resistant to both isoniazid and rifampin at the outset (multidrug-resistant cases) are essentially incurable. The cost to cure such patients is very high and the medications required are not available. It is essential to utilize at least two medications at all times to which the organisms are likely to be susceptible (essentially, two medications which have never previously been given alone). Finally, it must be ensured that there is a high level of compliance with the regimen. Combination preparations of proven bioavailability should be used whenever rifampin is administered or whenever only two medications are prescribed, such as in the continuation phase of initial treatment.

The regimen such as used in Tanzania have been shown to give results equivalent to any other regimens evaluated both in terms of immediate results (over 250,000 cases evaluated) and in terms of risk of relapse (47). For the majority of patients (and those with highest priority), a powerful, rifampin-containing regimen is provided with full protection of the use of rifampin and for a very low cost (medication costs are less than US$ 20 per case). In addition, operationally, the treatment becomes one of the shortest possible; the period of careful supervision is restricted to the initial 2 months, shorter even than the "short-course" treatment of 6 months using rifampin throughout (in which careful supervision must be maintained for the full 6 months and the treatment results are not better, under program conditions, than those obtained when rifampin is used only in the initial intensive phase). However, as noted previously, the regimen that includes thioacetazone in the continuation phase is applicable only in populations in which the drug is well tolerated.

8. The Prevention of Resistance to Antimycobacterial Drugs

As noted previously, an important objective of the model NTP is the achievement of a high degree of treatment success without the promotion of resistance to antituberculosis medications. How is drug resistance created? The most important cause of the development of drug resistance is medical or programmatic malpractice (48). The incorrect use of medications in the treatment of tuberculosis is the key determinant for the development of drug resistance. This is most serious in the case of resistance to the key medications, isoniazid and rifampin. A health worker or program that misuses medications may be responsible for the deaths of patients in whom drug-resistant organisms are produced and also for the spread of incurable disease to susceptible members of the community (including other health workers). Drug resistance is caused by treatment that consists of only a single drug to which the bacilli are sensitive or by programmatic malpractice, that is, the patient is not carefully monitored or inappropriate regimens are chosen. An example of the latter is the poorly supervised use of rifampin throughout a treatment regimen when there is a high level of primary resistance to isoniazid. An example of such a situation occurred in Djibouti, where, in 1989, it was decided to change the continuation phase of the treatment regimen from 6 months of isoniazid and ethambutol to 4 months of isoniazid and rifampin, presumably a stronger regimen (49). Patients failing to convert their sputum smears to negative were routinely monitored with culture and sensitivity tests. Prior to 1989, one-quarter of the patients who failed to convert their sputum had organisms that were resistant to both isoniazid and rifampin, whereas, after 1989, three-quarters were resistant to both medications (Fig. 6.5).

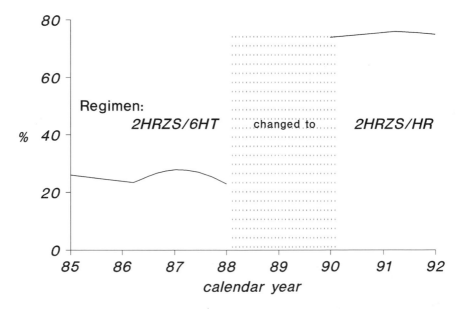

Figure 6.5. Occurrence of multidrug resistance (to isoniazid and rifampin) in Djibouti.

One of the most important achievements of the "model" programs has been the expansion of diagnosis and treatment throughout a country without the promotion of resistance to medications as shown in Figure 6.6 (41). This has been achieved in two ways: by the careful development of a complementary retreatment regimen which can ensure the cure of virtually all cases who fail on the first treatment and by the strict supervision of the use of rifampin and restriction of its use to the initial intensive phase of treatment, using the "disposable" drug thioacetazone as the companion of isoniazid in the continuation phase.

What can be done if resistance to medications is already well established in the community as a result of anarchic treatment? This problem has been studied in two settings in, what were then, low-income countries (45,50). In both Algeria and Korea, treatment of tuberculosis in an unstandardized manner, without careful attention to follow-up, had resulted in a high level of both primary and acquired resistance to both isoniazid and streptomycin. In both countries, rifampin had not been used widely and combined resistance to both isoniazid and rifampin was uncommon. With a strengthening of the program structure, supervision, and standardization of approach (51), along with the introduction and controlled use of a new powerful medication (rifampin) which had not previously been used to any great extent, it was possible to reduce rates of both primary and acquired resistance (Fig. 6.7).

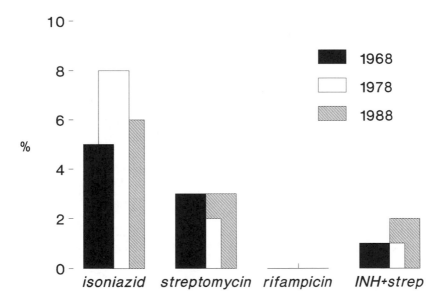

Figure 6.6. Trend in primary resistance to antituberculosis medications in Tanzania from 1968 to 1988.

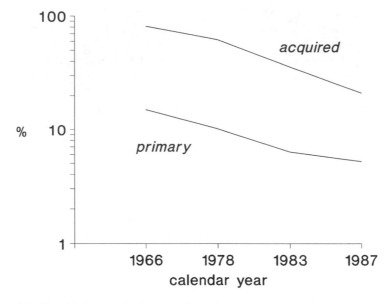

Figure 6.7. Trend in levels of primary and acquired resistance in Korea.

9. The Impact of HIV on Treatment of Tuberculosis in Low-Income Countries

The pandemic of infection with the human immunodeficiency virus (HIV) has led to a rise in the case rates of tuberculosis in a number of countries, particularly sub-Saharan Africa. This has led some people to feel that the situation is hopeless for containing tuberculosis in such countries (52). How does HIV infection affect the treatment of tuberculosis in low-income countries?

Several studies from Africa have shown clearly that the results of treatment of smear-positive patients infected with HIV with regimens not containing rifampin are not as good as in patients who are not infected with HIV (53,54). There is a higher fatality rate among those who are infected with HIV, although the cause of death is usually a disease other than tuberculosis (55). Possibly as many as one-third of all such patients die while being on treatment for their tuberculosis (56). Moreover, there is a continuing trend in fatality such that the life expectancy of such patients is markedly reduced with a median survival after diagnosis possibly as low as 18 months—death being related to HIV infection rather than tuberculosis itself.

Among patients who survive, the results of treatment are similar in HIV-infected and uninfected persons when rifampin-containing regimens are used

(56,57); the proportions of patients who convert their sputum smears to negative, of those who do not convert their smears, and of those who default from treatment (Fig. 6.4), are similar. The likelihood of relapse after cure following treatment with the regimens recommended for use in low-income countries by the IUATLD has not been studied. The contribution of such relapse cases to the spread of tuberculosis is unlikely to be high, as life expectancy of such cases is substantially reduced.

10. The Future of Treatment Programs in Low-Income Countries

What of the future of efforts to control tuberculosis, especially in low-income countries? Since the recognition of the cost-effectiveness of the model programs and their adoption by the WHO, a large number of the countries of Africa have adopted them as the basis for their NTP and many have obtained donor support. The extension of such programs to include all countries where tuberculosis is common is the most important next step. The contribution of the World Bank to the establishment of such programs in large countries such as China, India, and Bangladesh has been an important development.

References

1. World Bank (1993) World Development Report 1993: Investing in health. Oxford: World Bank/Oxford University Press.

2. Sudre P, ten Dam G, Kochi A (1992) Tuberculosis: a global overview of the situation today. Bull WHO 70:149–159.

3. Kochi A (1991) The global tuberculosis situation and the new control strategy of the World Health Organization. Tubercle 72:1–6.

4. Shaw JB, Wynn-Williams N (1954) Infectivity of pulmonary tuberculosis in relation to sputum status. Am Rev Tuberc 69:724–732.

5. Loudon RG, Williamson J, Johnson JM (1958) An analysis of 3,485 tuberculosis contacts in the city of Edinburgh during 1954–1955. Am Rev Tuberc 77:623–643.

6. Grzybowski S, Barnett GD, Styblo K (1975) Contacts of cases of active pulmonary tuberculosis. Bull Int Union Tuberc 50:90–106.

7. Nsanzumuhire H, Lukwago EW, Edwards EA, Stott H, Fox W, Sutherland I (1977) A study of the use of community leaders in case-finding for pulmonary tuberculosis in the Machakos District of Kenya. Tubercle 58:117–128.

8. Grigg ERN (1958) The arcana of tuberculosis. Am Rev Tuberc 78:151–172.

9. Cummins SL (1920) Tuberculosis in primitive tribes and its bearing on tuberculosis of civilized communities. Int J Publ Hlth 1:10–171.

10. Rutledge CJA, Crouch JB (1919) The ultimate results in 1694 cases of tuberculosis treated at the Modern Woodmen of America Sanatorium. Am Rev Tuberc 2:755–763.

11. Springett VH (1971) Ten-year results during the introduction of chemotherapy for tuberculosis. Tubercle 52:73–87.

12. Krebs W. Die Fälle von Lungentuberkulose in der Aangahischen Heilstätte Barmelwell aus der jahren 1912–27. Beitr Klin Tbk 74:345–379.

13. National Tuberculosis Institute (1974) Tuberculosis in a rural population of South India: a five year epidemiological study. Bull WHO 51:473–488.

14. Rouillon A, Perdrizet S, Parrot R (1976) Transmission of tubercle bacilli: the effects of chemotherapy. Tubercle 57:275–299.

15. Enarson D, Rieder H, Arnadottir T (1991) Tuberculosis Guide for Low Income Countries, 3rd Ed. Paris: Les Editions de l'Aulne.

16. Ryan F (1992) Tuberculosis: The Greatest Story Never Told, pp. 148–170. Bromsgrove U.K.: Swift Publishers.

17. Collins JJ (1982) The contribution of medical measures to the decline of mortality from respiratory tuberculosis: an age-period-cohort model. Demography 19:409–427.

18. Crofton J (1960) Tuberculosis undefeated. Br Med J i:679.

19. Kakou M, and Kakihara H (1992) Analysis of chronic excretions of *Mycobacterium tuberculosis* in Aichi Perfecture. Kekkaku 67:29–44.

20. Canetti G (1955) The Tubercle Bacillus in the Pulmonary Lesions of Man. New York: Springer-Verlag.

21. Ryoken (1979) A study on prevalence of resistance to primary and secondary drugs among newly admitted pulmonary tuberculosis patients in 1977. Kekkaku 54:549–555.

22. Schulzer M, Enarson DA, Grzybowski S, Hong JP, Lim TP (1987) An analysis of pulmonary tuberculosis data from Taiwan and Korea. Int J Epidemiol 16:584–589.

23. Centers for Disease Control (1992) Transmission of multidrug-resistant *Mycobacterium tuberculosis* among immunocompromised persons in a correctional system—New York 1991. Morbid Mortal Wkly Rep 41:507–509.

24. World Health Organization (1974) WHO Expert Committee on Tuberculosis, Ninth Report. Geneva: World Health Organization.

25. Bannerji D, Anderson S (1963) A sociological study of awareness of symptoms among persons with pulmonary tuberculosis. Bull WHO 29:665–693.

26. Frimodt-Möller J (1962) Changes in tuberculosis prevalence in a south Indian rural community following a tuberculosis control programme over a seven years period (a preliminary report). Indian J Tuberc 9:187–191.

27. Hopewell PC, Sanchez-Hernandez M, Baron RB, Ganter B (1984) Operational evaluation of treatment for tuberculosis; results of a "standard" 12-month regimen in Peru. Am Rev Respir Dis 129:439–443.

28. Enarson DA (1990) The tuberculosis control programme of international child care in Haiti. Unpublished report to the Canadian Lung Association.

29. Korean National Tuberculosis Association (1975) Report on the Third Tuberculosis Prevalence Survey in Korea. Seoul: Korean National Tuberculosis Association.

30. Aoki M (1984) Comments on the Results of the Sixth Tuberculosis Prevalence Survey in Taiwan Area. Tokyo: Research Institute of Tuberculosis Japan Anti-Tuberculosis Association.

31. Nationwide random survey for the epidemiology of pulmonary tuberculosis conducted in 1979. Chin J Respir Dis 5:67–71.

32. Datta M, Radhmani MP, Selvaraj R, Paramasivan CN, Gopalan BN, Sudeendra CR, Prabhakar R (1993) Critical assessment of smear-positive pulmonary tuberculosis patients after chemotherapy under the district tuberculosis programme. Tubercle Lung Dis 74:180–186.

33. Grzybowski S (1993) Drugs are not enough; failure of short-course chemotherapy in a district in India. Tubercle Lung Dis 74:145–146.

34. Iseman MD, Cohn DL and Sbarbaro JA (1993) Directly observed treatment of tuberculosis We can't afford not to. New Engl J Med 328:576–578.

35. Enarson DA (1991) Principles of IUATLD Collaborative National Tuberculosis Programmes. Bull Int Union Tuberc Lung Dis 66:195–200.

36. Murray CJ, DeJonghe E, Chum HJ, Nyangulu DS, Salomao A, Styblo K (1991) Cost-effectiveness of chemotherapy for pulmonary tuberculosis in three sub-Saharan African countries. Lancet 338:1305–1308.

37. Kochi A (1991) The global tuberculosis situation and the new control strategy of the World Health Organization. Tubercle 72:1–6.

38. Manalo F, Tan F, Sbarbaro JA, Iseman MD (1990) Community-based short-course treatment of pulmonary tuberculosis in a developing nation. Am Redi Respir Dis 142:1301–1305.

39. Fox W (1990) Drug combinations and the bioavailability of rifampicin. (Discussion paper). Tubercle 71:241–245.

40. Acocella G (1989) Studies of bioavailability in man. Bull Int Union Tuberc Lung Dis 64:40–42.

41. Fox W, Mitchison DA (1975) Short-course chemotherapy for pulmonary tuberculosis. Am Rev Respir Dis 111:325–353.

42. International Union Against Tuberculosis and Lung Disease (1988) Antituberculosis regimens of chemotherapy. Recommendations from the Committee on Treatment of the IUATLD. Bull Int Union Tuberc Lung Dis 63:60–64.

43. World Health Organization Tuberculosis Programme (1993) Treatment of Tuberculosis: Guidelines for National Programmes. Geneva: World Health Organization.

44. Costello HD, Caras GJ, Snider DE (1980) Drug resistance among previously treated tuberculosis patients, a brief report. Am Rev Respir Dis 121:313–316.

45. Boulahbal F, Khaled S, Tazir M (1989) The interest of follow-up of resistance of the tubercle bacillus in the evaluation of a programme. Bull Int Union Tuberc Lung Dis 64:23–25.

46. Chonde TM (1989) The role of bacteriological services in the National Tuberculosis and Leprosy Programme in Tanzania. Bull Int Union Tuberc Lung Dis 64:37–39.

47. Garcia FD, Mayorga H (1990) Relapse rate after short-course chemotherapy in smear-positive patients with pulmonary tuberculosis under routine conditions. Am Rev Respir Dis 141:A896.

48. Sbarbaro J (1981) The nature of man and physician (editorial). Am Rev Respir Dis 123:147.

49. Rodier G, Gravier P, Sevre J-P, Binson G, Omar CS (1993) Multidrug-resistant tuberculosis in the Horn of Africa. J Infect Dis 168:523–524.

50. Kim SJ, Kim SC, Bai GH (1982) Drug resistance to *Mycobacterium tuberculosis* isolated from patients with pulmonary tuberculosis discovered in the fourth nation-wide tuberculosis survey in Korea. Tubercle Lung Dis 29:1–10.

51. Jin BW, Kim SC, Mori T, Shimao T (1993) The impact of intensified supervisory activities on tuberculosis treatment. Tubercle Lung Dis 74:267–72.

52. Stanford JL, Grange JM, Pozniak A (1991) Is Africa lost? Lancet 338:557–558.

53. Perriens JH, Colebunders RL, Karahunga C, Willame JC, Jengmons J, Kaboto M, Mukadi Y, Panwels P, Ryder RW, Prignol J et al. (1991) Increased mortality and tuberculosis treatment failure rate among human immunodeficiency virus (HIV) seropositive compared with HIV seronegative patients with pulmonary tuberculosis treated with "standard" chemotherapy in Kinshasa, Zaire. Am Rev Respir Dis 144:750–755.

54. Hawken M, Nunn PP, Gathua S, Brindle R, Godfrey-Faussett P, Githui W, Odhiambo J, Batchelor B, Gilks C, Morris J, McAdam K (1993) Increased recurrence of tuberculosis in HIV-1 infected patients in Kenya. Lancet 342:332–337.

55. Mohamed A, Lwechunguru S, Chum HJ, Styblo K, Broekmans JF (1990) Excess fatality, caused—to a great extent—by HIV infection, in smear-positive patients enrolled on short-course chemotherapy after sputum conversion. Am Rev Respir Dis 141:A267.

56. Enarson DA, Chum HJ, Gninafon M, Nyangulu DS, Salomao MA (1993) Tuberculosis and human immunodeficiency virus infection in Africa. In: Reichman LB, Hershfield ES, eds. Tuberculosis: A Comprehensive International Approach, pp. 395–412. New York: Marcel Dekker.

57. Small PM, Schechter GF, Goodman PC, Sande MA, Chaisson RE, Hopewell PC (1991) Treatment of tuberculosis in patients with advanced human immunodeficiency virus infection. N Engl J Med 324:289–294.

7

Chemotherapy of Drug-Resistant Tuberculosis In the Context of Developed and Developing Countries

Tadao Shimao

1. Introduction

Tuberculosis, once a dreadful disease, has become a curable disease mainly due to marked progress in chemotherapy. Therefore, the emergence of drug resistance is one of the major problems of current tuberculosis control programs. Drug-resistant tuberculosis is produced by poor case management and the cost of treatment of drug-resistant cases is much higher than that for drug-sensitive cases. In addition, its efficacy is generally poorer. Thus, the priority of a national tuberculosis control program should be to achieve a high cure rate of newly detected tuberculosis by improving case management through various measures, such as directly observed therapy.

The principle of the treatment of drug-resistant tuberculosis is the use of three to four, or even five drugs to which the strain is sensitive, and in cases who fail to convert to culture negative, surgery may be considered. However, the application of this principle is quite different in developed and developing countries. In the former, treatment can be based on the results of drug-susceptibility tests. In the latter, there are difficulties such as a lack of laboratory facilities for routine drug-susceptibility tests, a shortage of financial resources to procure sufficient amounts of the more expensive secondary drugs and a shortage of trained staff in the peripheral health facilities. The application of a standard retreatment regimen is therefore inevitable.

In this chapter, chemotherapy of drug-resistant tuberculosis in developed countries is discussed first and then the same topic in developing countries.

2. Chemotherapy for Drug-Resistant Tuberculosis in Developed Countries

2.1. Basic Concept

The basic concept of treating drug-resistant tuberculosis in developed countries is to prescribe a regimen based on the results of drug-susceptibility tests. The number of drugs used is normally three to four, but sometimes five drugs may be used. For multidrug-resistant tuberculosis, chest surgery may be considered for suitable cases, as the efficacy of chemotherapy is limited.

Close supervision of patients during treatment is particularly important, as the planned chemotherapy could be the last chance to cure patients by chemotherapy alone and drugs used for drug-resistant tuberculosis are generally speaking more toxic than drugs used for the original treatment.

2.2. Grouping of Antituberculosis Drugs Used in Chemotherapy for Drug-resistant Tuberculosis

As the efficacy of chemotherapy depends on the potency of combined drugs, it is worthwhile to group those currently available into the following four groups (Table 7.1).

Class A drugs are bactericidal and less toxic. They are rifampicin (R), isoniazid (INH or H), pyrazinamide (PZA), and streptomycin (SM). They are the first-choice drugs even in the treatment of drug-resistant tuberculosis if they are previously unused and the strain is sensitive.

Class B drugs are bacteriostatic and less toxic or bactericidal but more toxic. They are ethambutol (EMB), kanamycin (KM), and capreomycin (CPM).

Class C drugs are less potent and/or more toxic and are used in combination with Class A or B drugs to prevent the emergence of drug resistance. They are

Table 7.1. Grouping of TB Drugs

Class A (Bactericidal, less toxic)
    Rifampicin (R) or its derivatives such as rifabutin (RBT), isoniazid (INH),
        pyrazinamaide (PZA), streptomycin (SM)
Class B (Bacteriostatic and less toxic or bactericidal but more toxic)
    Ethambutol (EMB) kanamycin (KM), capreomycin (CPM)
Class C (Substitutes, less potent and/or more toxic)
    Ethionamide or prothionamide (1314Th, 1321Th, Th), thiacetazone (T), para-
        aminosalicyclic acid (PAS), cycloserine (CS), viomycin (VM), or enviomycin
        (EVM)
Class D (Experimental drugs)
    New Derivatives of R such as KRM1648, quinolones and macrolide compounds

---

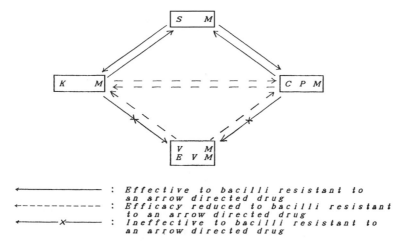

Figure 7.1. Cross-resistance among SM, KM, CPM, VM, and EVM.

ethionamide or prothionamide (1314 Th, 1321 Th, or Th), thiacetazone (T), para-aminosalysilic acid (PAS), cycloserine (CS), viomycin (VM) and enviomycin (EVM).

Class D drugs are under experiment. They are new derivatives of rifampicin RFP such as KRM 1648 (rifabutin), quinolones and macrolide compounds. Some of them, in particular KRM 1648, have given promising results in both *in vitro* and *in vivo* experiments and clinical trials are expected.

In determining the regimen, priority should be given to Class A drugs, then Class B drugs, and priority is lowest with Class C drugs. It is, of course, necessary to avoid a combination of drugs which are given by injection, as there is cross-resistance between SM, KM, CPM, and VM (EVM) as shown in Figure 7.1. SM should be used first, then KM or CPM, and finally VM or EVM.

2.3. Regimens Before the Results of Drug-Susceptibility Tests Are Known

Investigations are in progress on rapid tests to detect and identify mycobacteria and to determine drug susceptibility. Using currently available methods, it takes up to 2 months to obtain these results. Treatment, therefore, has to be started, without knowing the results of drug-susceptibility tests.

In determining the appropriate regimen for these cases the results of previous treatment are important. If previous treatment was successful and the patient converted to culture negative, completed treatment, and was declared "cured" by the attending physician, but subsequently relapsed, chemotherapy can be restarted using the standard regimen for new smear-positive pulmonary tuberculosis,

Cured (Bacilli converted to negative and treatment was
       terminated according to physician's instruction)

Regimen of Chemotherapy : 2~3 H R Z S (E)

Failed or Results Unknown

Combined Use of Previously Unused Drugs

No. of previously unused class A drugs	Regimen of chemotherapy	Examples
3	Combined use of 3 A drugs	R Z S
2	Combined use of 2 A & 1 B drugs	R Z E, R S E S Z E
1	Combined use of 1 A & 2~3 B & C drugs	R E KM, R KMTh S E Th, Z E KMTh
0	Combined use of 3~4 B & C drugs	E KMTh, KMThCS KMTh T CS

Figure 7.2. Regimen of chemotherapy for possible drug-resistant cases before pretreatment drug-sensitivity tests results are available; based on results of previous treatment.

namely 2HRZS (E) and 4HR, as tubercle bacilli are likely to still be sensitive to drugs used in the previous treatment.

If previous chemotherapy failed or the results were unknown, the regimen should be a combination of previously unused drugs on the assumption that the strain is likely to be resistant to previously used drugs, and this regimen will depend on how many Class A drugs were previously unused. As INH is used in most cases, it was excluded from the regimens in Figure 7.2. If two or more Class A drugs are available, chemotherapy is started with a combination of two or three class A drugs with the addition of a Class B drug. If only one or no Class A drug is available, a combination of three to four drugs is needed.

2.4. Regimens When the Results of Pretreatment Drug-Susceptibility Tests Are Known

Recommended regimens for drug-resistant cases when the results of pretreatment drug-susceptibility tests are known are presented in Figure 7.3. In determining the regimens, the number of Class A drugs to which the strain is sensitive is important and examples of regimens and the expected efficacy are shown in the figure.

STRAINS SUSCEPTIBLE TO ALL FOUR CLASS A DRUGS

If the strain is sensitive to all Class A drugs, the regimen is the same as that used in the original treatment of new smear positive pulmonary tuberculosis, namely 2HRZS (E)/4HR, and the expected efficacy is excellent.

No. of sensi- tive class A Drugs	Regimen of Chemotherapy	Examples	Expected Efficacy
4	Combined use of 3~4A drugs	$2HRZS\ (E) \to 6$~$7HR$ or 6~$7H_3\ R_3$	(+++)
3	Combined use of 3A drugs	$2HRZ \to 6$~$7HR$ $3HRS \to 6HR$ $2RZS \to 7RSE$	(+++) (+++) (+++)
2	Combined use of 2A and 1B drugs	$3HRE \to 6$~$9HR$ $3HSE \to 9HE$ $3RZTh \to 9RZ$ $3RSE \to 9RE$	(+++) (++) (++) (++)
1	Combined use of 1A and 2~3B and C drugs	3~$4HEKM \to 9$~$8HE$ 3~$4REKM \to 9$~$8RE$ 3~$4SETh \to 9$~$8SE$ 3~$4ZKMTh \to 9$~$8ZTh$	(++) (++) (+) ~ (++) (+) ~ (++)
0	Combined use of 3~4B and C drugs	$12KM \cdot Th \cdot CS$ $12CPM \cdot Th \cdot CS \cdot T$ $12CPM \cdot Th \cdot PAS \cdot T$	(+) (+) (±) ~ (+)

Notes (+++) : *Excellent results*   (++) : *Good results*   (+) : *Limited results*   (±) : *Poor results*

Figure 7.3. Regimen of chemotherapy for drug-resistant cases of pulmonary TB after pretreatment drug-sensitivity test results are available.

STRAINS SENSITIVE TO THREE CLASS A DRUGS

If the strain is sensitive to three Class A drugs, then these should be combined in the initial phase of 2–3 months, and the total duration of chemotherapy should be 8–9 months, including the continuation phase. The expected efficacy is excellent.

THE STRAIN IS SENSITIVE TO TWO CLASS A DRUGS

If the strain is sensitive to two Class A drugs, they should be combined with EMB, provided the drug is sensitive to this drug. If bacilli are resistant to EMB, either ethionamide (Th) or kanamycin (KM), capreomycin (CPM), or some other Class C drug must be substituted. The duration of the initial phase should be 3 months. Excellent efficacy can be expected if 3HRE/6–9HR is used. In other regimens, expected efficacy might be slightly less but is generally good.

THE STRAIN IS SENSITIVE TO ONE CLASS A DRUG

A strain sensitive to a single class A drug should be treated with a combination of two or three Class B or C drugs, and the initial phase should be 3–4 months. Expected efficacy is worse than the above three categories; however, cure by chemotherapy alone is still high if the strain is sensitive to INH or R. As Class B or C drugs are more toxic, close supervision is needed. Moreover, this treatment may be the last chance of cure by chemotherapy, and the best results can be expected by only full adherence to treatment. Taking into account all these factors, it is recommended that patients be admitted to the hospital where there are tuberculosis experts.

THE STRAIN IS RESISTANT TO ALL CLASS A DRUGS

Three to four Class B or C drugs have to be combined. As the efficacy of chemotherapy is limited, it is necessary to consider surgery. The incidence of side effects will be high, and this can be difficult to cope with in the outpatient department. It is recommended that patients in this category be treated in hospital where there are tuberculosis experts.

GENERAL CONSIDERATIONS

Regimens of chemotherapy shown in Figure 7.3 are examples and may need to be modified taking into account the response and the incidence and severity of side effects.

In the initial phase, fully supervised drug-taking either by hospitalization or directly observed in the outpatient department is strongly recommended. Even in the continuation phase, treatment may be better supervised by the use of an intermittent thrice- or twice-weekly, regimen.

2.5. Efficacy of Chemotherapy for Drug-Resistant Tuberculosis

Clinical efficacy of chemotherapy for drug-resistant tuberculosis depends mainly on the number of Class A drugs to which the strain is sensitive. Regimens are tailored to individual cases and the number of combined drugs and the duration of the initial as well as of the continuation phase should be determined case by case, taking into account factors such as severity of the disease, response to treatment, severity of side effects, and so forth.

EFFICACY OF INITIAL TREATMENT OF DRUG-RESISTANT TUBERCULOSIS

Several controlled clinical trials of short-course (6–8 months) chemotherapy have been carried out by the British Medical Research Council in collaboration with East African countries, Hong Kong, and Singapore. (2–12) In these studies, results were reported separately for patients whose strains pretreatment were sensitive and those with resistant strains. The results for those with sensitive strains were, in general, excellent. The efficacy of a 6–8-month regimen for patients with pulmonary tuberculosis due to resistant strains is shown in Table 7.2 according

Table 7.2. Efficacy of short-course chemotherapy of 6–8 months for initially resistant cases of pulmonary TB by the number of sensitive class A drugs

No. of Class A drugs	Regimen of Chemotherapy in the Initial Phase							
	With PZA				Without PZA			
	Total	Failed	Total	Relapse	Total	Failed	Total	Relapse
2	40	2 5.0%	36	4 11.1%	104	1 1.0%	89	22 24.7%
1	66	8 12.1%	53	3 5.7%	68	13 19.1%	46	8 17.4%
0	100	13 13.0%	71	17 23.9%	52	16 30.8%	29	13 44.8%

Notes: The efficacy of 6–8 months' short-course chemotherapy for initially resistant cases to INH and/or SM were summarized from the studies carried out by the British Medical Research Council (Refs. 2–12). The results were tabulated by the number of Class A drugs in the regimen.

to the number of Class A drugs in the regimen and whether or not PZA was used in the regimen in the initial phase.

Excellent results can be expected if two Class A drugs are sensitive; however, the duration of the continuation should be extended, particularly when PZA is not included in the initial phase as the relapse rate is slightly higher in this group.

If only one Class A drug is sensitive, treatment failure is higher than cases treated with two Class A drugs, and efficacy depends on whether the strain is susceptible to RFP or INH.

If all Class A drugs are resistant, results are much worse than the above two groups, and the results illustrate the difficulty of treating multidrug-resistant tuberculosis.

EFFICACY OF RETREATMENT REGIMENS FOR DRUG-RESISTANT TUBERCULOSIS

Studies on retreatment regimens using secondary drugs have been carried out for chemotherapy failures when drugs such as INH, SM, and PAS have been used. Some of these are presented in Table 7.3. (13–20)

The treatment committee of the International Union Against Tuberculosis and Lung Disease (IUATLD) have carried out collaborative studies on the combined use of PZA, Th, and CS for the treatment of patients who had failed chemotherapy with primary drugs. (13) Twenty-eight hospitals from 14 European countries participated. Patients were given three drugs for 4 months. They were then divided into three groups; those continuing the same three drugs, and those given Th and CS, or PZA and Th for another 8 months. Conversion to culture negativity was good in those who completed treatment in all three groups; however, the conversion rate in all cases was low due to the high dropout rate because of side effects.

If RFP and EMB were used, the results were much better. A study in Hong Kong revealed that 87% of cases responded favorably when given RFP and EMB daily for 18 months or 16 months once weekly after both drugs had been given daily for 2 months. (14) This study also showed that the results of 6-months' use of PZA, Th, and CS followed by 12 months of PZA and Th were very good if the strains were sensitive to Th.

Table 7.3. Efficacy of retreatment regimens of chemotherapy for drug-resistant pulmonary TB

Author (Ref.)	Resistant to	Regimen of chemotherapy		No. of cases	Favorable		Relapse
					To whole cases	To treatment completed	
IUAT, Second	INH	12ZThCS		43	94%	67%	
Study, Series	and	4ZThCS→8ThCS		41	87%	63%	
II (13)	S M	8ZTh		42	85%	67%	
Hong Kong TB		18RE		91		87%	
Treatment	INH	2RE→16R_1E_1		62		87%	
Service/BMRC (13)		18R_2E_2		84		79%	
		18R_1E_1		53		71%	
		6ZThCS→12ZTh		59		93%	
		(Th: sensitive)					
			RFP (mg)				
Zierski (15)	INH	3RE→9R_1E_1	600	38		95%	} 7%
Poland		9R_2E_2	600	40		95%	
		9R_1E_1	1200	35		97%	} 0%
		9R_2E_2	1200	30		93%	
		15R_1E_1	1200	35		97%	} 2%
		15R_2E_2	1200	35		97%	
		21R_1E_1	1200	32		100%	} 0%
		21R_2E_2	1200	36		100%	
Goble (16) (Denver,	INH and	Individually tailored				65%	
USA)	RFP	regimen		134			14%[a]
RYOKEN (17,18) (Japan)	INH	12KMThCS		64		53%	15%[2]
	and	12 Previously Used Drugs		67		11%	20%
	SM						
		I. 6R + Previously Used				47%[3]	
RYOKEN (19,20) (Japan)	INH, SM	Drug		121			31%
	KM,Th,CS	IIA. 6RE		67		72%	10%
		IIB. 6R_2E_2		65		80%	8%

[a]Out of a total 171 of cases resistant to INH and RFP, 63 (37%) died, among them 37 (22%) from tuberculosis.
[b]Fatality during 2 years follow-up was 20% in the KMThCS group and 21% in the control group. Out of bacilli-positive cases at 12 months, none converted to negative later in the KMThCS group and 9% in the control; all of them were surgically treated.
[c]In approximately two-thirds of cases, RFP was further continued, and the negativity conversion to culture from positive cases at 6 months were seen in 2% in the group I, in 35% of the group IIA, and in 25% of the group IIB. Adding these cases, the negative conversion rate by culture at 12 months was 48%, 82%, and 85% respectively in groups I, IIA, and IIB.

A study in Poland showed excellent results of retreatment by the combined use of RFP and EMB when treatment with primary drugs had failed (15). If the dose of RFP in the intermittent regimen was increased to 1200 mg and was continued for 12 months or longer, nearly all cases converted to culture negative and very few relapses were observed.

2.6. Case Management During Treatment

DIFFICULTIES IN REGULAR DRUG-TAKING DURING TREATMENT

Regular drug-taking is the key to the success of treatment. However, there are difficulties in maintaining regular drug-taking during treatment.

They are summarized as follows:

1. *Poor awareness of the disease by patients themselves*: Tuberculosis patients often complain of persisting cough and sputum, tiredness, weight loss, and so forth; however, the more severe symptoms such as high fever or pain are rarely seen. Cough and sputum will disappear and patients notice an increase in weight soon after the start of chemotherapy. They feel well long before they are actually cured. To continue regular drug-taking after they are well is not easy. Nowadays short-course regimens are widely used, but 6–8 months' treatment is still long for most tuberculosis patients, and it is difficult to continue regular drug-taking for this length of time.
2. *Roles played by patients*: Patients must take treatment until their disease is cured. At the same time, most are allowed to work either from the start of treatment, or during treatment but after their disease has improved. Patients who are busy sometimes forget that they are patients and do not take their drugs.
3. *Poor approach of health workers including physicians to patients*: In treating tuberculosis, patients must understand their disease. They should be instructed thoroughly in how it may be cured. It is particularly important to stress the necessity of regular drug-taking during treatment. Lack of understanding on the part of health workers and physicians can lead to a lack of awareness of patients of their tuberculosis. Health workers often claim that patients do not take drugs regularly or default from treatment; however, in many cases, this is the responsibility of the health workers who failed to instruct their patients properly. This can result in drug resistance because the patients were doubtful of or lost confidence in the health services due to failure of previous treatment. However, some failures of treatment are also due to the patient. Every effort should be made to convince patients of the necessity of regular drug-taking during treatment.
4. *Concentration on vulnerable groups*: The decline in tuberculosis in developed

countries has taken place more rapidly in socially favorite groups but has been slow or not seen at all in the vulnerable groups such as the homeless, chronic alcoholics, drug users, migrants, refugees, and so on. It is well known that compliance with the chemotherapy for a long period of time is much more difficult in patients of these latter groups than in patients in more favorably social groups.

MEASURES TO ENSURE REGULAR DRUG-TAKING

Before the start of treatment, it is necessary to take time to instruct patients on the nature of the tuberculosis and how it may be cured. In particular, the importance of regular drug-taking until they have been declared as cured should be stressed. If physicians are too busy and can spend little time to talk to patients, tuberculosis nurses or some other health workers should take on the physician's role.

Patients may have problems which prevent them taking their drugs even though they are well motivated to be compliant. Among these are how to cover the cost of treatment, how to feed the family, who will take care of the children, concern at losing a job, and so forth. Health care workers need to support patients in cooperation with the social workers to help solve these problems. It is vital that health workers maintain contact with patients during treatment.

Regular drug-taking in the initial phase of chemotherapy is very important, as there are often large numbers of tubercle bacilli in the lesions and the risk of the emergence of drug resistance is high if drugs are not taken regularly. One measure to ensure regular drug-taking during this period is to admit patients to a hospital, where drugs are given directly by nurses.

If hospitalization is impossible, treatment has to be conducted at clinics, and drugs should be given by health workers to patients directly at least during the initial phase. In the latter case, thrice or twice weekly regimens are more convenient for directly observed therapy (DOT).

This is only feasible when patients live near the clinic. If they live too far away, they should be hospitalized until adherence to treatment has been established.

Action against defaulters should be taken immediately if patients do not come to the clinic on the appointed day. This is particularly important in cases receiving intermittent therapy, as the risk of the emergence of drug resistance is high when drugs are taken irregularly.

2.7. The Use of INH Alone for Patients with Strains Resistant to All Available Drugs

Chemotherapy for patients with drug-resistant tuberculosis who continue to discharge tubercle bacilli is difficult. We are reliant on treatments that hark back to the prechemotherapy era, namely bed rest and good nutrition in fresh air. The use of INH alone may be considered in such cases, as this attenuates the virulence of tubercle bacilli. The use of the newer drugs like the quinolones and the macrolides has yet to be established.

INCIDENCE OF TUBERCULOSIS AMONG CONTACTS OF CHRONIC CASES

In animal models it is recognized that the virulence of strains of *Mycobacterium tuberculosis* which are resistant to isoniazid is attenuated. However, it is not known if this is true for patients.

Results of a study in Japan on the incidence of tuberculosis among contacts of chronic cases are summarized in Table 7.4 (21). Chronic cases in this study were treatment failures in Tokyo and Okinawa who continued to discharge tubercle bacilli resistant to nearly all available drugs. The control groups were the contacts of sex- and age-matched new bacillary tuberculosis cases registered in the same health center in the same year as the chronic cases. These patients converted to culture negative following treatment. The ambulatory group were contacts of

Table 7.4. Incidence of tuberculosis among contacts of chronic tuberculosis cases—surveys in Tokyo and Okinawa

| | Tokyo | | Okinawa | | Chronics at JATA clinic |
	Chronics	Control	Chronics	Control	clinic
No. of cases	39	39	7	7	15
No. of contacts	129	127	27	32	58
Within 3 years after registration					
Observed person-years	340	355	61	70	140
New case of TB (per 100					
person years)	3	7	1	1	4
	0.88	1.97	1.63	1.42	2.86
3 yrs. or later after Registration					
Observed person years	798	468	190	200	237
New case of TB (per 100					
person years)	3	3	2	1	1
	0.38	0.64	1.05	0.50	0.42

Note: Chronics are chemotherapy failure cases discharging tubercle bacilli in sputum continuously, and the control is age- and sex-matched bacillary cases of tuberculosis, who converted to negative by chemotherapy and were registered at the same health center in the same year of chronics. Chronics at JATA clinic are cases treated ambulatorilly at the JATA clinics.

Table 7.5. Prognosis of chronic bacillary pulmonary patients in hospitals; Surveys by RYOKEN

Observation started in	1975	1981	1988
Survey was carried out in	1981	1984	1991
Observation period	5 years, 3 months	3 years	2 years, 2 months
No. of patients	513	366	182
Died	275 53.6%	143 39.1%	58 31.9%
Hospitalization continued	172 33.5%	185 50.5%	107 58.8%
Discharged alive	61 11.9%	36 9.8%	17 9.3%
Unknown	5 1.0%	2 0.5%	—

cases treated ambulatorily at the clinics of the Japan Anti-Tuberculosis Association in Tokyo. The incidence of tuberculosis was high in the initial 3 years after registration, and thereafter it fell. However, there was no significant difference in the incidence between the contacts of chronic cases and the control, both in the earlier and later period after registration.

In analyzing these data, the fact that many of the chronic cases were hospitalized must be taken into consideration. Out of 39 cases in Tokyo, 30 were hospitalized when they registered and 20 by the time of the survey, and out of 7 cases in Okinawa, 5 cases were hospitalized by the time of the survey as shown in Table 7.5. Thus, the chances of infection by contact with chronic cases were likely to be reduced, although family members often visited patients in the hospital and patients sometimes were allowed to go back home and, therefore, had contact with family members. However, the incidence of tuberculosis among contacts of chronic patients who lived with their families while receiving treatment was similar to that of the control groups in the above study. Of course, chronic cases knew their own disease status and took care to reduce direct contact with their family members. It can be assumed that the risk of chronically infected patients infecting family members is low.

The use of INH as monotherapy in chronic cases resistant to all available drugs is acceptable, as the virulence of *M. tuberculosis* highly resistant to INH is attenuated, it is inexpensive, and the incidence of serious side effects is low. (22–33)

FINDINGS SUPPORTING THE ATTENUATED VIRULENCE OF STRAINS HIGHLY RESISTANT TO ISONIAZID

Among chronic cases, there are some whose sputum is smear positive but on culture there is no growth. This suggests that the viability of the bacilli in these cases is impaired and the number of viable bacilli in sputum is much less than in pretreatment smear positive sputa.

Ryoken repeatedly made surveys of the prevalence of initial as well as acquired resistance among newly admitted tuberculosis patients in specific hospitals. Fig. 7.4 indicates the correlation between the prevalence of drug resistance and its grade to INH, SM and PAS in previously untreated and treated cases in 1966 and 1972. (22,23) The prevalence of previously treated cases represents a part of drug resistance pattern among the source of tuberculosis infection in a community, and the prevalence of drug resistance among previously untreated cases indicates the resistance pattern of those primarily infected from the source of infection. There was a good positive correlation between the prevalence of drug resistance and its grade to INH, SM and PAS among previously untreated and treated cases, except the prevalence of cases resistant to INH 5 mcg/ml completely. The prevalence of strains which are highly resistant to INH was high among previously treated cases, while it was disproportionately low in previously untreated cases, and the same fact was seen both in the 1966 and 1972 surveys. The low prevalence of cases highly resistant to INH among the previously untreated cases in contrast to the high prevalence of such cases among previously treated cases could be explained by the attenuated virulence of INH highly resistant tubercle bacilli, through which either infection itself or the break down of the disease from the primarily infected persons is prevented.

The prognosis of chronic cases investigated by RYOKEN is presented in Table 7.5. (24–26) In the prechemotherapy era, approximately half of smear positive pulmonary tuberculosis patients died within two years after the detection of the

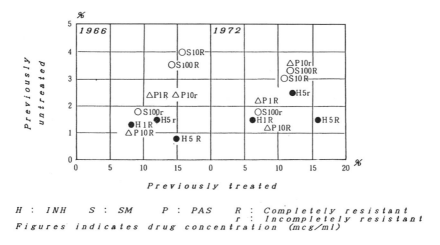

H : INH   S : SM   P : PAS   R : Completely resistant
                                r : Incompletely resistant
Figures indicates drug concentration (mcg/ml)

Figure 7.4. Correlation between the prevalence of drug resistance and its grade to INH, SM, and PAS in previously treated and untreated pulmonary TB studies of RYOKEN in 1966 and 1972.

Table 7.6. Tuberculous complications in patients admitted to national sanatoria and died of pulmonary tuberculosis

	Survey year						
	1959	1964	1969	1974	1979	1984	1989
No. died of pulmonary TB	1838	1922	1363	1042	902	728	551
Cases with TB complications	339 18.4%	195 10.1%	157 11.5%	54 5.2%	46 5.1%	39 5.4%	31 5.6%
Miliary TB	4 0.2%	5 0.3%	5 0.4%	2 0.2%	4 0.4%	4 0.5%	3 0.5%
Meningitis	1 0.1%	1 0.1%	1 0.1%	—	7 0.8%	0.7%	0.7%
Intestinal TB	93 5.1%	37 1.9%	23 1.7%	10 1.0%	11 1.2%	7 1.0%	4 0.7%
Laryngeal TB	43 2.3%	22 1.1%	17 1.2%	4 0.4%	—	1 0.1%	—
Bone and joint TB	46 2.5%	26 1.4%	20 1.5%	16 1.5%	17 1.9%	10 1.4%	4 0.7%
Urogenital TB	36 2.0%	10 0.5%	10 0.7%	4 0.4%	1 0.1%	5 0.7%	1 0.2%

disease, while the fatality of chronic cases in these studies was lower than that of the prechemotherapy era, though the prognosis of chronics was in general rather unfavourable. The difference in the fatality could be caused by the difference in the virulence of bacilli.

In the prechemotherapy era, most cases of tuberculosis died not only of pulmonary tuberculosis but also of complicated intestinal and laryngeal tuberculosis which led patients to emaciation. Studies on causes of deaths among tuberculosis patients admitted to national sanatoria in the whole of Japan were carried out every 5 years since 1959, and the prevalence of tuberculous complications among cases died of pulmonary tuberculosis by the time of their death was presented in Table 7.6. (27–33) The prevalence of complicated intestinal and laryngeal tuberculosis was low throughout the repeatedly carried out surveys, and the fact indicates certain changes in the disease progression pattern of tuberculosis in the pre- and post-chemotherapy era, which could be explained by the attenuated virulence of tubercle bacilli of INH highly resistant strains.

Recent nosocomial infection of tuberculosis with multidrug resistant tubercle bacilli among HIV positives in the U.S. taught us that INH resistant tubercle bacilli could cause tuberculosis in HIV positives. In persons with normal immunological status, however, the risk of break down of tuberculosis might be lower and the progression of the disease might be slower among cases highly resistant to INH.

The use of INH as monotherapy in chronic cases resistant to all available drugs is acceptable as the virulence of *M. tuberculosis* highly resistant to INH is attenuated, it is cheap and the incidence of serious side-effects is low.

2.8. Adverse Reaction to Drugs

The treatment of drug-resistant tuberculosis inevitably involves the use of Class B and/or C drugs, which are more toxic than Class A drugs. As the incidence of side effects is high when using Class B and/or C drugs, patients must be supervised closely not only for compliance but also for possible side effects.

The major side effects of antituberculosis drugs are presented in Table 7.7.

Before starting and during chemotherapy, patients should be instructed about the possible side effects of drugs, and if they detect any abnormality, they must immediately report to an attending physician. Many side effects are detected by patients complaints as shown in Table 7.8, methods of coping with them is also indicated in the table. The most common side effects are gastrointestinal, which are encountered often when Th, PAS, and R are used. In the case of R symptoms are often improved by taking it after food. A good principle for coping with the gastrointestinal disturbances is to give drugs in divided doses.

Vertigo caused by SM can be eased by reducing the daily dose to 0.5 g/day in the elderly patients. For central nervous system disturbances caused by CS or Th and arthralgia caused by PZA, drugs should be discontinued and substituted by other drugs.

The detection of side effects requires the laboratory examinations shown in Table 7.9. How to cope with them is also indicated in the table. Irreversible side effects are hearing impairment caused by SM, KM, CPM, VM, and EVM; hence,

Table 7.7. Commonly seen side effects of anti-tuberculosis drugs

	INH	R	PZA	S M	EMB	K M	CPM	T h	T	PAS	C S	V M
Hearing impairment				◎		◎	◎			◎		
Visual disturbances					◎							
Peripheral nerve dist.	○				○			○				
Central nervous system dist.	○							○			◎	
Allergic reactions	○	○		○				○	◎			
Skin reaction, eruption, etc.	○	○		○				○	◎	◎	○	
Hepatotoxicity	○	◎	◎					◎	◎	○		
Nephrotoxicity						○	○					○
Hematopoietic organ dist.	○	○							◎			
Gastrointestinal truct dis.	○	○	○		○			◎		◎	○	
Others			◎	◎					○	○		
			(1)	(2)					(3)	(4)		

Note: ◎: Frequently seen side effects ○: Rather rare side effects

(1): Arthralgia (2): Paresthesia around mouth, rigidness in face
(3): Fall off of hair (4): Thyroid gland dysfunction

Table 7.8. Side effects of TB drugs detected by patient complaints

Side effects	Symptoms	Possible causing agents	Measures needed
Gastrointestinal truct disturbance	Loss of appetite, nausea, vomiting, constipation, diarrhea	Th, PAS	Daily dose divided into 2 to 3 dosages. Take drugs after meal.
Allergic reactions	Fever, exanthem, thrombocytopenia, shock, flu syndrome	PAS, R	Desensitization; use of steroid
Peripheral nerve dist.	Paresthesia in the extremities	INH, EMB	Use of vitamin B6
Vestibular disturbance	Vertigo	SM	Reduction of SM dose
Visual disturbance	Decline of visual acuity, color sense abnormality, dimness of sight	EMB	Discontinue EMB
Central nervous system dist.	Irritation, uneasiness, cramp, epilepsia	CS, Th	Discontinue drug(s)
Articular dist.	Arthralgia of large joint	PZA	Discontinue PZA

Table 7.9. Side effects of TB drugs detected by laboratory examinations

Side effects	Laboratory examinations indicated	Possible causing agents	Measures needed
Hearing impairment	Audiometry	SM, KM, CPM, VM, EVM	Discontinue injection
Hepatotoxicity	GOT, GPT, ALP, etc.	PZA, Th, R	Check GOT, GPT at short interval when levels exceed 40; Stop drugs when levels exceed 100
Hematopoietic	Blood cell count	R, T	Discontinue drugs
Visual disturbance	Visual acuity, color sense	EMB	Discontinue EMB
Articular disturbance	Uric acid	PZA	Check uric acid at short interval when its level exceeds 6
Nephrotoxicity	Urinalysis (protein, BUN, sediments)	KM, CPM, VM, EVM	Discontinue drug

pretreatment audiometry is necessary. Hepatotoxicity is severe if it is due to PZA or Th, and the use of the drug(s) should be discontinued. In the case of R elevated GOT and/or GPT levels often decrease spontaneously whilst continuing R. If GOT and/or GPT levels exceed 100, R should be discontinued temporarily. However, in many cases R can be used again after GOT and/or GPT levels return to normal.

Cases who have severe side effects are impossible to cope with in the out patient department, and hospitalization is recommended.

2.9. Surgical Treatment for Drug-Resistant Cases of Pulmonary Tuberculosis

As mentioned before, the efficacy of treatment for drug-resistant pulmonary tuberculosis depends on the number of Class A drugs to which the strain is sensitive, in particular the sensitivity to INH and RFP. If a strain is sensitive to only one Class A drug, or to SM or PZA, or resistant to all Class A drugs, a high cure rate by chemotherapy alone cannot be expected and surgery is indicated, if this is possible.

The results of surgical treatment at the author's institute between 1976 and 1981 are shown in Table 7.10. They are presented separately for those with positive preoperative culture results, including the grade of drug resistance and those with negative cultures.

The results of chest surgery were worse in patients with preoperative positive cultures. However, in multidrug-resistant cases, 83.6% converted to culture negative and 80.1% were rehabilitated socially. Hence, chest surgery played an important role in curing drug resistant cases.

Using a combination of Class B and/or Class C drugs, the cure rate by chemotherapy alone is low, however, conversion to culture negative for a short period of time after their initial use can be expected, and if surgery is used during this period, the results are much better. For good results, the following conditions are indispensable:

1. Major foci can be resected by pneumonectomy, lobectomy or combined resection, or be collapsed by thoracoplasty.
2. Remaining foci should heal spontaneously or be cured by the use of a regimen not containing INH and RFP.
3. Postoperative lung function must be good enough for social rehabilitation.
   (A practical index is to maintain the postoperative forced expiratory volume (FEV)/Predicted Vital Capacity above 40%)
4. At least one or two drugs to which the strain is sensitive even if these are Class B or C, are reserved to cover preoperative and postoperative periods, in particular for pulmonary resection.

Table 7.10. Results of surgical treatment for drug-resistant pulmonary tuberculosis—Ryoken, 1976–1981

Drug resistance	No. of cases	Bacilli (−)	Bacilli (+)	Died	Socially rehabil.	With complic.
Bacilli (+) within 2 months before surgery						
Multidrug resistance	146	83.6	13.7	2.7	80.1	10.4
R resist. + at least						
1 another	162	91.4	6.2	2.5	87.7	6.2
Unknown	15	93.3	—	6.7	93.3	73.3
Total	323	87.9	9.3	2.8	84.5	11.1
Bacilli (−) within 2 months before surgery						
Total	621	98.6	1.1	0.3	96.8	1.4

Note: Multidrug resistant cases are resistant to R, INH and/or SM, EMB and/or KM; namely resistant to at least three important drugs including R. Final results of surgery are divided into bacilli (−), (+) and died, and socially rehabilitated and cases with complications are relisted.

If the patient continued to discharge bacilli in spite of chemotherapy but fulfilled the above conditions, the use of chest surgery should be strongly considered.

3. Chemotherapy of Drug-Resistant Tuberculosis in Developing Countries

3.1. Background

Chemotherapy for drug-resistant tuberculosis in developing countries is quite different from that of developed countries. The difficulties can be summarized as follows:

1. *Weak health infrastructure.* The national tuberculosis control programme (NTP) should be integrated into the primary health care system. However, the quality and quantity of the peripheral health care facilities are often poor and many drug-resistant cases are produced due to irregular drug taking and defaulting during treatment. There are few if any tuberculosis experts at the district level and it is difficult to prescribe individual regimens or to change regimens if side effects occur. It is a fact that side-effects of drugs used in drug-resistant tuberculosis are seen more frequently than in cases treated with standard short-course chemotherapy (SCC).
2. *Lack of shortage of laboratory facilities for culture and drug-sensitivity tests.* In most developing countries, sputum culture and drug-sensitivity tests are not done routinely. It is rarely possible, therefore, to prescribe regimens according to laboratory results.

3. *Cost of secondary drugs.* Most developing countries suffer from a shortage of financial resources to procure antituberculosis drugs even for new smear-positive pulmonary cases. In addition, the cost of the drugs used in the treatment of drug-resistant tuberculosis is generally, much higher than that of the drugs used in the SCC, and it is rarely possible to cover the cost of secondary drugs unless external financial resources are available.

Taking into account these conditions, it is difficult to define precisely drug resistant cases in most developing countries. Therefore the regimen for relapses, treatment failure and chronic cases is the same as that for drug-resistant cases as the prevalence of drug resistance is presumed to be high in these patients.

The definition of relapse, treatment failure, and chronic disease is as follows: (35)

Relapse. Originally treated for tuberculosis and declared as cured and currently having at least two sputum specimens smear positive for acid-fast bacilli (AFB) and a chest X-ray showing findings consistent with pulmonary tuberculosis.

Treatment failure. Sputum smear remains positive for AFB 5 months or more after the start of chemotherapy, or the treatment is interrupted 1–5 months after the start of chemotherapy and a subsequent sputum is smear positive for AFB.

Chronic cases. Sputum smear positive for AFB in spite of completing a retreatment regimen under supervision.

Of the above three categories, the chronic cases should be dealt with separately, as their strains are most likely to be resistant to all routinely available drugs.

In most developing countries, a standardized regimen irrespective of the results of drug-sensitivity tests is used for the treatment of relapses and treatment failures. However, the World Health Organization (WHO) and the International Union Against Tuberculosis and Lung Disease (IUATLD) have recommended regimens for the treatment of tuberculosis, and among these are regimens for relapses and treatment failures (see below). Table 7.11.

Under certain circumstances where laboratory facilities for routine culture and drug-sensitivity tests are available, and the treatment of tuberculosis patients is carried out under the supervision of tuberculosis experts, some modification of the standard regimen of chemotherapy for relapses or treatment failures can be considered. However, even in such cases, modification of the regimen is often limited due to shortage of financial resources and in the majority of cases regimens recommended by WHO and IUATLD have to be followed.

3.2. Recommended Regimen of Chemotherapy for Relapses or Treatment Failures

WHO and IUATLD have recommended the following regimens for relapses and treatment failures under conditions where facilities for drug-sensitivity tests are either (1) routinely not available or (2) routinely available.

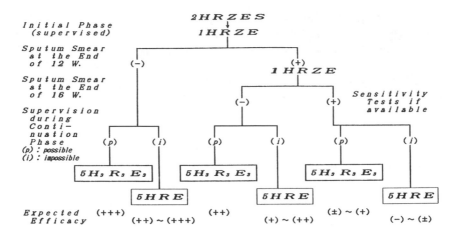

Figure 7.5. Recommended regimen for relapse and treatment failure cases; drug-sensitivity tests routinely not available.

UNDER CONDITIONS WHERE ROUTINE DRUG-SENSITIVITY TESTS ARE NOT AVAILABLE

The flowchart of the recommended regimen is presented in Figure 7.5. Treatment is started with the combination of five drugs: INH, R, PZA, EMB and SM daily for 2 months, followed by a four drug combination of INH, R, PZA and EMB for 1 month (2HRZES/1HRZE). As the risk of emergence of multiple-drug resistance is very high in these patients, treatment during this initial intensive phase should be fully supervised.

If the sputum is still smear positive, all drugs are stopped for 2–3 days, and a sputum specimen sent to the laboratory for culture and sensitivity tests if these facilities are available. A continuation phase is then started, consisting of INH, R, and EMB given three times a week for 5 months (5H3R3E3) under supervision. If supervision is impossible, treatment is continued with the above three drugs daily for 5 months (5HRE).

If the sputum is still smear positive at the end of the continuation phase, the case is no longer eligible for the retreatment regimen, and is classified as a chronic case.

If the sputum smear is negative at the end of the third month, the continuation phase is started and continued for 5 months (5H3R3E3) if supervision is possible or 5HRE if it not.

UNDER CONDITIONS WHERE DRUG-SENSITIVITY TESTS ARE AVAILABLE

The flowchart of the recommended regimen for sensitive and resistant cases is shown in Figure 7.6. If isolates are fully sensitive to all drugs including INH and

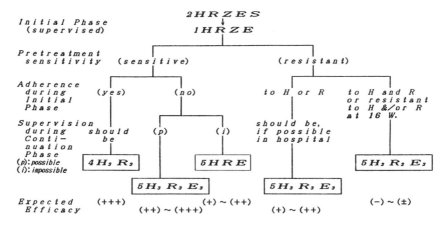

Figure 7.6. Recommended regimen for relapse and treatment failure cases; drug sensitivity tests available.

RFP, the patient adhered to the initial phase of treatment, and there is evidence of a favorable response such as conversion of sputum to smear-negative, treatment is continued with INH and RFB three times a week for 4 months (4H3R3) under supervision. It is particularly important to supervise closely during the entire period to ensure full adherence to the treatment.

If the pretreatment sensitivity tests showed resistance to either INH or RFP, the continuation phase regimen (5H3R3E3) is started under close supervision, preferably in a referral hospital.

If the pretreatment sensitivity tests showed resistance to both INH and RFP or a strain isolated at the end of 16 weeks is resistant to either INH or RFP or to both, the continuation phase regimen (5H3R3E3) is continued. However, the chance of attaining sputum-negative status is limited.

Dosages of the drugs in the recommended regimens for relapses and treatment failures are presented in Table 7.11. (35)

CHEMOTHERAPY FOR CHRONIC CASES

Under current conditions in most developing countries, the sole method for treating chronic cases is to continue INH alone. The theoretical basis for this approach has already been presented.

4. Future Problems of Chemotherapy for Drug-Resistant Tuberculosis

As mentioned above, the treatment of drug-resistant tuberculosis, in particular, multidrug-resistant tuberculosis is very difficult and, moreover, new drug-resistant

Table 7.11. Treatment regimens of chemotherapy for relapse and treatment failure cases—Recommended by WHO and IUATLD

	Initial Phase (3 months)				Continuation Phase (5 months)		
Pretreatment body weight (in kg)	INH + R(HR) tablet 100 mg + 150 mg 150 mg + 300 mg	PZA(Z), tablet 500 mg	EMB(E), tablet 400 mg	SM(S)[a] powder for injection 1 g base in vial	INH + R(HR), tablet 100 mg + 150 mg	INH(H), tablet 300 mg	EMB(E),[b] tablet 400 mg
<33	2 tablets (100 mg + 150 mg) daily	2 tablets daily	2 tablets daily	500 mg daily	2 tablets, 3 times weekly	1 tablet, 3 times weekly	2 tablets, 3 times weekly
33–50	3 tablets (100 mg + 150 mg) daily	3 tablets daily	2 tablets daily	750 mg daily	3 tablets, 3 times weekly	1 tablet, 3 times weekly	3 tablets, 3 times weekly
>50	2 tablets (150 mg + 300 mg) daily	4 tablets daily	3 tablets daily	750 mg daily	4 tablets, 3 times weekly	1 tablet, 3 times weekly	4 tablets, 3 times weekly

Note: [a]Administration of SM is restricted to the first 2 months of the initial treatment phase.
[b]When INH + RFP and EMB are given daily in the continuation phase, the drug doses are the same as in the initial phase.
Source: Reproduced, by permission, from Treatment of Tuberculosis: Guidelines for National Programmes, Geneva, World Health Organization, Table 3, p. 11.

cases will be produced if we fail to prevent new primary infection from drug-resistant sources of infection. Hence, problems of drug-resistant tuberculosis give and will continue to give a heavy burden for the national tuberculosis programs in countries where the prevalence of drug resistance is high. How does one cope with this serious problem?

First, it is most important not to produce new cases of drug-resistant tuberculosis and this can be achieved by improving case management during treatment. The priority of the tuberculosis programme of WHO is to raise the cure rate of new smear-positive pulmonary tuberculosis to a minimum of 85% in developing countries and 95% in developed countries by the year 2000 by ensuring adherence to treatment. "Prevention prevails on treatment" is true also in the management of drug-resistant tuberculosis.

Second, prevention of primary infection of contacts of drug-resistant cases is also important to halt the production of more new cases of drug-resistant tuberculosis. Measures to achieve this are discussed earlier.

Third, development of new drugs which are effective against drug-resistant tubercle bacilli is indispensable to cure multidrug-resistant tuberculosis. As shown in Table 7.1, some promising new drugs are now under investigation, such as new derivatives of rifampicin, quinolones, and macrolide compounds. *In vitro* and *in vivo* activities of some of them have been confirmed. However, clinical trials of these drugs are not easy. They are very expensive and the benefit to manufacturers is limited by the small number of patients in developed countries where expensive drugs are affordable and the lack of finance in most developing countries where there are large numbers of patients. Measures such as the orphan drug policy to promote the development of new drugs by governments are needed.

Fourth, even if some new drug(s) are developed, the combined use of at least two drugs to which the strain is sensitive is needed, otherwise single use will produce resistant tubercle bacilli.

Fifth, studies on immunomodulation of host defense mechanisms should be encouraged. Even if no drugs are available, the approach is expected to increase the probability of spontaneous cure by ameliorating the host defense mechanisms.

Sixth, the intensification of international cooperation is needed to cope globally with drug-resistant tuberculosis. Currently available short-course chemotherapy is potent enough to cure nearly all new cases of tuberculosis. However, its application is still limited in many developing countries mainly due to the shortage of financial resources. The cost to developing countries of existing and new antituberculosis drugs together with technical cooperation in the improvement of case management as a part of the international programs must be considered seriously. If it is not, chemotherapy, a powerful weapon we have against tuberculosis, will lose its potency through misuse before tuberculosis is brought under control. International cooperation in tuberculosis research to develop new rapid diagnostic methods and new drugs is necessary to solve drug-resistant tuberculosis, which is one of the major obstacles in the global fight against this disease.

References

1. Kudoh S (1975) Tubercle bacilli. In: Shimao T, ed. A New Introduction to Phthisiology, pp. 75–95. Tokyo: Japan Anti-Tuberculosis Association.

2. East African/British Medical Research Council (1973) Controlled clinical trial of four short-course (6-month) regimens of chemotherapy for treatment of pulmonary tuberculosis. Second report. Lancet 7816:1331–1339.

3. East African/British Medical Research Council (1974) Controlled clinical trial of four short-course (6-month) regimens of chemotherapy for treatment of pulmonary tuberculosis. Third report. Lancet 7875:237–240.

4. Second East African/British Medical Research Council Study (1974) Controlled clinical trial of four short-course (6-month) regimens of chemotherapy for treatment of pulmonary tuberculosis. Lancet 7889:1100–1115.

5. Second East African/British Medical Research Council Study (1976) Controlled clinical trial of four short-course (6-month) regimens of chemotherapy for treatment of pulmonary tuberculosis. Second report. Am Rev Resp Dis 114:471–475.

6. Third East African/British Medical Research Council Study (1978) Controlled clinical trial of four short-course regimens of chemotherapy for two durations in the treatment of pulmonary tuberculosis. First report. Am Rev Resp Dis 118:39–48.

7. Third East African/British Medical Research Council Study (1980) Controlled clinical trial of four short-course regimens of chemotherapy for two durations in the treatment of pulmonary tuberculosis. Second report. Tubercle 61:59–69.

8. East and Central African/British Medical Research Council Fifth Collaborative Study (1986) Controlled clinical trial of 4 short-course regimens of chemotherapy (three 6-month and one 8-month) for pulmonary tuberculosis. Final report. Tubercle 67:5–15.

9. Hong Kong Chest Service/British Medical Research Council (1978) Controlled trial of 6-month and 8-month regimens in the treatment of pulmonary tuberculosis. First report. Am Rev Resp Dis 118:219–227.

10. Hong Kong Chest Service/British Medical Research Council (1979) Controlled trial of 6-month and 8-month regimens in the treatment of pulmonary tuberculosis. The results up to 24 months. Tubercle 60:201–210.

11. Tanzania/British Medical Research Council Study (1985) Controlled clinical trial of two 6-month regimens of chemotherapy in the treatment of pulmonary tuberculosis. Am Rev Resp Dis 131:727–731.

12. Singapore Tuberculosis Service/British Medical Research Council (1985) Clinical trial of three 6-month regimens of chemotherapy given intermittently in the continuation phase in the treatment of pulmonary tuberculosis. Am Rev Resp Dis 132:374–378.

13. The Committee on Treatment of the International Union Against Tuberculosis (1969)

A comparison of regimens of ethionamide, pyrazinamide and cycloserine in the retreatment of patients with pulmonary tuberculosis. Bull IUAT 42:7–57.

14. Hong Kong Tuberculosis Treatment Services/British Medical Research Council (1975) A controlled trial of daily and intermittent rifampicin plus ethambutol in the re-treatment of patients with pulmonary tuberculosis: Results up to 30 months. Tubercle 56:179–189.

15. Zierski M (1977) Prospects of retreatment of chronic resistant pulmonary tuberculosis patients. A critical review. Lung 154:91–102.

16. Goble M, Iseman M, Madsen LA, et al. (1993) Treatment of 171 patients with pulmonary tuberculosis resistant to isoniazid and rifampin. N Engl J Med 328:527–532.

17. Tuberculosis Research Committee, RYOKEN/Oka H, Oomori K (1962) A study on clinical efficacy of the combination of KM, Th and CS for severe pulmonary tuberculosis resistant to primary drugs. Results of treatment for 12 months. Nihon Iji Shimpo (Japan Med J) 1997:3–14.

18. Tuberculosis Research Committee, RYOKEN/Oka H, Oomori K (1963) Follow-up results of patients with severe pulmonary tuberculosis treated with the combination of KM, Th and CS. Nihon Iji Shimpo (Japan Med J) 2061:14–18.

19. Tuberculosis Research Committee, RYOKEN (1970) Therapeutic effect of rifampicin on re-treatment cases of pulmonary tuberculosis. Kekkaku (Tuberculosis) 45:227–235.

20. Tuberculosis Research Committee, RYOKEN (1974) Three-year follow-up results of the first clinical study on the daily and twice-weekly administration of rifampicin, investigated by the Tuberculosis Research Committee, RYOKEN. Kekkaku (Tuberculosis) 49:107–112.

21. Saitoh M (1983) Infectiousness of chronic bacillary cases of pulmonary tuberculosis and problems on contacts control. Hokenfu no Kekkaku Tenbou (Rev Tuber Publ Hlth Nurses) 22:48–58.

22. Tuberculosis Research Committee, RYOKEN/Oka H, Gomi J, Chiba Y, et al. (1969) Prevalence of drug resistance among pulmonary tuberculosis patients newly admitted to member's institutions. The results in 1966 compared with those of 1957, 1959, 1961 and 1963. Nihon Iji Shimpo (Japan Med J) 2355:3–8.

23. Tuberculosis Research Committee, RYOKEN/Oka H, Gomi J, Chiba Y, et al. (1975) Prevalence of drug resistance among pulmonary tuberculosis patients newly admitted to member's institutions. Part 1. The results in 1972 compared with those of previous 6 surveys. Kekkaku (Tuberculosis) 50:1–8.

24. Tuberculosis Research Committee, RYOKEN (1985) Follow-up study of patients with tuberculosis staying at hospitals for more than five years by the time of survey in 1975 (Part 1). Results of the survey in 1981, with special reference to the dead and discharged patients. Kekkaku (Tuberculosis) 60:351–403.

25. Tuberculosis Research Committee, RYOKEN (1984) Three years' follow-up study of patients with tuberculosis staying at hospitals for more than three years in 1981. Annual Report of RYOKEN for 1984, pp 4–16.

26. Tuberculosis Research Committee, RYOKEN (1991) Two years' follow-up study of patients with tuberculosis staying at hospitals for more than two years in 1988. Annual Report of RYOKEN for 1991, pp. 5–13.

27. Kino C (1961) A survey on deaths in 1959 among tuberculosis patients admitted to national sanatoria in whole Japan. Kekkaku Kenkyu no Shimpo (Adv Tuber Res) 30:170–214.

28. TB Death Survey Teams of National Sanatoria and of Japan Anti-Tuberculosis Association (JATA) (1967) A survey on deaths in 1964 among tuberculosis patients admitted to national sanatoria in whole Japan. Kekkaku to Kokyuki Shikkan Bunken no Shouroku Sokuhou (Abstr Curr Liter Tuberc Respir Dis) 18:42–60.

29. TB Death Survey Teams of National Sanatoria and of JATA (1971) A survey on deaths in 1969 among tuberculosis patients admitted to national sanatoria in whole Japan. Kekkaku to Kokyuki Shikkan Bunken no Shouroku Sokuhou (Abst Curr Liter Tuberc Respir Dis) 22:571–590.

30. TB Death Survey Teams of National Sanatoria and of JATA (1976) A survey on deaths in 1974 among tuberculosis patients admitted to national sanatoria in whole Japan. Kokyuki Shikkan to Kekkaku Bunken no Shouroku Sokuhou (Abstr Curr Liter Tuberc Respir Dis) 27:575–590.

31. TB Death Survey Teams of National Sanatoria and of JATA (1981) A survey on deaths in 1979 among tuberculosis patients admitted to national sanatoria in whole Japan. Kokyuki Shikkan to Kekkaku Bunken no Shouroku Sokuhou (Abst Curr Liter Tuberc Respir Dis) 32:393–401.

32. TB Death Survey Teams of National Sanatoria and of JATA (1987) A survey on deaths in 1984 among tuberculosis patients admitted to national sanatoria in whole Japan. Kokyuki Shikkan to Kekkaku Bunken no Shouroku Sokuhou (Abstr Curr Liter Tuberc Respir Dis) 38:57–87.

33. TB Death Survey Team of National Sanatoria (1992) A survey on deaths in 1989 among tuberculosis patients admitted to national sanatoria in whole Japan. Shiryo to Tebou (Inform Rev Tuberc Respir Dis Res) 1:59–86.

34. Tuberculosis Research Committee, RYOKEN and Its Subcommittee on Surgical Treatment (1984) Changes in surgical treatment for pulmonary tuberculosis in recent 6 years. A comparative study between preoperative culture positive and negative cases. Kekkaku (Tuberculosis) 59:467–474.

35. World Health Organization (1993) Treatment of Tuberculosis. Guidelines for National Programmes, pp. 1–43. Geneva: World Health Organization.

8

Tuberculosis and the Acquired Immunodeficiency Syndrome

R. Shaw and R.J. Coker

1. Introduction

The World Health Organization (WHO) estimates that there are 1.9 billion people infected with tuberculosis worldwide, 14 million infected with human immunodeficiency virus (HIV), and 5.6 million coinfected with tuberculosis and HIV. The overlap between at-risk populations especially in the developing world is increasing, which suggests that these figures will increase rather than improve. Alone, these two pathogens confer an enormous disease burden on mankind. In combined infection, both organisms have a synergistic effect, increasing morbidity and mortality.

2. The Interaction Between Mycobacterium Tuberculosis and HIV

The presence of HIV infection increases mortality from tuberculosis (1) and there is an accelerated course of HIV infection after tuberculosis (2). The mechanism of the interaction between HIV and *M. tuberculosis* infection is the subject of intense research. Both organisms cause a degree of immunosuppression and facilitate replication of each other (3). HIV causes a reduction in CD4 lymphocytes and prevents the development of cell-mediated immunity needed to kill *M. tuberculosis* following phagocytosis by macrophages (4). Similarly, tuberculosis also reduces the number of CD4 cells (5,6). Phagocytosis of *M. tuberculosis* enhances HIV gene expression (7), and HIV infection of antigen-presenting cells alters cytokine production and microbicidal activity (8). The impairment of lymphocyte function by HIV is complex. There is debate as to whether HIV infection results in an imbalance between the Th1 and Th2 subtypes (9,10), which promote a cell-mediated or an allergic immune response, respectively. Whatever the mechanism, there is a clear epidemiological association between HIV infection and tuberculosis in both developed and developing countries.

3. The Problem in Developed Countries

The relationship between the increase in the number of cases of tuberculosis and HIV infection has been most clearly described in the United States (11). This relationship in developed countries may relate in part to the reduced resistance to tuberculosis associated with HIV infection but may also relate to socioeconomic factors which predispose to both infections as well as to a period of reduced funding for tuberculosis control programs (12). Thus, in the United States in 1990, 4.3% of AIDS cases were infected with tuberculosis. Among U.S.-born tuberculosis patients, 11% were HIV seropositive, with some clinics reporting seroprevalance rates as high as 57% (13). Patients with tuberculosis and AIDS have been predominantly young men and a high proportion have been black or Hispanic. By contrast in Britain, the recent increase in notifications of tuberculosis is linked more to socioeconomic factors, immigration from countries where tuberculosis is endemic, and possibly improved notification practices than to HIV infection alone (14,15). In Britain, HIV infection seems to have had a relatively small impact on tuberculosis (16). This may reflect the lack of overlap between the populations affected by the two infections. This could change if the HIV epidemic in India increases, as the overlap between HIV infection and tuberculosis in immigrant groups will increase.

One of the central questions in the control of tuberculosis in HIV-infected individuals has been whether the immunosuppression causes a reactivation of previously acquired disease or whether immunosuppression and life-style factors render individuals more prone to acquire new infection. The use of restriction fragment-length polymorphism to type organisms collected from patients in New York and San Francisco indicated that nearly two-thirds of patients infected with HIV or who had AIDS were in clusters of cases infected by an identical *M. tuberculosis* strain (17,18). This indicates that the majority of tuberculosis in these HIV-infected individuals results from new infection, not reactivation of old disease.

The complexity of the relationship between HIV and tuberculosis has been compounded by the increase in the incidence of multidrug-resistant tuberculosis (that is, tuberculosis resulting from infection with *M. tuberculosis* resistant to rifampicin and isoniazid with or without resistance to other drugs as well.) This has been most marked in the large cities in the United States. For example, in

New York, isolates examined during the month of April 1991 revealed that 33% of patients had isolates resistant to at least one drug and 19% of isolates were resistant to isoniazid and rifampicin (1). Previously treated patients, those infected with HIV, and intravenous drug abusers were at increased risk of drug resistance. The introduction of programs of "directly observed therapy" may have started to reduce the number of new cases to multidrug-resistant tuberculosis caused by poor compliance with therapy.

By contrast in Britain, drug resistance, even in inner-city areas, has been low. In east London over the period of 1984–1992, 3.4% of isolates were resistant to one drug and 3.1% to more than one drug (19). Over a similar period at the national level, drug resistance to one drug had a prevalence of 3–6%, and a multidrug resistance of 0.8% (20,21).

4. The Problem in Developing Countries

Tuberculosis is the most important opportunistic disease associated with HIV infection in the developing world. For example, an autopsy study in Abidjan in West Africa revealed that tuberculosis was the cause of death in 32% of 247 HIV-positive patients dying of HIV disease (22). In up to half the patients dying of the HIV wasting disease "slim disease," the pathology was tuberculosis (23). The importance of tuberculosis in HIV disease in the developing world relates both to the high prevalence of tuberculosis, such that most adults will have been exposed, as well as to the fact that tuberculosis affects individuals at an early stage of immunocompromise when the CD4 count is relatively preserved. Thus, in Kinshasa, the median peripheral blood lymphocyte count in HIV-positive tuberculous patients was 316 per microliter (24). The magnitude of the problem in Africa is illustrated by the estimate that by the end of the century in Kampala, Uganda, 2% per year of the population will develop tuberculosis (25).

It is feared that the spread of the HIV epidemic through other parts of the world such as Southeast Asia, India, and South America where tuberculosis is already endemic will result in a similar pattern to that currently observed in sub-Saharan Africa.

5. Nosocomial Spread

Transmission of tuberculosis is a function of the number and virulence of live bacilli produced by the source case, intensity and duration of exposure, and immune status of the contact. In a hospital or institutional setting, patients immunocompromised by HIV are vulnerable to nosocomial infection with tuberculosis. This has caused greatest concern when there has been nosocomial spread of multidrug-resistant disease (26,27). The technique of restriction fragment-length polymorphism has confirmed that the same organism may infect many individuals living in shelters for the homeless and hospitals (28–30). Studies monitoring immune status to tuberculosis by tuberculin testing have also shown that transmission can occur between prison inmates and staff (31). The evidence that nosocomial spread can occur has lead to a series of recommendations which emphasize the need to isolate smear-positive cases, attend to ventilation and airflow of isolation rooms, as well as the wearing of masks by staff and visitors (13). There has been much debate on issues such as which type of mask is required. The situation is further complicated with moral and ethical issues becoming highlighted, where patients with AIDS, who may have only a short life expectancy, require isolation for long periods for treatment of multidrug-resistant tuberculosis.

6. Clinical Features

The hallmark of infection with the human immunodeficiency virus (HIV) is the development of progressive immunodeficiency that is primarily cell mediated. Because cell-mediated immunity plays a key role in containing mycobacterial infection, clinical features of tuberculosis in this population depend on the stage of HIV-induced immunodeficiency (11,32–37).

There are, therefore, two consequences of the association between tuberculosis and HIV infection (38). First, the potential that an HIV-seropositive person who is infected with *M. tuberculosis* will subsequently develop clinical tuberculosis is much greater than that of the general population. About 10% of individuals without underlying HIV disease who have *M. tuberculosis* infection will subsequently develop active disease at some time during their lifetime. The risk of a *M. tuberculosis*-infected individual with HIV disease developing active tuberculosis is estimated at about 100 times greater, and in patients with AIDS, the risk is perhaps 170 times greater (39,40). In the United States, 5–10% of symptom-free HIV-seropositive individuals develop active tuberculosis within a year of HIV diagnosis (41).

Second, in addition to tuberculosis occurring more frequently in patients with HIV disease, it also occurs more rapidly after infection with *M. tuberculosis* than it does in the immunocompetent host (38). DNA fingerprinting of *M. tuberculosis* isolates from tuberculosis outbreaks has showed that up to 37% of HIV-positive individuals developed primary tuberculosis within 5 months of exposure to a source patient (28,42) compared historically to between only 2% and 4% of immunocompetent household contacts of tuberculosis developing tuberculosis within 12 months (43).

Because *M. tuberculosis* is a more virulent organism than many of the other opportunistic infections associated with AIDS, reactivation of tuberculosis commonly occurs at an earlier stage of HIV disease. HIV-infected individuals are also vulnerable to exogenous tuberculous infection, with rapid progression to active tuberculosis. Until recently, it was thought that approximately 90% of active cases of tuberculosis arose from reactivation of foci of infection acquired years earlier (41,44). Recent reports have suggested, however, that a third or more cases of tuberculosis can result from recent person-to-person transmission rather than reactivation of latent infection (17,18) and that, in individuals infected with HIV, clustering can occur so that much of HIV-associated tuberculosis, at least in the United States, arises from recent transmission.

7. Variation in Clinical Features According to Degree of Immunocompetence

The clinical presentation of tuberculosis depends on the stage at which infection with *M. tuberculosis* occurs and the site involved. In HIV-seropositive individuals without AIDS and with less advanced HIV infection, extrapulmonary infection is uncommon and the chest x-ray usually shows changes suggestive of reactivation tuberculosis. Symptoms in individuals who manifest these features are likely to be typical (45,46).

Tuberculosis in patients with AIDS already or who are severely immunocompromised through HIV infection commonly presents with extrapulmonary tuberculosis and frequently with chest x-ray findings suggestive of progressive primary tuberculosis. In these cases, symptoms are commonly nonspecific, can occur in association with other conditions associated with AIDS, and are often present for many weeks before the diagnosis is made (47–50).

The lungs are involved in approximately 70% of HIV-infected patients with coexistent tuberculosis (51). In those who are relatively immunocompetent, typical features develop which include weakness and tiredness, associated with anorexia and weight loss, persistent cough and hemoptysis. The chest x-ray typically shows upper-lobe infiltrates, often with cavitation (45,46,52,53). This picture is rarely seen in those with advanced HIV disease. Indeed, apical infiltrates in the absence of hilar or mediastinal lymphadenopathy occurs in less than 10% of AIDS-associated cases (41,47,54). The radiographic picture is more often atypical. Hilar and mediastinal lymphadenopathy, lower-lobe infiltrates, miliary infiltrates, and pleural effusions or a normal chest x-ray are more common in severely immunosuppressed patients (34,45,47,54–58) (Figs. 8.1 and 8.2).

Extrapulmonary disease occurs in up to 70% of HIV-infected individuals with tuberculosis, often with lymphatic or disseminated forms (miliary TB and/or two or more tuberculous sites) (34,41,47,51,58). In addition, unusual clinical presentations of tuberculosis can occur. Extrapulmonary tuberculosis frequently coexists with pulmonary tuberculosis, occurring in about a third of cases and can be more frequent in those with more advanced disease (34,51,59).

Peripheral lymph nodes and bone marrow are the most frequently involved extrapulmonary sites. Other extrapulmonary sites include urine, blood, bone, joint, cerebrospinal fluid, brain and spinal cord, liver, prostate, testicle, pancreas, spleen, skin, pericardium, gastrointestinal mucosa, and ascitic fluid (47,53,58–72).

With the increased use of steroids in *Pneumocystis carinii* pneumonia (PCP) (73,74), concern has been expressed about the risk of reactivation of tuberculosis

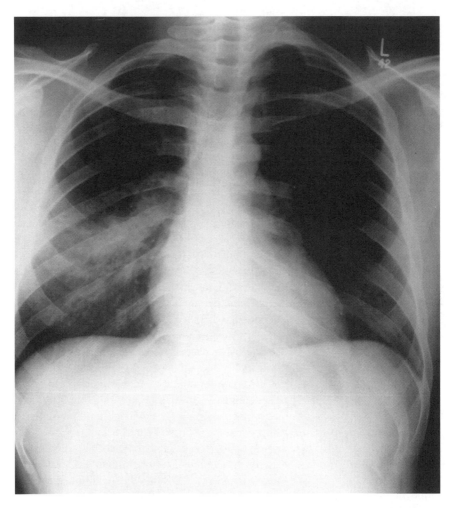

Figure 8.1. Chest x-ray showing right middle-lobe consolidation from tuberculosis in a moderately severely immunocompromised patient.

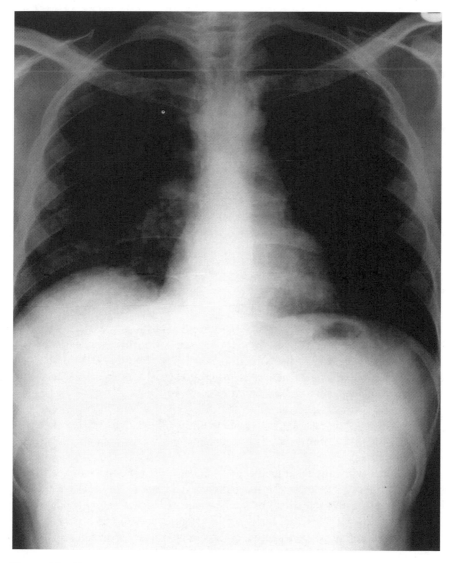

Figure 8.2. Chest x-ray showing miliary infiltrates in a severely immunocompromised patient.

or exacerbating undiagnosed tuberculosis (75–79). Allowing for the fact that much of tuberculosis occurring in association with HIV may be newly acquired, use of corticosteroids for the treatment of PCP probably does not increase its rate (80).

Multidrug-resistant tuberculosis (i.e., tuberculosis arising as a result of organisms resistant to rifampicin and isoniazid with or without resistance to other drugs occurs more frequently in patients with AIDS (particularly injecting drug users) as opposed to those with no previous AIDS-defining diagnosis, where it occurs in association with severe immunosuppression, and is frequently associated with recent infection with *M. tuberculosis* rather than reactivation (1,26,28,81–87). In New York in April 1991, a third of patients with tuberculosis had isolates resistant to one or more antituberculous drugs, with 19% of isolates resistant to both isoniazid and rifampicin (1).

In drug-resistant disease, dissemination involving both pulmonary and extrapulmonary sites is up to three times more likely, probably reflecting the host's more profound immunosupression. Fever, cough, dyspnea, and night sweats are the most common presenting symptoms. Cough is probably more likely to be unproductive. Lymphadenopathy is less common in matched controls with tuberculosis sensitive to all standard drugs. On chest radiograph, alveolar, reticular interstitial, and miliary infiltrates are more frequently seen in patients with multidrug-resistant tuberculosis (27).

8. Diagnosis

A high index of suspicion and an aggressive diagnostic approach is needed to diagnose tuberculosis associated with HIV infection. Conversely, in patients with proven tuberculosis, a high index of suspicion for concomitant HIV infection is required, particularly if the patient is in an AIDS risk group or presents with atypical pulmonary or extrapulmonary tuberculosis.

Because of immunodeficiency, the tuberculin skin test is frequently negative in patients with HIV disease and active tuberculosis (47,88). HIV-seropositive patients may then be at significant risk for tuberculosis if they are either tuberculin positive or anergic (89).

In a patient with suspected tuberculosis, at least three sputum samples should be sent for mycobacterial smear and culture examination. Several investigators have reported on the sensitivity of sputum smears in patients with pulmonary tuberculosis and HIV disease. Some studies suggest there is no significant difference in the rate of positive sputum smears in pulmonary tuberculosis between patients with and without HIV infection (45,58,90–92). Supporting these data, a recent study has suggested that tuberculosis associated with HIV infection is no more infectious than tuberculosis in the absence of HIV infection (93). Conflicting studies have suggested, however, that the rate of positive smears is less frequent in HIV-seropositive individuals (55,94–96). Nevertheless, about 50% of HIV-infected patients with pulmonary tuberculosis have positive sputum smears. The absence of cavitary disease does not appear to reduce the frequency of positive sputum smears (58,91,92). Indeed, sputum and bronchoalveolar lavage fluid may be positive for *M. tuberculosis* despite the chest x-ray being normal (47,92,97). Fiber-optic bronchoscopy with bronchoalveolar lavage and transbronchial lung biopsy may be necessary if either sputum samples cannot be obtained or sputum smears are nondiagnostic. The yield of bronchoscopy for the diagnosis of pulmonary tuberculosis in patients with HIV infection is similar to that in patients without HIV infection (98). Transbronchial biopsy provides a higher yield for an immediate diagnosis (98,99).

Tuberculous bacteremia is a frequent consequence of tuberculosis in the setting of HIV infection and blood cultures performed with the lysis–centrifugation system are positive in between 26% and 42% of patients, especially if they have miliary tuberculosis and elevated serum alkaline phosphatase or lactate dehydrogenase (49,50,100).

Mycobacteria other than *M. tuberculosis* may be seen in clinical specimens from HIV-infected patients, particularly those who are severely immunocompromised The high incidence of *M. avium-intracellulare* and other atypical mycobacteria in this group lead to confusion about the diagnosis before the results of culture are known (101). When extrapulmonary or disseminated tuberculosis is considered possible, sites suspected of *M. tuberculosis* involvement should be either aspirated or biopsied. With advanced immunosuppression, the host response becomes more severely impaired and poorly formed and sometimes absent granulomas without caseation may be seen histologically.

Delays in diagnosis and recognition of drug resistance have contributed to nosocomial spread of multidrug-resistant tuberculosis. A reduction in these delays would allow a more rapid response in implementing proper infection control procedures and appropriate chemotherapy. Even with newer and faster culture techniques, however, it can take as long as 3 weeks to establish the diagnosis and define drug sensitivity (102,103). The BACTEC system is a liquid culture method which detects radioactive CO_2 generated when mycobacteria are grown in the presence of a ^{14}C-labeled carbon source. Selective inhibition of members of *M. tuberculosis* by *p*-nitro-acetyl-amino-β-hydroxypropiophenone (NAP) allows this organism to be differentiated from other mycobacteria (103).

A novel approach, using firefly luciferase expressed by genetically altered mycobacteriophage could potentially reduce the time to establish drug sensitivity (104). Luciferase expression occurs only after recombinant bacteriophage infection of growing *M. tuberculosis* and the production of light, detected and assayed by a luminometer, enables both detection and drug sensitivity to be reported within hours rather than weeks.

9. Treatment

In HIV-seropositive patients, the clinical response generally occurs early, as in HIV-negative patients if organisms are sensitive to antituberculous agents (51,52,60,105,106). Similarly, sputum conversion occurs early and cultures are usually negative at a median of 10 weeks (51). Failure of treatment and relapse rates of tuberculosis chemotherapy are similar to those in HIV-negative patients (105).

Although tuberculosis might respond to conventional treatment the disease is associated with high fatality. This, in part, reflects the fact that tuberculosis occurs at different stages of HIV disease; for some, it is diagnosed before any other AIDS-defining illness, simultaneously for others, and after for some. In one study, the median survival of patients who developed tuberculosis after they had AIDS was only 10.6 months. In this same study, patients in whom tuberculosis was the AIDS-defining diagnosis had a median mortality of 18.4 months and this was most often unrelated to tuberculosis (51).

The bacterial population in cavitary pulmonary tuberculosis is estimated at between 10^7 and 10^9 organisms, and spontaneous mutations leading to drug resistance occur with a frequency of approximately 1 in 10^6–10^8, replications depending on the drug (107). As a consequence, there is a high probability of resistant mutations emerging and becoming predominant due to selection pressure if single-agent therapy is used. Sites for resistance are chromosomally located and are not linked; therefore, the likelihood of an organism spontaneously developing resistance to two agents is in the order of 1 in 10^{14}. Use of at least two drugs to which the organism is sensitive, therefore, means that the probability of resistance developing becomes negligible. Standard drug regimens used for the treatment of tuberculosis in HIV-negative patients are used for the treatment of tuberculosis in HIV-positive patients and patients respond well. In a retrospective study, 132 HIV-seropositive patients with tuberculosis sputum samples became clear of acid-fast bacilli after a median of 10 weeks' therapy. There was only one treatment failure and this was in a poorly compliant individual infected with multidrug-resistant organisms. Mortality from tuberculosis among treated patients in this study was 6% (51). However, because of increasing resistance, drug-susceptibility testing should always be done on all *M. tuberculosis* isolates. In industrialized countries, including the United Kingdom and the United States, when drug-resistant tuberculosis is infrequent, the first-line drugs are isoniazid, rifampicin, pyrazinamide, and ethambutol. In the United Kingdom rifampicin, isoniazid, pyrazinamide and ethambutol are advocated for the initial 2 months. Rifampicin and isoniazid are then continued in both HIV-positive and -negative individuals for at least a further 4 months (108,109). In the United States, in individuals infected with HIV, this phase of treatment is continued for 7 months (11,110). In the United Kingdom, and increasingly in the United States, compliance is improved by the use of multidrug preparations; thus "Rifater®" contains rifampicin, isoniazid, and pyrazinamide, and "Rifinah®" contains rifampicin and isoniazid. Compliance with treatment is the most important factor in the prevention of drug resistance. Compliance is markedly increased by using directly observed therapy (DOT) or Supervised therapy, in which the patient is observed taking therapy and compliance is documented (111).

Drug resistance, particularly to isoniazid and rifampicin, portends badly. In a series of 171 individuals, all of whom were presumed HIV seronegative, 37% died, with nearly 60% of these deaths attributed to tuberculosis. The response rate in patients in whom there was sufficient follow-up data was only 65% (87/134). Thirty-five percent of patients (47/134) showed no response, as indicated by continually positive cultures. Several patients who initially responded to treatment relapsed subsequently and the overall response rate was only 56% (112). An unfavorable response was seen in male patients in contrast to female patients and in those who had received a large number of antituberculous drugs before the study. In HIV-seropositive individuals, multidrug-resistant tuberculosis is associated with devastating consequences. In 62 patients, the median survival was only 2.1 months in one study compared to 14.6 months for controls with tuberculosis caused by single-drug resistant or susceptible bacilli (27). In another study, patients with AIDS and multidrug-resistant tuberculosis were 1.7 times more likely to die than those with drug-susceptible organisms. Ninety-one percent of AIDS patients infected with organisms resistant to both isoniazid and rifampicin died, with a median survival of about 3 months (1). More recently, however, it has been shown that early treatments with at least two antituberculous drugs which the patient has not been given before can markedly improve the outcome. (112a)

Adverse drug reactions to conventional antituberculous drugs occur in about 4% of non-HIV-infected individuals (105). In the HIV-seropositive population, adverse events are more frequent. In a retrospective study, 18% of HIV-seropositive patients receiving both isoniazid and rifampicin developed adverse drug reactions (51). Adverse reactions to rifampicin are especially common, occurring in up to 12% of patients (51,58). Ninety percent develop within the first 2 months and they usually take the form of rash or hepatitis, although life-threatening anaphylaxis has been reported (113,114). A recent study from Uganda compared a rifampicin-containing regimen (isoniazid, rifampicin, and pyrazinamide for 2 months, followed by isoniazid and rifampicin for 7 months) with a thiacetazone-containing regimen (streptomycin, thiacetazone, and isoniazid for 2 months, followed by thiacetazone and isoniazid for 10 months) (115). Of 101 patients receiving the rifampicin-containing regimen however, only one developed an adverse drug reaction (cutaneous hypersensitivity).

Patients at increased risk for drug-resistant *M. tuberculosis* (and therefore perhaps all HIV-infected individuals) should receive ethambutol in addition, until drug-susceptibility results are available (108,116,117).

If isoniazid resistance in the HIV-positive individual is shown, then a regimen of pyrazinamide, rifampicin, and ethambutol for 18 months is recommended (11,110). In cases of drug resistance to both isoniazid and rifampicin, this regimen is insufficient. Failure rates in HIV-negative individuals, for example, with rifampicin-resistant tuberculosis may be between 40% and 70% (118–120). In these cases, treatment should be planned on an individual basis and, while awaiting sensitivity results, include at least two effective drugs to which the patient has never been exposed (112a). Subsequently treatment should include at least three drugs to which the organism is sensitive. Lung resection may be helpful. In a series of 99 patients with multidrug-resistant pulmonary tuberculosis, 29 had surgical resection in addition to chemotherapy. Pneumonectomy or lobectomy were the surgical techniques performed in patients with sufficiently localized disease, and of the 27 survivors, 25 remained sputum culture negative for an average of 3 years (121).

Alternative newer antituberculous agents should be reserved for multidrug-resistant tuberculosis. Potential candidates include the quinolones, ofloxacin, ciprofloxacin, and sparfloxacin (122–124). The rifamycins, rifabutin and rifapentine, are active both in vitro and in vivo, but cross-resistance with rifampicin is usual (125).

In many sub-Saharan African countries, standard regimens for treating pulmonary tuberculosis often contain thiacetazone instead of rifampicin because of thiacetazone's low cost and ease of administration (126). The drug was widely used before the HIV epidemic, despite the known associated adverse drug reactions and risks of recurrent disease (127–129). Severe cutaneous hypersensitivity reactions and recurrent disease are recognized in association with HIV-related tuberculosis (115,130–133). In common with other drugs used to treat HIV-infected individuals, antituberculous treatment is associated with a greater frequency of adverse cutaneous reactions. Of 111 HIV-seropositive patients receiving an antituberculous regimen containing thiocetazone, 22 (20%) developed cutaneous hypersensitivity, an 18 times greater risk than in HIV-seronegative population. In this study, three patients, all HIV seropositive, developed toxic epidermal necrolysis and died as a result of this reaction (131). The cutaneous hypersensitivity associated with thiacetazone use appears to be related to increasing immunosuppression and is assumed to have an immunological basis (131), although earlier studies have suggested that it is dose related (i.e., is toxic) rather than immunological in origin (134). In addition to its toxicity, thiacetazone-containing antituberculous regimens are less effective at sterilizing sputum within 2 months than rifampicin-containing regimens (115).

10. Prevention of Tuberculosis

On a communitywide basis, reduction in poverty, alcoholism, and drug abuse, as well as improvements in general health, housing, nutrition, and so on are the main ways tuberculosis can be reduced. Vaccination with Bacille–Calmette–Guérin (BCG) has served as the main method to provide protection from tuberculosis for individuals. Comparison of BCG efficacy in nonimmunocompromised subjects reveals a wide range of protection efficacy from >50% in some studies to 0% in others (92). The protective effect, when present, probably lasts less than 20 years. The exact extent to which any protective immunity from BCG is carried over following the development of immunocompromise is unknown. This may relate to the extent of the immunocompromise and the time interval since vaccination. The risk of systemic infection with BCG in patients with HIV-induced immunodeficiency (135) has led to a concern that BCG should not be given to individuals immunocompromised by HIV. This may cause difficulty in planning communitywide BCG programs in areas of high seroprevalence for HIV.

11. Preventive Therapy

Isoniazid preventive therapy decreases the occurrence of active tuberculosis by between 60% and 98% in non-HIV-infected tuberculin skin-test-positive individuals (110,136–138). About 10% of patients with positive tuberculin skin tests who do not receive preventive therapy develop tuberculosis each year (88). The benefits of isoniazid prophylaxis in the overall population have been evaluated in several cost–benefit analyses (139). These rely on estimates of hepatitis mortality and tuberculosis prevalence. These benefits have been questioned recently because, it is suggested, projections of reactivation and mortality have been exaggerated in the population at large and the risks of death from isoniazid hepatitis underestimated (140,141).

Preventive therapy for drug-sensitive *M. tuberculosis* has been shown to be effective in patients infected with HIV and in patients with AIDS (142). Chemoprophylaxis with isoniazid decreases the likelihood of progression of tuberculosis infection to disease and, therefore, the use of prophylaxis is based on three factors: first, the likelihood of tuberculous infection; second, the probability that infection will progress to disease; and third, the risk of hepatotoxicity from isoniazid. The likelihood of infection is estimated from the patient's history of exposure to tuberculosis and response to the tuberculin skin test. The greater the degree of cellular immune deficiency, the greater the risk of progression of tuberculous infection. Hepatotoxicity from isoniazid increases with age and may be increased in women particularly if they are pregnant or in the postpartum period (116,143–145). Preventive therapy with isoniazid may, in addition, delay the onset of HIV-related disease and death in symptom-free individuals from areas of high tuberculosis prevalence (136).

Several international health organizations have recently issued guidelines, although these are still widely debated, for the use of tuberculosis preventive therapy in individuals dually infected (146,147) which reflect those issued from national committees. Those from the United States (116,137,148) have suggested that prophylaxis using isoniazid should be offered, after active tuberculosis has been excluded, to HIV-seropositive patients who are likely to be infected with *M. tuberculosis,* irrespective of their tuberculin skin test response, to those who have had previously positive tuberculin skin tests, and to those with chest x-rays that show fibrotic lesions likely to represent old healed tuberculosis. In the United Kingdom, similarly, guidelines suggest that isoniazid prophylaxis should be given to patients with a positive tuberculin test response who have no history of BCG vaccination (109). In those who have been vaccinated, a strongly positive tuberculin skin test (Heaf grade 3–4 or Mantoux >10 mm) indicates that prophylaxis should be given. Negative tuberculin skin tests are, however, only interpretable if skin tests for at least two other recall antigens are positive. Twelve months is the recommended duration of prophylaxis with isoniazid (109,116). Whether prophylaxis with isoniazid should be recommended for tuberculin-negative HIV-infected individuals is unclear. It has been suggested that it should be offered because such patients have a more accelerated course of infection by *M. tuberculosis* (85), although there are only limited data to support this (148).

The regimen of isoniazid for adults is a daily dose of 5 mg/kg up to a maximum of 300 mg for 12 months. Patients should be seen monthly and assessed for compliance, drug toxicity, and signs of active tuberculosis. If supervision is indicated, then isoniazid should be given twice weekly at a dose of 15 mg/kg (109,116,146). Patients who interrupt preventive treatment should restart with the aim of providing at least 6 months of preventive therapy over a year.

For multidrug-resistant tuberculosis, although recommendations have been made, it is uncertain whether prophylaxis is effective. The Centers for Disease Control have recommended that people exposed to multidrug-resistant TB should receive prophylaxis with pyrazinamide and ethambutol or pyrazinamide and a fluoroquinolone for 12 months (39). Ofloxacin, ciprofloxacin, and sparfloxacin have, in order of increasing efficacy, all been shown to have in vitro antimycobacterial activity (122–124). Aminoglycosides such as streptomycin, kanamycin, or amikacin should also be considered. In addition, patients should be reviewed every 3 months both medically and radiographically for the subsequent 2 years to detect active disease.

12. The Role of Molecular Techniques

Routine microbiological techniques can take 2–3 months to provide evidence of mycobacterial infection, its subtype, and drug-sensitivity pattern. In patients with tuberculosis and AIDS, the median time to death may be less than this period (1). This long delay and uncertainty about whether appropriate drugs have been used also causes difficulty with issues such as the need for prolonged isolation of a patient with a short life expectancy due to AIDS. In an attempt to overcome these problems, there is interest in the use of molecular biology techniques which have the potential to diagnose tuberculosis on the basis of the presence of *M. tuberculosis* DNA. The polymerase chain reaction (PCR) technique amplifies selected sequences of DNA 10^9–10^{12} times, allowing visualization on a gel of DNA derived from 1–100 original copies. Many studies have suggested that this sensitive detection technique can be used to identify *M. tuberculosis* DNA in clinical samples (149). Studies have used a variety of target sequences in the *M. tuberculosis* genome as targets for amplification. The sequence IS6110 has been most frequently used because it is specific for a small group of mycobacteria, including *M. tuberculosis,* and because it is present in more than one copy per genome thus giving the technique greater sensitivity. Early studies suggested that the technique was not very reproducible among laboratories (149), and there was some evidence that there was not a 100% correlation with active disease (150). More recently, a commercially available PCR test using a different target has been developed (Amplicor™, Roche Diagnostics). Reports from other centers suggest that the this has a high sensitivity and specificity in the diagnosis of active tuberculosis (151). It seems likely that in the near future the optimum system will be defined and that a PCR-based diagnostic system will be available for tuberculosis.

The possibility of using PCR to diagnose drug resistance was suggested by the identification in the mycobacterial genome of genes which confer susceptibility or resistance to specific drugs. Thus, mutations in the *katG* gene of *M. tuberculosis* are associated with the development of resistance to isoniazid (152). Similarly, the *INHA* gene is associated with isoniazid and ethionamide sensitivity/resistance (153,154), the *rpoB* with rifampicin sensitivity (155,156), and the *rpsL* gene with streptomycin sensitivity (153). It was initially hoped that identifying mutations in the *katG* would allow the presence of isoniazid resistance to be predicted. Unfortunately, the number of bases over which mutations occur in this gene is large, there are many mutations which may have no functional significance, and a proportion of isoniazid resistance occurs as a result of mutations in other genes. By contrast, the mutations thought to confer rifampicin resistance are nearly all confined to a 100 base-pair region of the *rpoB* gene which encodes the beta subunit of RNA polymerase. This observation opens the possibility of screening clinical samples for mutations in PCR-amplified products of this 100 base-pair region. At least at the research level, the technique of PCR combined with single-stranded conformational polymorphism has been used to provide a rapid diagnostic test for drug resistance (157). It is hoped that this will lead to the development of a technique suitable for the routine laboratory.

Another use of molecular biology is to study the epidemiology of tuberculosis. A number of studies have taken advantage of subtle differences in the DNA sequence of individual *M. tuberculosis* strains and used the technique of restriction fragment-length polymorphism (RFLP) to identify the spread of individual organisms in the community (158,159). Two studies in the United States studied large numbers of patients and were able to identify epidemiological factors associated with recent infection as opposed to reactivation (17,18). These factors included younger age, poverty, and membership of different ethnic groups. A similar study using RFLP was able to identify the movement of clones of *M. tuberculosis* from Denmark to Greenland (160). All these studies relied on a prior culture of the organism. Other techniques have been suggested which have the potential to type *M. tuberculosis* directly from sputum (161).

Although at an early stage, these technical developments are likely to offer rapid diagnosis of *M. tuberculosis* and identify drug-sensitivity patterns as well as define epidemiological factors predisposing to transmission. These techniques are, however, relatively expensive and their use is likely to be restricted to more affluent countries.

13. Conclusion

Combined infection with *M. tuberculosis* and HIV poses an enormous disease burden on mankind. There is every prospect that the number of cases and the proportion of the global population affected will increase. The presence of combined infection alters the clinical features and the prognosis of both tuberculosis and AIDS. Furthermore, as a measure of overall tuberculosis control HIV highlights weaknesses in our control programs (162). Thus, it is likely that over the next few decades, clinicians will spend an increasing proportion of their effort addressing this clinical problem.

References

1. Frieden TR, Sterling T, Pablos-Mendez A, Kilburn JO, Cauthen GM, Dooley SW (1993) The emergence of drug-resistant tuberculosis in New York City. N Engl J Med 328:521–526.

2. Whalen C, Horsburgh CR, Hom D, Lahart C, Simberkoff, Ellner J (1995) Accelerated course of human immunodeficiency virus infection after tuberculosis. J Respir Crit Care Med 151:129–135.

3. Zhang Y, Nakata K, Weiden M, Rom WN (1995) *Mycobacterium tuberculosis* enhances human immunodeficiency virus-1 replication by transcriptional activation at the long terminal repeat. J Clin Invest 95:2324–2331.

4. Dannenberg AJ (1991) Delayed-type hypersensitivity and cell-mediated immunity in the pathogenesis of tuberculosis. Immunol Today 12:228–233.

5. Dube M, Holtom F, Larsen R (1992) Tuberculous meningitis in patients with and without human immunodeficiency virus infection. Am J Med 93:520–524.

6. Ho WZ, Cherukuri R, Douglas SD (1994) The macrophage and HIV-1. Immunol Series 60:569–587.

7. Shattoch RJ, Friedland JS, Griffin GE (1994) Phagocytosis of *Mycobacterium tuberculosis* modulates human immunodeficiency virus replication in human monocytic cells. J Gen Virol 75:849–856.

8. Meyaard L, Schuitemaker H, Miedema F (1993) T-cell dysfunction in HIV infection: anergy due to defective antigen-presenting cell function. Immunol Today 14:161–164.

9. Clerici M, Shearer G (1994) The Th1–Th2 hypothesis of HIV infection: new insights. Immunol Today 15:575–581.

10. Graziosi C, Pantaleo G, Gantt KR, Fortin JP, Demarest JF, Cohen OJ, Sekaly RP, Fauci AS (1994) Lack of evidence for the dichotomy of the TH1 and TH2 predominance in HIV infected individuals. Science 265:248–252.

11. Barnes PF, Bloch AB, Davidson PT, Snider DE Jr (1991) Tuberculosis in patients with human immunodeficiency virus infection. N Engl J Med 324:1644–1650.

12. Brudney K, Dobkin J (1991) Resurgent tuberculosis in New York City: human immunodeficiency virus, homelessness and the decline of tuberculosis control programmes. Am Rev Respir Dis 144:745–749.

13. American Thoracic Society (1992) Control of tuberculosis in the United States. Am Rev Repir Dis 146:1623–1633.

14. Bhatti N, Law MR, Halliday R, Moore-Gillon J (1995) Increasing incidence of tuberculosis in England and Wales: a study of the likely causes. Br Med J 310:967–969.

15. Darbyshire JE (1995) Tuberculosis: old reasons for a new increase? Br Med J 310:954–955.

16. Watson JM, Meredith SK, Whitmore-Overton E, Bannister B, Darbyshire JH (1993) Tuberculosis and HIV: estimates of the overlap in England and Wales. Thorax 48:199–203.

17. Alland D, Kalkut GE, Moss AR, et al. (1994) Transmission of tuberculosis in New York City—an analysis by DNA fingerprinting and conventional epidemiologic methods. N Engl J Med 330:1710–1716.

18. Small PM, Hopewell PC, Singh SP, et al. (1994) The epidemiology of tuberculosis in San Francisco—a population-cased study using conventional and molecular methods. N Engl J Med 330:1703–1709.

19. El Jarad N, Parastatides S, Paul EA, Sheldon CD, Gaya H, Rudd RM, Empey DW (1992) Characteristics of patients with drug resistant and drug sensitive tuberculosis in East London between 1984 and 1992. Thorax 49:808–810.

20. Warburton ARE, Jenkins PA, Waight PA, Watson JM (1993) Drug resistance in initial isolates of *Mycobacterium tuberculosis* in England and Wales, 1982–1991. Communic Dis Rep (CDR review) 13:175–179.

21. Medical Research Council Cardiothoracic Epidemiology Group (1992) National survey of notifications of tuberculosis in England and Wales in 1988. Thorax 74:770–775.

22. Lucas SB, Hounnou A, Peacock C, et al. (1993) The mortality and pathology of HIV infection in a West African city. AIDS 7:1569–1579.

23. Lucas SB, de Cock KM, Hounnou A, et al. (1994) Slim disease in Africa: the contribution of tuberculosis. Br Med J 308:1531–1533.

24. Mukadi Y, Perriens JH, St. Louis ME, et al. (1993) Spectrum of immunodeficiency in HIV-1 infected outpatients with pulmonary tuberculosis in Zaire. Lancet 342:143–146.

25. Schulzer M, Fitzgerald JM, Enarson DA, Grzybowski S (1992) An estimate of the future size of the tuberculosis problem in sub-Saharan Africa resulting from HIV infection. Tuberc Lung Dis 73:52–58.

26. Edlin BR, Tokars JI, Grieco MH, et al. (1992) An outbreak of multidrug-resistant tuberculosis among hospitalized patients with the acquired immunodeficiency syndrome. N Engl J Med 326:1514–1521.

27. Fischl MA, Daikos GL, Uttamchandani RB, et al. (1992) Clinical presentation and outcome of patients with HIV infection and tuberculosis caused by multiple-drug-resistant bacilli. Ann Intern Med 117:184–190.

28. Daley CL, Small PM, Schecter GF, et al. (1992) An outbreak of tuberculosis with accelerated progression among persons infected with the human immunodeficiency syndrome. N Engl J Med 326:231–235.

29. Dwyer B, Jackson K, Raios K, Sievers A, Wilshire E, Ross B (1993) DNA restriction fragment analysis to define an extended cluster of tuberculosis in homeless men and their associates. J Infect Dis 167:490–494.

30. Geneweiun A, Telenti A, Bernasconi C, et al. (1993) Molecular approach to identifying route of transmission of tuberculosis in the community. Lancet 342:841–844.

31. Centers for Disease Control (1993) Probable transmission of multidrug resistant tuberculosis in a correctional facility—California. Morbid Mortal Wkly Rep 42:48–51.

32. Hopewell PC (1989) Tuberculosis and human immunodeficiency virus infection. Semin Respir Infect 4:111–122.

33. Chaisson RE, Slutkin G (1989) Tuberculosis and human immunodeficiency virus infection. J Infect Dis 159:96–100.

34. Pitchenik AE, Fertel D (1992) Tuberculosis and non tuberculous mycobacterial disease. Med Clin North Am 76:121–171.

35. de Cock KM, Soro B, Coulibaly IM, Lucas SB (1992) Tuberculosis and HIV infection in sub-Saharan Africa. J Am Med Assoc 268:1581–1587.

36. Jones BE, Young SM, Antonikis D, Davidson PT, Kramer F, Barnes PF (1993) Relationship of the manifestations of tuberculosis to CD4 cell counts in patients with human immunodeficiency virus infection. Am Rev Respir Dis 148:1292–1297.

37. Shafer RW, Chirgwein KD, Glatt AE, Dahdouh MA, Landesmam SH, Suster B (1991) HIV prevalence, immunosuppression, and drug resistance in patients with tuberculosis in an area epidemic for AIDS. AIDS 5:399–405.

38. Hopewell PC (1992) Impact of human immunodeficiency virus infection on the epidemiology, clinical features, management and control of tuberculosis. Clin Infect Dis 15:540–547.

39. Centers for Disease Control (1992) National action plan to combat multidrug-resistant tuberculosis. Meeting the challenge of multidrug-resistant tuberculosis: summary of a conference. Management of persons exposed to multidrug-resistant tuberculosis. Morbid Mortal Wkly Rep 41:5–71.

40. Rieder HL, Cauthen GM, Comstock GW, Snider DE Jr (1989) Epidemiology of tuberculosis in the United States. Epidemiol Rev 11:79–98.

41. Selwyn PA, Hartel D, Lewis VA, Schoenbaum EE, Vermund SH, Klein RS, Walker AT, Friedland GH (1989) A prospective study of the risk of tuberculosis among intravenous drug users with human immunodeficiency virus infection. N Engl J Med 320:545–550.

42. Dooley SW, Villarino ME, Lawrence M, et al. (1991) Nosocomial transmission of tuberculosis in a hospital unit for HIV-infected patients. J Am Med Assoc 267:2632–2635.

43. Bailey WC, Albert RK, Davidson PT, et al. (1983) Treatment of tuberculosis and other mycobacterial diseases: an official statement of the American Thoracic Society. Am Rev Respir Dis 127:790–796.

44. Horwitz O, Edwards PQ, Lowell AM (1973) National tuberculosis control program in Denmark and the United States. Hlth Serv Rep 88:493–498.

45. Pitchenik AE, Burr J, Suarez M, Fertel D, Gonzalez G, Moas C (1987) Human T-cell lymphotropic virus-III (HTLV-III) seropositivity and related disease among 71 consecutive patients in whom tuberculosis was diagnosed. A prospective study. Am Rev Respir Dis 135:875–879.

46. Theuer CP, Hopewell PC, Elias D, Schecter GF, Rutherford G, Chaisson RE (1990) Human immunodeficiency virus infection in tuberculosis patients. J Infect Dis 162:8–12.

47. Sunderam G, McDonald RJ, Maniatis T, Oleske J, Kapila R, Reichman LB (1986) Tuberculosis as a manifestation of the acquired immunodeficiency syndrome (AIDS). J Am Med Assoc 256:362–366.

48. Pitchenik AE (1990) Tuberculosis control and the AIDS epidemic in developing countries. Ann Intern Med 113:89–91.

49. Modilevsky T, Sattler FR, Barnes PF (1989) Mycobacterial disease in patients with human immunodeficiency virus infection. Arch Intern Med 149:2201–2205.

50. Kramer F, Modilevsky T, Waliany AR, Leedom JM, Bares PF (1990) Delayed diagnosis of tuberculosis in patients with human immunodeficiency virus infection. Am J Med 89:451–456.

51. Small PM, Schecter GF, Goodman PC, Sande MA, Chaisson RE, Hopewell PC (1991) Treatment of tuberculosis in patients with advanced human immunodeficiency virus infection. N Engl J Med 324:289–294.

52. Colebunders RL, Ryder RW, Nzilambi N (1989) HIV infection in patients with tuberculosis in Kinshasa, Zaire. Am Rev Respir Dis 139:1082–1085.

53. Given MJ, Khan MA, Reichman LB (1994) Tuberculosis among patients with AIDS and a control group in an inner-city community. Arch Intern Med 154:640–645.

54. Pitchenik AE, Rubinson HA (1985) The radiographic appearance of tuberculosis in patients with the acquired immunodeficiency syndrome (AIDS) and pre-AIDS. Am Rev Respir Dis 131:393–396.

55. Long R, Maycher B, Scalcini M, Manfeda J (1991) The chest roentgenogram in pulmonary tuberculosis patients seropositive for human immunodeficiency virus type 1. Chest 99:123–127.

56. Saks AM, Posner R (1992) Tuberculosis in HIV positive patients in South Africa: a comparative radiological study with HIV negative patients. Clin Radiol 46:387–390.

57. Goldman KP (1987) AIDS and tuberculosis. Br Med J 295:511–512.

58. Chaisson RE, Schecter GF, Theuer CP, Rutherford GW, Echenberg DF, Hopewell PC (1987) Tuberculosis in patients with the acquired immunodeficiency syndrome. Am Rev Respir Dis 136:570–574.

59. Pitchenik AE, Fertel D, Bloch AB (1988) Mycobacterial disease: epidemiology, diagnosis, treatment and prevention. Clin Chest Med 9:425–441.

60. Duncanson FP, Hewlett D Jr, Maayan S, Estepan H, Perla EN, McLean T, Rodriguez A, Miller SN, Lenox T, Wormser GP (1986) *Mycobacterium tuberculosis* infection in the acquired immunodeficiency syndrome. A review of 14 patients. Tubercle 67:295–302.

61. Handwerger S, Mildvan D, Senie R, et al. (1987) Tuberculosis and the acquired immunodeficiency syndrome at a New York City hospital: 1978–1985. Chest 91:176–180.

62. Louie E, Rice LB, Holzman RS (1986) Tuberculosis in non-Haitian patients with acquired immunodeficiency syndrome. Chest 90:542–545.

63. Bishburg E, Sunderam G, Reichman LB, Kapila R (1986) Central nervous system tuberculosis with the acquired immunodeficiency syndrome and its related complex. Ann Intern Med 195:210–213.

64. Fischl MA, Pitchenik AE, Spira TJ (1985) Tuberculous brain abscess and toxoplasma encephalitis in a patient with the acquired immunodeficiency syndrome. J Am Med Assoc 253:3428–3430.

65. Doll DC, Yarbro JW, Phillips K, Lott C (1987) Mycobacterial spinal cord abscess with an ascending polyneuropathy. Ann Intern Med 106:333–334.

66. Woolsey RM, Chambers TJ, Chung HD, McGarry JD (1988) Mycobacterial meningomyelitis associated with human immunodeficiency virus infection. Arch Neurol 45:691–693.

67. Dalli E, Quesada A, Juan G, Navaro R, Paya R, Tormo V (1987) Tuberculous pericarditis as the first manifestation of acquired immunodeficiency syndrome. Am Heart J 114:905–906.

68. D'Cruz IA, Sengupta EE, Abrahams C, Reddy HK, Turlapati RV (1986) Cardiac involvement, including tuberculous pericardial effusion, complicating acquired immune deficiency syndrome. Am Heart J 112:1100–1102.

69. de Silva R, Stoopack PM, Raufman JP (1990) Esophageal fistulas associated with mycobacterial infection in patients at risk for AIDS. Radiology 175:449–453.

70. Wasser LS, Shaw GW, Talavera W (1988) Endobronchial tuberculosis in the acquired immunodeficiency syndrome. Chest 94:1240–1244.

71. Moreno S, Pacho E, Lopez-Herce JA, Rodriguez-Creixems M, Scapa MC, Bourza E (1988) *Mycobacterium tuberculosis* visceral abscesses in the acquired immunodeficiency syndrome (AIDS). Ann Intern Med 109:437.

72. Freed JA, Pervez NK, Chen V, Damasker B (1987) Cutaneous mycobacteriosis: occurrence and significance in two patients with the acquired immunodeficiency syndrome. Arch Dermatol 123:1601–1603.

73. The National Institutes of Health/University of California Expert Panel for Corticosteroids as Adjunctive Therapy for Pneumocystis Pneumonia (1990) Consensus statement on the use of corticosteroids as adjunctive therapy for pneumocystis pneumonia in the acquired immunodeficiency syndrome. N Engl J Med 323:1500–1504.

74. Bozzete SA, Sattler FR, Chiu J, Wu AW, Gluckstein D, Kemper C, Bartok A, Niosi J, Abraham I, Coffman J et al. (1990) A controlled trial of early adjunctive treatment with corticosteroids for *Pneumocystis carinii* pneumonia. N Engl J Med 323:1451–1457.

75. Sattler FR (1991) Who should receive corticosteroids as adjunctive treatment for *Pneumocystis carinii* pneumonia? Chest 99:1058–1061.

76. Alzeer AH, FitzGerald JM (1993) Corticosteroids and tuberculosis: risks and use as adjunct therapy. Tuberc Lung Dis 74:6–11.

77. Haanaes OC, Bergman A (1983) Tuberculosis emerging in patients treated with corticosteroids. Eur J Respir Dis 64:294–297.

78. Clumeck N, Hermans P (1991) Corticosteroids as adjunctive therapy for pneumocystis pneumonia in patients with AIDS. N Engl J Med 324:1666–1667.

79. El-Sadr W, Milder J, Capps L, Sivalapan V (1991) Corticosteroids as adjunctive therapy for pneumocystis pneumonia in patients with AIDS. N Engl J Med 324:1667.

80. Jones BE, Taikwel EK, Mercado AL, Sian SU, Barnes PF (1994) Tuberculosis in patients with HIV infection who receive corticosteroids for presumed *Pneumocystis carinii* pneumonia. Am J Respir Crit Care Med 149:1686–1688.

81. Pitchenik AE, Burr J, Laufer M, Miller G, Cacciatore R, Bigler WJ, Witte JJ, Cleary T (1990) Outbreaks of drug-resistant tuberculosis at AIDS centers. Lancet 336:440–441.

82. Centers for Disease Control (1988) Mycobacterium tuberculosis transmission in a health clinic: Florida, 1988. Morbid Mortal Wkly Rep 38:256–264.

83. Centers for Disease Control (1990) Nosocomial transmission of multidrug-resistant tuberculosis to health care workers and HIV-infected patients in an urban hospital—Florida. Morbid Mortal Wkly Rep 39:718–722.

84. Coronado VG, Beck-Sague CM, Hutton MD, et al. (1993) Transmission of multidrug-resistant *Mycobacterium tuberculosis* among persons with human

immunodeficiency virus infection in an urban hospital: epidemiologic and restriction fragment length polymorphism analysis. J Infect Dis 168:1052–1055.

85. Tabet SR, Goldbaum GM, Hooton MD, Eisenach KD, Cave MD, Nolan CM (1994) Restriction fragment length polymorphism analysis detecting a community-based tuberculosis outbreak among persons infected with human immunodeficiency virus. J Infect Dis 169:189–192.

86. Beck-Sague C, Dooley SW, Hutton MD, et al. (1992) Hospital outbreak of multidrug-resistant *Mycobacterium tuberculosis* infections—factors in transmission to staff and HIV-infected patients. J Am Med Assoc 268:1280–1286.

87. Pearson ML, Jereb JA, Frieden TR, et al. (1992) Nosocomial transmission of multidrug-resistant *Mycobacterium tuberculosis*—a risk to patients and health workers. Ann Intern Med 117:191–196.

88. Selwyn PA, Sckell BM, Alcabes P, Friedland GH, Klein RS, Schonbaum EE (1992) High risk of active tuberculosis in HIV-infected drug users with cutaneous anergy. J Am Med Assoc 268:504–509.

89. Moreno S, Baraia-Etxaburu J, Bouza E, Parras F, Perez-Tascon M, Miralles P, Vicente T, Alberdi JC, Cosin J, Lopez-Gay D (1993) Risk for developing tuberculosis among anergic patients infected with HIV. Ann Intern Med 119:194–198.

90. Long R, Scalcini M, Manfreda J, Jean-Baptiste M, Hershfield E (1991) The impact of HIV on the usefulness of sputum smears for the diagnosis of tuberculosis. Am J Publ Hlth 81:1326–1328.

91. Fournier AM, Dickenson GM, Erdfrocht IR, Cleary T, Fischl MA (1988) Tuberculosis and nontuberculous mycobacteriosis in patients with AIDS. Chest 93:772–775.

92. Smith RL, Yew K, Berkowitz KA, Aranda CP (1994) Factors affecting the yield of acid-fast sputum smears in patients with HIV and tuberculosis. Chest 106:684–686.

93. Nunn P, Mungai M, Nyamwaya J, et al. (1994) The effect of human immunodeficiency virus type-1 on the infectiousness of tuberculosis. Tuberc Lung Dis 75:25–32.

94. Helbert M, Robinson D, Buchanan D, et al. (1990) Mycobacterial infection in patients infected with the human immunodeficiency virus. Thorax 45:45–48.

95. Elliott AM, Namaambo K, Allen BW, et al. (1993) Negative sputum smear results in HIV-positive patients with pulmonary tuberculosis in Lusaka, Zambia. Tuberc Lung Dis 74:191–194.

96. Klein N, Dunanson F, Lenox T, Pitta A, Cohen S, Wormser G (1989) Use of mycobacterial smears in the diagnosis of pulmonary tuberculosis in AIDS/ARC patients. Chest 95:1190–1192.

97. Fertel D, Pitchenik AE (1989) Tuberculosis in acquired immune deficiency syndrome. Semin Respir Infect 4:198–205.

98. Kennedy DJ, Lewis WP, Barnes PF (1992) Yield of bronchoscopy for the diagnosis of tuberculosis in patients with human immunodeficiency virus infection. Chest 102:1040–1044.

99. Salzman SH, Schindel ML, Aranda CP, Smith RL, Lewis ML (1992) The role of bronchoscopy in the diagnosis of pulmonary tuberculosis in patients at risk of HIV infection. Chest 102:143–146.

100. Shafer R, Goldberg R, Sierra M, Glatt AE (1989) Frequency of *Mycobacterium tuberculosis* bacteremia in patients with tuberculosis in an area endemic for AIDS. Am Rev Respir Dis 140:1611–1613.

101. Mitchell DM, Miller RF (1992) Recent developments of the pulmonary complications of HIV disease. AIDS and the lung update. Thorax 47:381–390.

102. Middlebrook G, Reggiardo Z, Tigertt WD (1977) Automatable radiometric detection of growth of *M. tuberculosis* in selective media. Am Rev Respir Dis 115:1066–1069.

103. Roberts GD, Goodman NL, Heifets L, Larsh HW, Lindner TH, McClatchy JK, McGinnis MR, Siddigi SH, Wright P (1983) Evaluation of BACTEC radiometric method for recovery of mycobacteria and drug susceptibility testing of *M. tuberculosis* from acid-fast smear-positive specimens. J Clin Microbiol 18:689–696.

104. Jacobs WR, Udani R, Barletta R, et al. (1993) Rapid assessment of drug susceptibilities of *Mycobacterium tuberculosis* using luciferase reporter phage. Science 260:814–822.

105. Slutkin G, Schecter GF, Hopewell PC (1988) The results of a 9-month isoniazid–rifampin therapy for pulmonary tuberculosis under program conditions in San Francisco. Am Rev Respir Dis 138:1622–1624.

106. FitzGerald J, Grzybowski S, Allen EA (1991) The impact of human immunodeficiency virus infection on tuberculosis and its control. Chest 100:191–200.

107. Vareldzis BP, Grosset J, de Kantor I, Crafton J, Laszlo A, Felten M, Raviglione MC, Kochi A (1994) Drug-resistant tuberculosis: laboratory issues. World Health Organization recommendations. Tuberc Lung Dis 75:1–7.

108. Joint Tuberculosis Committee of the British Thoracic Society (1990) Chemotherapy and management of tuberculosis in the United Kingdom: recommendations of the Joint Tuberculosis Committee of the British Thoracic Society. Thorax 45:403–408.

109. Subcommittee of the Joint Tuberculosis Committee of the British Thoracic Society (1992) Guidelines on the management of tuberculosis and HIV infection in the United Kingdom. Br Med J 304:1231–1233.

110. Centers for Disease Control (1989) Tuberculosis and human immunodeficiency virus infection: recommendations of the Advisory Committee for the Elimination of Tuberculosis (ACET). Morbid Mortal Wkly Rep 38:236–238.

111. Weis SE, Slocum PC, Blais FX, King B, Nunn M, Matney GB, Gomez E, Foresman BH (1994) The effect of directly observed therapy on the rates of drug resistance and relapse in tuberculosis. N Engl J Med 330:1179–1184.

112. Goble M, Iseman MD, Madsen LA (1994) Treatment of 171 patients with pulmonary tuberculosis resistant to isoniazid and rifampicin. N Engl J Med 328:527–532.

112a. Turett GS, Telzak EE, Torian LV, et al. (1995) Improved outcomes for patients with multidrug-resistant tuberculosis. Clin Inf Dis 21:1238–1244.

113. Wurtz RM, Abrams D, Becker S, Jacobson MA, Mass MM, Marks SH (1989)

Anaphylactoid drug reactions due to ciprofloxacin and rifampicin in HIV-infected patients. Lancet 1:955–956.

114. Dukes CS, Sugarman J, Cegielski JP, Lallinger GJ, Mwakyusa DH (1992) Severe cutaneous hypersensitivity reactions during treatment of tuberculosis in patients with HIV infection in Tanzania. Trop Geogr Med 44:308–311.

115. Okwera A, Whalen C, Byekwaso F, Vjecha M, Johnson J, Huebner RJ, Mugerwa R, Ellner J (1994) Randomised trial of thiacetazone and rifampicin-containing regimens for pulmonary tuberculosis in HIV-infected Ugandans. Lancet 344:1323–1328.

116. American Thoracic Society (ATS) (1986) Treatment of tuberculosis and tuberculosis infection in adults and children. Am Rev Respir Dis 134:355–363.

117. Davidson PT (1987) Drug resistance and the selection of therapy for tuberculosis. Am Rev Respir Dis 136:255–257.

118. Iseman MD, Madsen LA (1989) Drug-resistant tuberculosis. Clin Chest Med 10:341–353.

119. Mitchison DA, Nunn AJ (1986) Influence of initial drug resistance on response to short-course chemotherapy of pulmonary tuberculosis. Am Rev Respir Dis 133:423–430.

120. Hong Kong Chest Service/British Medical Research Council (1991) Controlled trial of 2, 4 and 6 months pyrazinamide in 6-month, three-times-weekly regimens for smear-positive pulmonary tuberculosis, including an assessment of a combined preparation of isoniazid, rifampicin, and pyrazinamide. Am Rev Respir Dis 143:700–706.

121. Iseman MD, Madsen L, Goble M, Pomerantz M (1990) Surgical intervention in the treatment of pulmonary disease caused by drug-resistant *Mycobacterium tuberculosis*. Am Rev Respir Dis 141:623–625.

122. Gorzynski EA, Gutman SI, Allen W (1989) Comparative antimycobacterial activities of difloxacin, temafloxacin, enofloxacin, pefloxacin, reference quinolones, and a new macrolide, clarithromycin. Antimicrob Agents Chemother 33:591–592.

123. Rastogi N, Goh KS (1991) In vitro activity of the new difluorinated quinolone sparfloxacin (AT-4140) against *Mycobacterium tuberculosis* compared with activities of ofloxacin and ciprofloxacin. Antimicrob Agents Chemother 35:1933–1935.

124. Leysen DC, Haemers A, Pattyn SR (1989) Mycobacteria and the new quinolones. Antimicrob Agents Chemother 33:1–5.

125. Baohong J, Truffot-Pernot C, Lacroix C, Raviglione MC, O'Brien RJ, Olliaro P, Roscigno G, Grosset J (1993) Effectiveness of rifampicin, rifabutin and rifapentine for preventive therapy of tuberculosis in mice. Am Rev Respir Dis 148:1541–1546.

126. Keers RY (1978) Pulmonary Tuberculosis. A Journey Down the Centuries. London: Cassell Ltd.

127. Miller AB, Fox W, Tall R (1966) An international co-operative investigation into thiacetazone (thiocetazone) side-effects. Tubercle 47:33–73.

128. Ferguson GC, Nunn AJ, Fox W, Miller AB, Robinson DK, Tall RA (1971) A second

international co-operative investigation into thiacetazone side-effects. Tubercle 52:166–181.

129. A Cooperative Study in East African Hospitals, Clinics and Laboratories with the Collaboration of the East African and British Medical Research Councils (1973) Isoniazid with thiacetazone (thiocetazone) in the treatment of pulmonary tuberculosis in East Africa. Third report of fifth investigation. Tubercle 54:169–179.

130. Hira SK, Wadhawan D, Kamanga J, et al. (1988) Cutaneous manifestations of human immunodeficiency virus in Lusaka, Zambia. J Am Acad Dermatol 19:451–457.

131. Nunn P, Kibuga D, Gathua S, et al. (1991) Cutaneous hypersensitivity reactions due to thiacetazone in HIV-1 seropositive patients treated for tuberculosis. Lancet 337:627–630.

132. World Health Organization (1992) Severe hypersensitivity reactions among HIV-seropositive patients with tuberculosis treated with thiacetazone. Geneva: World Health Organization, Weekly Epidemil Rec 67:1–3.

133. Perriens JH, Colebunders RL, Karahunga C, Williame JC, Jeugmans J, Kaboto M, Mukadi Y, Pauwels P, Ryder RW, Prinot I, et al. (1991) Increased mortality and tuberculosis treatment failure rate among human immunodeficiency virus (HIV) seropositive compared with HIV seronegative patients with pulmonary tuberculosis treated with "standard" chemotherapy in Kinshasa, Zaire. Am Rev Respir Dis 144:750–755.

134. Sen P, Chatterjee R, Saha J, Roy H (1974) Thiacetazone concentration in blood related to grouping of tubercular patients, its treatment, results and toxicity. Indian J Med 62:557–564.

135. Centers for Disease Control (1985) Disseminated *Mycobacterium bovis* infection from BCG vaccination of a patient with acquired immunodeficiency syndrome. Morbid Mortal Wkly Rep 34:227.

136. International Union Against Tuberculosis Committee on Prophylaxis (1982) Efficacy of various durations of isoniazid preventive therapy for tuberculosis: five year follow-up in the IUAT trial. Bull WHO 60:555–564.

137. Centers for Disease Control (1990) The use of preventive therapy for tuberculosis infection in the United States. Recommendations of the Advisory Committee for the Elimination of Tuberculosis. Morbid Mortal Wkly Rep 39:9–12.

138. Stead WW, To T, Harrison RW, Abraham JH (1987) Benefit–risk considerations in preventive treatment for tuberculosis in elderly persons. Ann Intern Med 107:843–845.

139. Colice GL (1990) Decision analysis, public health policy and isoniazid prophylaxis for young adult tuberculin skin reactors. Arch Intern Med 150:2517–2522.

140. Tsevat J, Taylor WC, Wong JB, Pauker SG (1988) Isoniazid from the tuberculin reactor: take it or leave it. Am Rev Respir Dis 137:215–215.

141. Israel HL (1993) Chemoprophylaxis for tuberculosis. Res Med 87:81–83.

142. Pape JW, Jean SS, Ho JL, Hafner A, Johnson WD Jr (1993) Effect of isoniazid

prophylaxis on incidence of active tuberculosis and progression of HIV infection. Lancet 342:268–272.

143. Hamadeh MA, Glassroth J (1992) Tuberculosis in pregnancy. Chest 101:114–120.

144. Snider DE Jr, Caras GJ (1992) Isoniazid-associated deaths: a review of available information. Am Rev Respir Dis 145:494–497.

145. American Thoracic Society (1990) Diagnostic standards and classification of tuberculosis. Am Rev Respir Dis 142:725–735.

146. World Health Organization (1994) Tuberculosis preventive therapy in HIV-infected individuals (news). Bull World Health Organ (WHO) 72:305–307.

147. (1994) Tuberculosis preventive therapy in HIV-infected individuals. A joint statement of the International Union Against Tuberculosis and Lung Disease (IUATLD) and the Global Programme on AIDS and the Tuberculosis Programme of the World Health Organization (WHO). Tuberc Lung Dis 75:96–98.

148. Centers for Disease Control (1991) Purified protein derivative (PPD)—tuberculin anergy and HIV infection: guidelines for anergy testing and management of anergic persons at risk of tuberculosis. Morbid Mortal Wkly Rep 40:27–33.

149. Noordhoek GT, Kolk AHJ, Bjune G, Catty D, Dale JW, Fine PE, Godfrey-Faussett P, Cho SN, Shinnick T, Svenson SB, et al. (1994) Sensitivity and specificity of PCR for detection of *Mycobacterium tuberculosis:* a blind comparison study among seven laboratories. J Gen Virol 75:849–856.

150. Walker DA, Taylor IK, Mitchell DM, Shaw RJ (1992) Comparison of polymerase chain reaction (PCR) amplification of two mycobacterial DNA sequences, IS6110 and the 65kD antigen gene, in the diagnosis of tuberculosis. Thorax 7:198–206.

151. D'Amato RF, Wallman AA, Hochstein LH, Colaninno PM, Scardamaglia M, Ardila E, Ghouri M, Kim K, Patel RC, Miller A (1994) Rapid laboratory diagnosis of pulmonary tuberculosis using Amplicor *Mycobacterium tuberculosis* test. Am J Resp Rev Crit Care Med 149:A45.

152. Zhang Y, Heym B, Allen B, Yung D, Cole S (1992) The catalase–peroxidase gene and isoniazid resistance of *Mycobacterium tuberculosis.* Nature 358:591–593.

153. Nair J, Rouse DA, Bai G-H, Morris SL (1993) The *rpsL* gene and streptomycin resistance in single and multiple drug-resistant strains of *Mycobacterium tuberculosis.* Mol Microbiol 10:521–527.

154. Banerjee A, Dubnau E, Quemard A, et al. (1994) *inhA,* a gene encoding a target for isoniazid and ethionamide in *Mycobacterium tuberculosis.* Science 263:227–230.

155. Talenti A, Imboden P, Marchesi F, et al. (1993) Detection of rifampicin-resistance mutations in *Mycobacterium tuberculosis.* Lancet 341:647–650.

156. Donnabella V, Martiniuk F, Kinney D, et al. (1994) Isolation of the gene for the beta subunit of RNA polymerase from rifampicin-resistant *Mycobacterium tuberculosis* and identification of new mutations. Am J Respir Cell Mol Biol 11:639–643.

157. Heym B, Honore N, Truffot-Perrot C, et al. (1994) Implications of multidrug resistance for the future of short course chemotherapy of tuberculosis: a molecular study. Lancet 344:293–298.

158. Godfrey-Fausset P, Mortimer PR, Jenkins PA, Stoker NG (1992) Evidence of transmission of tuberculosis by DNA fingerprinting. Br Med J 305:221–223.

159. Goyal M, Omerod LP, Shaw RJ (1994) Epidemiology of an outbreak of drug resistant tuberculosis in the UK using restriction fragment length polymorphism. Clin Sci 86:749–751.

160. Yang ZH, de Haas PEW, van Soolingen D, van Embden JDA, Anderson AB (1994) Restriction fragment length polymorphism of *Mycobacterium tuberculosis* strains isolated from Greeland during 1992: evidence of tuberculosis transmission between Greenland and Denmark. J Clin Microbiol 32:3018–3025.

161. Goyal M, Young D, Zhang Y, Jenkins PA, Shaw RJ (1994) Polymerase chain reaction amplification of a variable sequence upstream of *katG* gene to subdivide strains of *Mycobacterium tuberculosis* complex. J Clin Microbiol 32:3070–3071.

162. Coker R, Miller R (1997) HIV-associated tuberculosis: a barometer for wider tuberculosis control and prevention. Brit Med J 314:1847.

9

The Management of Extrapulmonary Tuberculosis

L.P. Ormerod

1. Introduction

Although not as frequent numerically as pulmonary and intrathoracic tuberculosis (TB), extrapulmonary forms of TB make up an important proportion of all forms of TB. Indeed, there is some evidence in developing countries that the level of extrapulmonary TB has stayed fairly constant, in the presence of falling levels of respiratory TB, thereby causing the proportion of TB at extrapulmonary sites to rise relative to respiratory TB (1).

In England and Wales, in the most recently published national survey of TB notifications in 1988 (2), 32% of previously untreated patients had extrapulmonary disease. There were important ethnic differences, with 57% of all nonrespiratory disease coming from the 3% of the population of Indian subcontinent (ISC) ethnic origin. Even within nonpulmonary sites, there were ethnic differences. In both white and ISC ethnic groups, lymph node disease was the commonest, with 37% and 52% of cases, respectively, but abdominal TB in the white ethnic group (6%) was under half that of the ISC ethnic group (14%), whereas, conversely, genitourinary TB was much commoner in the white ethnic group (28%) than in the ISC ethnic group (4%) (Table 9.1).

An earlier Medical Research Council (MRC) study in England and Wales in 1983 gave a more detailed breakdown of extrapulmonary sites and incidence (3) and showed that although the overall rate of TB in the ISC population was 25 times that of the white ethnic group, the rate was 51 times higher for extrapulmonary disease. Once again, there were considerable differences between extrapulmonary sites, with the difference in rates between the main ethnic groups

236

Table 9.1. Extrapulmonary disease in England and Wales in 1988

Site	Total		White		ISC[a]		Other	
	No.	%	No.	%	No.	%	No.	%
Lymph node	329	44	86	37	205	52	38	54
Bone and joint	87	12	31	13	51	13	5	7
Genitourinary	84	11	65	28	16	4	3	4
Abdominal	84	11	14	6	57	14	13	19
CNS	36	5	10	4	21	5	5	7
Miliary	51	7	19	8	27	7	5	7
Abscess	27	4	10	4	14	4	3	4
Other	54	7	18	8	32	8	4	6
Total sites	752		253		423		76	
Total patients[b]	698		233		395		70	

[a]Indian sub-continent (ISC).
[b]Some patients had disease at more than one site.
Source: Ref. 2.
"With permission from Thorax and from Dr. Janet Darbyshire of the MRC Cardiothoracic Epidemiology Group"

ranging from 72 times higher for lymph node disease but only 10 times higher for genitourinary disease.

A rise in the proportion of TB at extrapulmonary sites has not only been seen in the United Kingdom (1) but also in other developed countries. In the United States in 1964, 8% of reported TB was extrapulmonary, which increased to 15% in 1981 and 17.5% in 1986 (4,5). A rise in extrapulmonary TB has also been reported in Hong Kong (6) from 1.2% in 1967 to 6.6% in 1990.

One other factor which plays a part in the increasing proportion of extrapulmonary TB in some situations is HIV coinfection. Not only are such persons more likely to develop TB but also particularly extrapulmonary forms, which occur in over 50% of all such TB/HIV coinfected cases (7). There is also evidence that disease at some extrapulmonary sites such as the central nervous system or lymph nodes are more commonly affected in HIV-coinfected persons than in those without such coinfection (8).

The increase in the proportion of extrapulmonary TB in developed countries such as the United States and the United Kingdom is largely due to the effects of immigration from Third World countries. This trend together with HIV coinfection in the United States and HIV alone in Africa, Asia, and South America will make the diagnosis and management of extrapulmonary tuberculosis most important (9–13). The recognition of extrapulmonary TB has not been helped by the generally falling prevalence of TB, until recently, and hence declining clinical experience. The reduced physician experience, coupled with presentations that may be of gradual onset or atypical, may mean that TB is not considered in the differential diagnosis for some time during which further morbidity, or even death, can occur. In the less developed world, the problems of diagnosis are compounded by a lack of diagnostic resources, with few forms of extrapulmonary TB being positive on microscopy. Empirical treatment, or trials of treatment, will therefore more likely be given on clinical grounds only, without pathological and/or bacteriological support or confirmation.

2. Lymph Node Tuberculosis

Over 80% of lymph node TB is in cervical lymph nodes, with a small number of cases involving axillary, inguinal, and chest wall nodes (14). In England and Wales, it is most often seen in ISC immigrants in whom it accounts for over 50% of extrapulmonary disease (2). Lymph node disease accounts for 45% of extrapulmonary forms in Hong Kong (6) and 30% in the United States (15).

Mycobacterium bovis, which in the past accounted for a significant proportion of lymph node disease, is now less common, *M. tuberculosis* being the most frequent isolate (16). In young children in developed countries, particularly aged under 5, lymph node disease caused by *M. avium-intracellulare* can simulate TB histologically. If such atypical organisms are isolated, showing a nontuberculous mycobacterial lymphadenopathy, treatment is by surgical excision, not by drug treatment.

The source of infection in lymph node disease is usually by reactivation of disease originally disseminated during the initial primary airborne infection in the lung. Such reactivation occurs, as at other sites, when body defense mechanisms weaken, allowing local reactivation at previously contained sites. Primary lymph node infection and lymphatic spread from adjacent sites also occur. In the United Kingdom, 10% of cervical lymph node disease in ISC patients have associated mediastinal lymphadenopathy (2,3), suggesting retrograde spread from mediastinal to cervical nodes. A prospective study of the source of cervical lymph node TB infection (17), while showing one-third had evidence of current or previous lung TB suggesting earlier dissemination from a pulmonary source, also showed 6% had nasopharyngeal TB and that cervical nodes were part of the primary local infection.

In developed countries, the peak incidence of lymph node disease is between 20 and 40 years of age (18), but in high-prevalence countries, it is highest in childhood. In the ISC ethnic group, there is a female preponderance of this form of disease (19), for which deficiency of vitamin D has been invoked as an explanation.

3. Clinical Features

The lymph node enlargement in TB is usually gradual and painless, but it can occasionally be more rapid and painful. The individual nodes are firm and discrete (Fig. 9.1) but may later become matted together and fluctuate. There is seldom any accompanying erythema or warmth, the so-called "cold abscess." Unless treatment is begun at this stage, the nodes may proceed to discharge with resultant sinuses, superficial abscesses, and scarring. In immunosuppressed patients, the

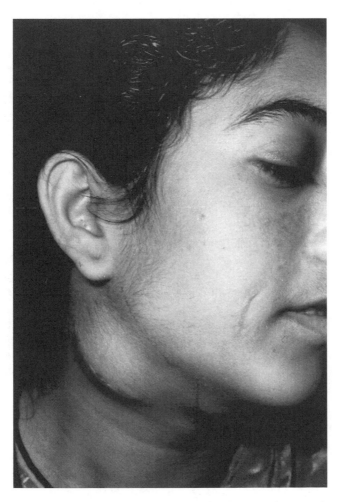

Figure 9.1. Cervical lymphadenopathy in an Asian female; caseating granulomata on biopsy, positive culture for *M. tuberculosis*.

presentation may mimic acute pyogenic infection with marked local pain, swelling, and erythema.

Constitutional features such as fever, weight loss, night sweats, and malaise are seen in a minority of patients, but they are usually absent. There may be clinical or x-ray evidence of tuberculosis elsewhere, usually with either pulmonary parenchymal or mediastinal lymph node involvement (2,3), the latter commonly in ISC or African ethnic patients.

4. Diagnosis

In many developing parts of the world, lymph node tuberculosis is diagnosed clinically from the typical features, sometimes supported by a strongly positive tuberculin skin test. Strictly, the diagnosis depends on the demonstration of *M. tuberculosis* or *M. bovis* in pus or aspirates from nodes. Acid-fast bacilli, however, are only seldom seen on direct smear from such samples, because the actual number of bacilli in infected nodes is small. The majority of the clinical features are not due to the bacterial infection per se but to the marked immunological response to mycobacterial antigens, mainly tuberculoproteins. Acid-fast bacilli are more often seen on smears from biopsy samples from lymph nodes (20), but the positive culture rate only reaches 50–70% (14,18,20,21).

Biopsy of glands shows a spectrum of histology from mild reactive hyperplasia and granulomas through extensive necrosis and caseation. Lymph node biopsy is sometimes carried out because TB is not suspected, but as a diagnostic procedure because of the clinical suspicion of lymphoma or secondary carcinoma. It has been shown that excision biopsy does not speed up healing or enable shorter treatment to be given (14,18). It should be borne in mind that granulomatous histology, particularly if the granulomas are noncaseating, can be caused by fungal infections, brucellosis and sarcoidosis, and, typically, "tuberculous histology" with acid-fast bacilli on microscopy by nontuberculous mycobacteria. Fine-needle aspiration cytology (FNAC) has been shown to have a high specificity (22), and showing granulomatous changes in between 71 and 83% (22–25) in combination with a positive tuberculin skin test (24) is an acceptable alternative to surgical biopsy.

5. Treatment

Several controlled prospective studies over the last 15 years have established the role of chemotherapy as the main treatment for lymph node tuberculosis. A study in the 1970s showed that 18 months' treatment with either isoniazid/rifampicin or isoniazid/ethambutol supplemented by 2 months of initial streptomycin gave good clinical results (26). This was followed by a British Thoracic Society study which compared 18-month and 9-month regimens of isoniazid/rifampicin, each supplemented by initial ethambutol for 2 months. The 9-month regimen performed just as well during treatment (27) and during 5 years of follow-up (28) when there were no clinical or microbiological relapses; good cosmetic results were obtained with both regimens.

On theoretical grounds, pyrazinamide should be superior to ethambutol in the initial phase of treatment (29), as it acts at intracellular pH, is bactericidal rather than bacteriostatic, and can reach bacteria sequestered inside macrophages or lymphocytes. A retrospective study which compared speed of radiological improvement in mediastinal lymph node tuberculosis with 12 months of isoniazid/rifampicin, supplemented by either ethambutol or pyrazinamide for the initial 2 months (30), showed that the pyrazinamide group responded more rapidly at 2, 5, and 7 months. Following this study, the British Thoracic Society Research Committee carried out a further controlled prospective study of short-course chemotherapy in lymph node tuberculosis. This third study (31) compared two 9-month regimens of isoniazid/rifampicin supplemented by either ethambutol or pyrazinamide for the initial 2 months, with a 6-month regimen of isoniazid/rifampicin supplemented by 2 months' initial pyrazinamide. During treatment, there were no differences among the three regimens in terms of resolution of lymph nodes, or the proportion with residual lymph nodes (31). However, repeat aspiration after commencement of treatment was more common in the ethambutol-treated patients ($p = .005$). Follow-up for 30 months from commencement of treatment (32) showed no differences in enlargement of glands, development of new glands or sinuses, or in the proportions with residual lymph nodes. In the follow-up period, there were nine cases where the clinician felt relapse had occurred, but no bacteriological confirmation was obtained in the five cases where material was cultured.

The most recent short-course regimen trial (32) gave prospective confirmation to the retrospective clinical series of McCarthy and Rudd (33) that a regimen of isoniazid/rifampicin supplemented by 2 months of pyrazinamide is satisfactory for lymph node tuberculosis with sensitive organisms. This regimen is now recommended as standard treatment in the United Kingdom (34). Short-course treatment, but with a different regimen, has also been shown to give a 97% success rate in children in India (35). The regimen used in this trial (34) was thrice-weekly supervised isoniazid/rifampicin/pyrazinamide/streptomycin followed by twice-weekly outpatient isoniazid/streptomycin for 4 months.

In all three of the British trials on lymph node tuberculosis, enlargement of existing nodes and development of new nodes were reported (27,31,36), as was the development of new glands after cessation of treatment or enlargement of persistent nodes which were residual at the end of treatment (26,28,32). The persistence of lymphadenopathy at the end of treatment, and particularly the development of lymph node enlargement or new lymphadenopathy during or after treatment, causes concern to physicians who are not experienced with treating lymph node disease. This might lead to unnecessary extension of treatment or the reintroduction of treatment on the basis of "relapse." Such events occur in a significant minority of treated patients and do not of themselves mean that progress is adverse. In the 1985 (27) and 1992 (31) studies 12% and 16–22%, respectively, developed new lymph nodes during treatment. After cessation of treatment, similar rates of persistent lymphadenopathy at the end of treatment and development of new nodes of 9% and 11% (28) and 15% and 5% (32) were reported. These nodes, if biopsied, are usually negative on culture (32), and although clinical "relapse" may be diagnosed, they are bacteriologically sterile. It is more than likely that such phenomena are immunologically mediated, being due to hypersensitivity to tuberculoprotein perhaps from disrupted macrophages (27) and may not indicate an unfavorable outcome.

Surgical excision or biopsy plays no part in the treatment of lymph node disease; patients without surgical intervention did just as well as those with such intervention (27,28). Surgical biopsy, however, may be carried out as a diagnostic procedure to obtain both material for histology and culture if aspiration or fine-needle aspiration cytology (22–25) is not used. Surgical biopsy may, however, have a role in obtaining adequate material for culture, and surgical excision is the treatment of choice for nontuberculous mycobacterial lymphadenopathy (37,38).

The success of short-course therapy (28,32) has been shown for fully sensitive organisms but may not apply to isolates with significant drug resistance (e.g., to isoniazid). In the U.K. studies, there was an increasing incidence of isoniazid resistance over time with 0/32 in 1977 (36), 0/29 in 1985 (27) but 13/108 (12%) in 1992 (31). The 6-month regimen (32) is therefore only applicable to fully sensitive organisms and may have to be modified for isoniazid-resistant organisms to a longer period of therapy with rifampicin/ethambutol (34).

6. Bone and Joint Tuberculosis

Bone and joint tuberculosis presents several years after the initial respiratory infection (39), tubercle bacilli becoming hematogenously spread at that time, with a predilection for the spine and the growing ends of long bones lying dormant until clinical disease occurs. In developed countries, orthopedic tuberculosis makes up some 15–20% of extrapulmonary sites, but with a substantially higher incidence in immigrant groups (3). This study (3) showed rates of 0.2/100,000 and 16/100,000 in white and ISC ethnic groups, respectively.

7. Spine

The spine is the commonest site of orthopedic tuberculosis (40,41). The usual presenting symptom is back pain which can have been present for months and occasionally longer. More unusual presentations with radicular pain mimicking abdominal conditions (42) or with referred neurological symptoms involving legs and sphincters due to spinal cord compression also occur. Local tenderness or slight kyphosis may be found, with progression to grosser kyphosis in advanced disease. Paraspinal abscesses not uncommonly accompany spinal tuberculosis; these can progress to psoas abscess which appears or discharges in the groin, being the presenting feature in some cases, with associated psoas spasm causing hip flexion.

7.1. Diagnosis

This can be delayed in low-prevalence groups because the condition is now very uncommon (41,43). Spinal infection usually commences in the intravertebral disk, and this discitis then spreads by means of the longitudinal and anterior spinal ligaments to involve the vertebral bodies above and below the disk. Thus, x-rays show erosion of the superior and inferior borders of the adjacent vertebrae and loss of disk space. The disease progresses with increasing destruction of the vertebrae, and loss of height and development of kyphosis at that level (Fig. 9.2). The lumbar and thoracic spine are the usual sites of involvement, with the cervical spine being less commonly involved. The disease usually involves a single intravertebral space, but multiple levels can be involved, in some cases with normal vertebrae between involved areas.

Developments in imaging have helped in the diagnosis and assessment of spinal tuberculosis. Computerized tomography (CT) may show involvement before the changes are apparent on plain x-ray (44) and can be better at defining the extent of involvement (44–46). Associated paravertebral and psoas abscesses are well demonstrated (Fig. 9.3). Nuclear magnetic resonance (NMR) imaging is also useful (47,48), with T_2-weighted images demonstrating epidural inflammation (48). The ability of NMR images to be reconstituted in both vertical and horizontal planes can enable full assessment of the extent of disease to be made (49).

Only a minority of cases have evidence of associated pulmonary tuberculosis, but when occurring together with classical spinal x-ray appearances, clinical diagnosis can be made without biopsy. Needle or open biopsy may be needed to make a diagnosis in isolated spinal disease. Unless tuberculosis is considered in the differential diagnosis, appropriate cultures may not be taken for both mycobacterial and pyogenic infections, and the appropriate treatment delayed (43). The main differential diagnosis is between acute pyogenic infection and metastatic spinal disease. With the latter, the radiological features usually are different, with erosion of the pedicles and vertebral body, but with preservation of the disk space, unlike either pyogenic or tuberculous infection. Pyogenic infection (e.g., with Staphylococci) mimic tuberculosis radiologically, but the onset is usually much

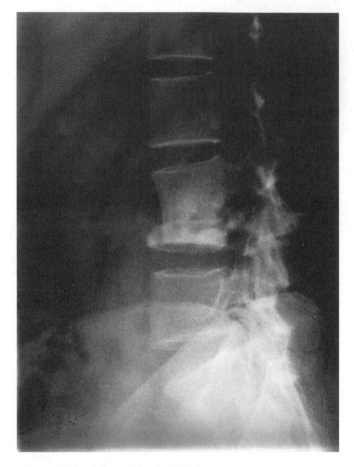

Figure 9.2. Tuberculosis of lumbar spine; L3/4 disk space lost with erosion of adjacent vertebral margins, and substantial destruction of the body of L4.

more acute, the pain can be severe, and there may be accompanying systemic features.

7.2. Treatment

The British Medical Research Council carried out a number of studies over an extended period which helped to define the relative roles of chemotherapy and surgery in the treatment of spinal tuberculosis.

Studies in Korea (50,51), Hong Kong (52), and Rhodesia (53) using isoniazid/para-aminosalicylic acid (PAS) for 18 months gave good results, with over 80%

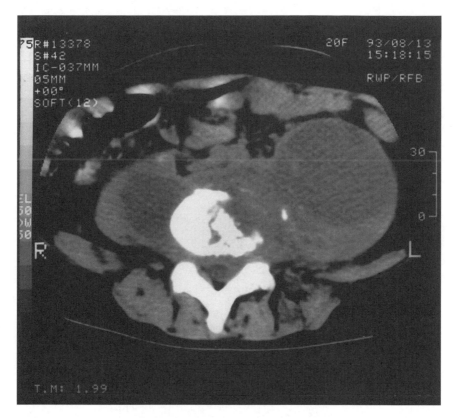

Figure 9.3. CT scan of 19-year-old Asian girl with left inguinal swelling. Erosion of lumbar vertebra with bilateral psoas abscesses (L>R).

of patients achieving favorable status at 3 years. They also showed no additional benefits from the addition of streptomycin for the initial 3 months of therapy, plaster jackets, bed rest for the initial 6 months of therapy, or debridement operations (50–53).

In Hong Kong (54), the so-called "Hong Kong operation," which involved excision of the spinal focus and bone grafting with anterior fusion, in combination with chemotherapy produced less residual deformity and more rapid bone fusion. A later comparison of those patients treated by the "Hong Kong operation" and those by chemotherapy alone showed no additional benefit accrued from surgery (55). These comparisons (56) also showed that 6 or 9 months of treatment with isoniazid/rifampicin supplemented by twice-weekly streptomycin was highly effective. Because pyrazinamide is more bactericidal than streptomycin, has good tissue penetration, but is only required for the initial 2 months of treatment, a regimen of isoniazid/rifampicin for 6 months with 2 months initial pyrazinamide can be recommended (34).

7.3. Other Bone/Joint Sites

Although the spine accounts for up to 50% of orthopedic disease, any bone or joint can be involved. Clinical series in ISC ethnic patients have shown a wide

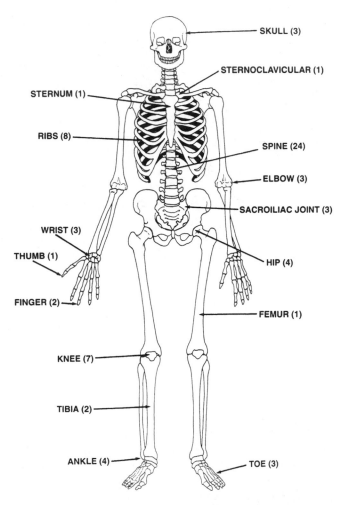

Figure 9.4. Distribution of sites of bone/joint tuberculosis seen in Asian patients updated from Ref. 41.

variety of sites (41) (Fig. 9.4). Tuberculosis should be included in the differential diagnosis of unusual joint lesions, particularly in a monoarthritis in an immigrant group, or there can be substantial delay in achieving a diagnosis (41). Both single (57) and multiple (58) joint presentations are described. There are occasional reports of cases with so many sites and of a cystic type that metastatic bone disease is simulated (59).

These nonspinal sites do not usually require surgical treatment, but surgery by either open biopsy or arthroscopy in the case of certain joints (e.g., knee or elbow) to obtain material for histology and culture is often required to make an initial diagnosis. The 6-month regimen of isoniazid/rifampicin with initial pyrazinamide is recommended (34). Occasionally, surgery is required after the completion of therapy if late presentation or extensive disease has caused major joint disease or instability. Arthrodesis of unstable joints may be necessary, and replacement of hip and knee joints, sometimes under antituberculosis drug cover, has been performed. Combined management of orthopedic tuberculosis, of whatever site, with a physician supervising the antituberculosis drug treatment (34) and the orthopedic surgeon managing the mechanical aspects of the disease is strongly recommended (41).

8. Genitourinary Tuberculosis

In developed countries, genitourinary tuberculosis is one of the commoner sites in white patients (2), with 28% of extrapulmonary cases in the white ethnic group, but only 4% in the ISC ethnic group being genitourinary. An earlier detailed analysis of sites of disease (3) showed rates of genitourinary tuberculosis of 0.4/100,000 and 4.0/100,000 in the white and ISC ethnic groups, respectively. The ratio of ISC : white rates at 10 times greater was the lowest of extrapulmonary sites, and this together with the numerical preponderance of white cases has led to discussions as to why genitourinary tuberculosis is relatively underrepresented in ISC patients (60). This may be an age-related phenomenon, most white patients being considerably older than the ISC ethnic patients, which would fit in with the likely natural history of genitourinary tuberculosis (*vide infra*). The same survey (3) also showed that in white patients, renal tract lesions predominated, but female genital disease predominated in the ISC ethnic group.

Although *M. tuberculosis* can sometimes be detected in the urine within a few months of the primary respiratory infection (39), proving hematogenous dissemination at this early stage, clinical disease usually presents many years after the initial infection (39), having lain dormant, often in the renal parenchyma, for that length of time. In areas where *M. bovis* has been eradicated from cattle for many years, the finding of *M. bovis* in genitourinary isolates (61,62) also supports the proposed natural history. Further support also comes from reactivation tuberculosis in transplanted kidneys, from presumed microscopic dormant foci, following immunosuppression (63).

9. Clinical

Renal tuberculosis is a relatively silent disease and can progress to unilateral renal destruction insidiously. Systemic symptoms of fever, weight loss, and nocturnal sweats are not common. As disease progresses, hematuria, dysuria, nocturia, and pain either in the loin or more anteriorly may occur. Loin or abdominal pain seem to be described more commonly in patients who present under the age of 25 years (64).

Renal tuberculosis may be found coincidentally during investigation of hypertension (65) but rarely presents as renal failure due to the destruction of renal parenchyma, or obstructive hydronephrosis from ureteric involvement (66). Renal tuberculosis can also present as a diffuse interstitial nephritis (67), an important diagnosis to be made, as corticosteroids in addition to antituberculosis drugs can significantly increase renal function and hence avoid progression to dialysis (68). Diffuse interstitial nehpritis has also been reported in transplanted kidneys (69).

Disease in the kidneys can progress to the ureter and then to the bladder by seeding of tubercle bacilli into the urine and implantation distally. Ureteric involvement can lead to irregular stenosis with consequent obstructive hydronephrosis and, occasionaly, to complete obstruction leading to a tuberculous pyonephrosis. Bladder involvement can lead initially to cystitis symptoms of dysuria and frequency. As bladder wall inflammation and associated fibrosis worsen, bladder capacity decreases and can become greatly diminished, leading to marked polyuria and nocturia. The urine usually shows hematuria and proteinuria on testing due to the cystitis, but a pyuria on microscopy which is negative on culture for standard bacterial pathogens. Such a sterile pyuria should routinely lead to the sending of early morning urine samples for mycobacterial culture.

9.2. Genital

The commonest genital sites are the prostate and epididymis with the testicle being less frequently involved (69). Although direct spread from adjacent foci in the genital tract or hematogenous spread can occur, antegrade infection from kidney or bladder are much commoner. Local symptoms of discharge or dysuria can mimic bacterial or chlamydial infections and tumors.

Female genital tuberculosis is commoner in the ISC ethnic group (3) and is spread either hematogenously or directly from tuberculous peritonitis. The fallopian tubes are almost invariably infected, with the endometrium in 90%, but ovarian involvement is reported only in 20%, and rarely in the cervix, vagina, and vulva (70). The commonest presentation is with infertility without associated features but pelvic pain or menorrhagia are reported in 20–25% with much smaller percentages having amenorrhea or postmenopausal bleeding (70).

9.3. Diagnosis

The diagnosis of urinary tract tuberculosis is still based on the intravenous urogram (IVU) and early morning urine cultures. There is a high percentage of abnormality of the IVU in renal disease. In the initial stages, there are just calyceal irregularities or clubbing, and there can be pelviureteric junction narrowing with associated pelvis dilatation. The latter can progress to pelvic obliteration and then to a small or nonfunctioning kidney. Calcification in renal tuberculosis is quite common and is an important pointer to the diagnosis (71) (Fig. 9.5). CT scanning

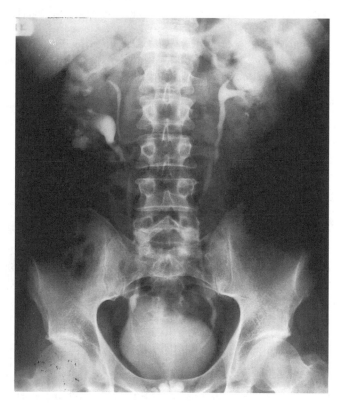

Figure 9.5. IVP showing clubbed and distorted calyces in the right kidney, with calcification in the right kidney and dense prostatic calcification. Early morning urine (EMU) positive for *M. tuberculosis.*

can also be useful with parenchymal retraction and calcification, low parenchymal density, and calycectasis being found in two-thirds of 20 reported cases (72). Pelvic contraction and ureteropelvic fibrosis or obstruction can also be well demonstrated on CT scanning (73). Perinephric abscesses can occur and may point in the groin as with a psoas abscess or in the loin.

Isotope renograms are sometimes of use in assessing differential renal function and may be the first definite indication of ureteric obstruction because of delay in excretion. If there is significant ureteric stenosis, serial renograms will show whether there is improvement in response to treatment.

The diagnosis may be made by biopsy done because of the clinical suspicion of tumor (e.g., renal or testicular), with tuberculosis not being considered until histology is received. Cultures of urine and tissue, if possible, should be done, particularly if an abscess is found in association with the kidney or epididymis.

Urine culture, best done by early morning urine on three consecutive mornings, should be carried out in all patients with urological tuberculosis. The positive yield on microscopy is small, the main positive results being on culture. The finding of a positive microscopy for acid-fast bacilli, which is unusual in genitourinary TB, should raise the question as to whether this is a false-positive one. Such false-positive microscopy can arise due to either contamination of laboratory reagents with environmental mycobacteria or from *Mycobacterium smegmatis,* a saprophytic mycobacteria which can occur in genital secretions. Because of the natural history of genitourinary tuberculosis (39), pulmonary tuberculosis is unusual in association with it. If there is a suspicious chest x-ray lesion present, sputum for acid-fast bacilli should be collected.

10. Management

10.1. Medical

All patients require medical treatment, even if all apparent disease has been removed surgically, because of the likelihood of residual bacilli leading to recurrence and the possibility of other foci elsewhere. Although there is not the support of prospective controlled trials, short-course 6-month regimens are recommended for genitourinary disease. The regimen of rifampicin/isoniazid/pryazinamide for 2 months followed by rifampicin/isoniazid for a further 4 months is recommended in the United Kingdom (34). As with other short-course regimens, it is the inclusion of pyrazinamide that allows the 6-months' treatment duration. If pyrazinamide is omitted or cannot be tolerated, 9 months total treatment should be given. Rifampicin is a particularly good drug for urological tuberculosis because urinary excretion of a significant proportion of the drug means urinary concentrations can reach 100 times that of the serum, and comfortably exceed the minimum inhibitory concentration for *M. tuberculosis.*

It is conventional to use corticosteroids to prevent progression of ureteric strictures. This is advocated by authors in some series (74) but has not been subjected to controlled trial. The addition of corticosteroids, however, has been shown to significantly improve renal function in tuberculous interstitial nephritis (75).

10.2. Surgery

Surgery may initially be to obtain a diagnostic biopsy but has significant roles in genitourinary disease together with effective drug treatment. Sometimes, surgery is excisional because of suspected tumor (e.g., testis) but may also be because of major organ damage, with removal being necessary to prevent possible complications of, for example, a destroyed nonfunctioning kidney. In Gow's major series of 1117 patients (74), 80 (7%) underwent partial nephrectomy because of tuberculous lesions in one pole causing persistent problems such as recurrent infections. In the same series (74), 30% had nephrectomies, 17% epididymectomies, and 4% orchidectomy.

Surgery may also be needed to overcome the mechanical problems of pelvi-ureteric obstruction, ureteric stricture, or reduced bladder capacity of chronic cystitis. In Gow's series (74), reimplantation of the ureter was performed in 6%, 5% had reconstructive bladder surgery, and 2% ureterocolonic transplantation.

Renal function can be monitored both quantitatively and qualitatively by isotope renograms to assess the significance of ureteric stenosis but may have to be repeated serially, and progression can sometimes be rapid (76). Ureteric strictures can be managed by endoscopic dilatation if at the lower end, or by stenting if in the upper ureter, with ureteric reimplantation or diversion reserved for cases where the lesser measures are not possible or fail (77). For the few patients with severe bladder volume reduction as a consequence of chronic tuberculous cystitis, procedures to increase the volume are needed but are best reserved until drug treatment has been completed. Augmentation with a ileal loop attached to the dome of the bladder can add 300–400 ml to bladder capacity.

11. Gastrointestinal Tuberculosis

In developed countries, this form of tuberculosis is uncommon in indigenous populations, being much commoner in immigrant groups. In the United Kingdom in 1983, the rate in the ISC ethnic group was 50 times that of the white ethnic group (3). In the Third World however, gastrointestinal tuberculosis is commonly reported in both HIV-negative and HIV-positive patients. In the pre-HIV era, one-quarter of all ascites was tuberculous in etiology (78), the proportion being over 40% in Lesotho by 1986 (79). In HIV-positive patients, both pulmonary and abdominal tuberculosis have recently been shown to make significant contributions to the wasting in "slim disease" seen in Africa (80). In such patients, intraabdominal lymphadenopathy is a predominant feature in abdominal disease.

11.1. Clinical Features and Presentation

The gastrointestinal tract can be involved anywhere throughout its length, with infection being due to either ingestion, hematogenous spread, or by local extension to peritoneum from nodes and gut. However, involvement of the upper gastrointestinal tract or perianal disease are uncommon, the former accounting for under 3% of a 500-patient series of surgically treated patients (81). Gastric (82) and duodenal tuberculous ulcers are described, which are not distinguishable from peptic ulcers other than on histology or by a positive culture of *M. tuberculosis* from the stomach or gastric washings. Esophageal disease is described, usually causing dysphagia, which could be caused by aspiration of bacilli or spread from contiguous glands in the mediastinum. At esophagoscopy, ulcerating tumor can be mimicked (83,84).

In a series of 109 patients (82), about one-third had an acute presentation, with the other two-thirds having a more gradual onset. Of the cases with an acute presentation, approximately half had acute right iliac fossa pain simulating appendicitis, and the other half had acute intestinal obstruction (82,85,86).

The commonest symptoms are abdominal pain, fever, malaise, and weight loss (82), being described in 60%, 72%, and 58%, respectively, in another series (87). Abdominal swelling, mainly due to ascites, is described in variable proportions from 10% (82) to 65% (88). Coexisting respiratory disease is found in approximately one-third of cases, 36% in one series (87), and 29% in another (82), of which 23% were active with positive cultures.

There are no diagnostic signs of abdominal tuberculosis (89) and the so-called classical "doughy" abdomen is not reported in large series (82). There may be tenderness in the right iliac fossa simulating appendicitis, or a right iliac fossa mass simulating carcinoma or appendix abscess. The ileocecal region is the commonest site of disease, with frequencies of between 24% and 80% in reported series (82,88,90,91). Here, the presentation may be with acute or subacute small-bowel obstruction with a distended abdomen and vomiting, with or without an abdominal mass (82,86). The colon (other than cecum) is involved in up to 10% (82) and may present with bleeding (92). Anal disease and tuberculous ischiorectal abscess (82) are occasionally described.

11.2. Diagnosis

Because of the nonspecific presentation in most cases, and the fact that two-thirds have normal chest x-rays, the diagnosis is not suspected in up to two-thirds of cases (86), other diseases such as carcinoma or inflammatory bowel disease being thought likelier.

Normochromic or hypochromic anemia, raised erythrocyte sedimentation rate (ESR), and reduced serum albumin ($<$ 35 g/L) are commonly found but are non-specific (82). The white blood count is usually normal (91). The tuberculin test is positive in most cases (82) but can be negative in undernourished, immuno-supressed, or HIV-positive individuals and in advanced disease.

The ascitic fluid in abdominal tuberculosis, in common with other serous membrane tuberculosis, is a lymphocyte-rich exudate (protein $>$ 35 g/L) and usually straw colored. Ziehl–Neelsen staining of the fluid is usually negative (93,94) and the percentage with a positive culture is not high. Acid-fast bacilli may be seen in gastric washings, particularly in children; sputum smear and culture should be done if there is a chest x-ray lesion.

Plain abdominal x-rays give no specific help but may show ascites or distended bowel loops, confirming bowel obstruction on an erect film. Barium meal is not helpful, but small-bowel studies and barium enema are. Small-bowel studies may show stricturing (Fig. 9.6), mucosal abnormalities, and even skip lesions and fistulas. These features, however, cannot differentiate tuberculosis from other inflammatory bowel disease on radiological appearance alone (82). Contrast from small-bowel studies often demonstrate ileocecal disease well, but this can also be shown by barium enema, with features of shortening of the ascending colon and vertical passage of the ileum into the colon being highly suggestive (85). In the colon, carcinoma can be mimicked with shouldering and annular lesions (82).

The noninvasive methods of ultrasound and CT give suggestive but not diagnostic features in both ascites and bowel involvement. The CT features are better described. The ascitic element is usually of high (15–30 Hounsfield units) density; the mesentery may be thickened and have a stellate appearance. Lymph node disease may be seen in the retroperitoneum or mesentery. Irregular soft tissue densities in the omentum or lymph nodes with a central, well-demarcated area of low density thought to represent caseation are very suggestive (95). This feature can be enhanced further, with intravenous contrast, the inflammatory rim becoming more predominant (96), but is described in other pathologies including lymphoma and carcinoma. Thickened bowel with nodularity of the wall may also be seen (96,98), and all the above features can be seen in combination within a poorly defined mass including bowel loops (98).

To make a definitive diagnosis requires either positive cultures and/or classical histological features from samples from the gastrointestinal tract or ascites. Laparotomy will give a definitive diagnosis in virtually all cases if adequate samples for histology and culture are collected. To avoid laparotomy, a number of less invasive techniques have been used. Levine (99) described the use of blind, percutaneous, peritoneal needle biopsy with a high yield and few complications. This level of positive results, however, was not reproduced by other series (100), and

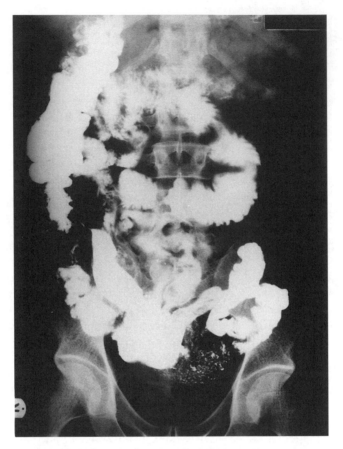

Figure 9.6. Small-bowel barium study showing retracted cecum and markedly narrowed terminal ileum due to tuberculosis. Two further strictures in jejunum and ileum were also present (not seen on this view). Patient required right hemicolectomy and two end-to-end small bowel resections due to intestinal obstruction.

open biopsy of the peritoneum is suggested by some (101) as a preferable alternative, with a lower risk of bowel perforation because the biopsy is taken under direct vision.

Laparoscopy is now the initial procedure of choice and has been shown to be safe, and to give a very high positive rate and few complications (91,94,102,103). Laparoscopy is safe, the risk of bowel perforation being lowest when ascites are present. The only time when open peritoneal biopsy is to be preferred is when there is intense plastic peritonitis present or when CT or ultrasound shows bowel loops adherent to the anterior abdominal wall, thereby increasing the risk of bowel perforation. Colonic lesions or even the ileocaecal valve area are accessible to the colonoscope, but adequate specimens are essential (104). Fine-needle aspiration in conjunction with colonoscopy has given reasonable preliminary results (105).

If positive cultures are not obtained, there may be histological difficulties in differentiating tuberculosis from Crohns disease. In the former, granulomas are more evident in the lymph nodes associated with the bowel than in the bowel wall itself, whereas in Crohn's disease, the pattern is reversed (106).

11.3. Treatment

In the prechemotherapy days, abdominal tuberculosis carried a high mortality (107,108). The results of published series with modern chemotherapy give a mortality of 5–7% (82,109), although some of the mortality was prior to the diagnosis being made. As with some other forms of nonpulmonary tuberculosis, there are no controlled prospective trials of short-course chemotherapy. Such treatment with a 6-month regimen, as for lymph node and bone disease, is recommended in the United Kingdom (34). Corticosteroids, although theoretically being useful in ascites, are not usually needed and are not recommended for routine clinical practice (82). Resection is only required if there is mechanical obstruction (82), and if resection is carried out, this should be by end-to-end anastomosis instead of by ileo-transverse anastomosis (82,85,110). Modern drug therapy has given good results on follow-up; of 103 patients followed up for 15 months, there was no recurrence of gastrointestinal problems (82), but 10% of female patients of childbearing age had either primary or secondary infertility after the treatment of their abdominal tuberculosis.

12. Miliary Tuberculosis

Miliary tuberculosis occurs when tubercle bacilli are spread acutely through the bloodstream. In high-prevalence areas, the majority of cases follow shortly after initial infection, but in low-prevalence areas, the majority of cases are in the elderly, representing reactivation. The lung is always involved, other organs variably so. Microscopically, the miliary lesions consist of Langerhans giant cells, epithelioid cells, and lymphocytes and contain acid-fast bacilli, sometimes with central caseation. In elderly or immunosuppressed patients, nonreactive pathological appearances are described with necrotic lesions containing no specific tuberculous features but teeming with acid-fast bacilli. In such cases, the diagnosis is usually made at postmortem (111). The symptoms are insidious of onset (112) and include anorexia, malaise, fever, and weight loss and occur in both the "acute" and cryptic forms. Miliary tuberculosis accounts for up to 5% of cases of extrapulmonary tuberculosis in the United Kingdom (3).

12.1. Acute Miliary Tuberculosis

In addition to the general symptoms, headache from coexistent tuberculous meningitis occurs frequently and should alert the clinician to perform a lumbar puncture. Cough, dyspnea, and hemoptysis are less common symptoms. Physical signs are few; the chest almost invariably sounds clear on auscultation. Enlargement of the liver, spleen, or lymph nodes may be found in a small number of cases (113). Involvement of the serosal surfaces can lead to the development of small pericardial or pleural effusions or slight ascites. Fundal examination should be carried out to detect choroidal tubercles which are more commonly seen in children. Skin lesions may also occur in the form of papules, macules, and purpuric lesions. These probably represent local vasculitic lesions caused by reaction to mycobacterial antigen.

The typical x-ray shows an even distribution of uniform-sized lesions 1–2 mm throughout all zones of the lung. Small bilateral pleural effusions may also be seen. An unusual variation with reticular shadowing due to lymphatic involvement had been described (114).

12.2. Cryptic Miliary Tuberculosis

As tuberculosis declines in incidence in developed countries, a form of miliary tuberculosis without typical x-ray shadowing, so-called "cryptic" miliary tuberculosis has been seen more frequently. This is usually seen in patients aged over 60 (115) but may be seen in young patients in some immigrant groups. The symptoms are usually insidious with weight loss, lethargy, and intermittent fever (116). Meningitis and choroidal tubercles are rarely found; mild hepatosplenomegaly may be found but physical signs are usually absent. Because of this, a high index of suspicion is required to reach a diagnosis, and commonly the diagnosis is not made until postmortem (117). The main differential diagnosis is with disseminated carcinoma. Table 9.2 contrasts the features of the classical and cryptic forms of miliary disease.

12.3. Diagnosis

The classical form of miliary disease is usually easy to diagnose because of the typical x-ray appearances which are only absent in the early stages. The tuberculin test is usually positive, and bacteriological confirmation may be obtained from sputum, urine, and cerebrospinal fluid (CSF). The diagnosis of the cryptic form rests initially in having clinical suspicion of the diagnosis, and then carrying out specific tests or monitoring response to a trial of antituberculosis drugs. Blood dyscrasias are not uncommonly seen in the cryptic form, pancytopenia (118,119) leukemoid reactions (120,121), and other granulocyte abnormal-

Table 9.2. Comparison of classical and cryptic forms of miliary TB

Feature	Cryptic	Classical
Age	Majority over 60 years	Majority under 40 years
TB history or contact	Up to 25%	Up to 33%
Malaise/weight loss	75%	75%
Fever	90%	75%
Choroidal tubercles	Absent	Up to 20%
Meningitis	Rare unless terminal	Up to 20%
Lymphadenopathy	Absent	Up to 20%
Miliary shadowing on x-ray	Rare	Usual except in early stages
Tuberculin test	Usually negative	Usually positive
Pancytopenia/leukemoid reaction	Common	Rare
Bacteriological confirmation	Urine; sputum; bone marrow	Sputum: CSF
Biopsy evidence	Liver up to 75%; bone marrow; lymph node	Seldom required

ities (122) have all been reported. Bone marrow aspiration may yield both granulomata on biopsy and acid-fast bacilli on culture, and should be considered if a blood dyscrasia is present. Liver biopsy has the highest diagnostic yield of granulomata which have been reported in up to 75% of biopsies. In cases where the patient is unwilling or where facilities for them do not exist, a clinical trial of antituberculosis drugs should be given. The fever usually responds within 7–10 days, followed by clinical improvement in 4–6 weeks.

12.4. Complications

Tuberculous meningitis may complicate miliary tuberculosis and is a manifestation of acute hematogenous spread. It occurs overtly in up to 20% of cases. Lumbar puncture should be performed if there any symptoms of meningism or headache. A positive microscopy for acid-fast bacilli from the CSF may be the most rapid way of confirming the clinical diagnosis of miliary tuberculosis. Adult respiratory distress syndrome (ARDS) can, rarely, be the presentation of miliary TB (123). In such cases, the breathlessness due to the ARDS can be dominant and the classical x-ray appearances obscured by diffuse confluent or ground-glass shadowing (124,125).

13. Central Nervous System (CNS) Tuberculosis

Although central nervous system (CNS) tuberculosis only makes up some 5% of notified cases in developed countries (3), its importance because of the disproportionate morbidity and mortality associated with this form of tuberculosis cannot be underestimated. In the developing world where difficulties in making a diagnosis and reduced care and treatment availability contribute, it is a major source of death or disability from TB. This is now compounded by an increased risk in HIV-infected individuals with rates for CNS tuberculosis of up to 2/100 in one series (8).

The great majority of cases are of tuberculous meningitis, but intracranial tuberculomata are seen not infrequently in association, and occasionally on their own. Occasional extradural abscesses are reported in association with skull vault bony lesions (41). Tuberculous infection of the meninges is almost always from a focus elsewhere. The method of spread is hematogenous, with up to 20% having overt miliary tuberculosis (Table 9.2). The bacilli usually gain access to the CSF not directly from the bloodstream but from small subpial tuberculomata (126). Symptoms and signs are sometimes as much caused by the intense inflammatory reaction which accompanies the infection as by the infection itself. The meninges look to be covered with gray, thickened exudate, which can be intense and occlude foramina, particularly in the posterior fossa. These meningeal changes lead to endarteritis obliterans which is a more frequent source of focal neurological changes than tuberculomata. The meningeal exudate can extend down the spinal cord onto spinal roots (127).

13.1. Clinical Findings

The initial symptoms are nonspecific, with malaise, anorexia, headache, and vomiting; in children, irritability, poor feeding, drowsiness, or altered behavior may be the dominant features. Unless there is evidence of tuberculosis elsewhere or in low-prevalence countries, the nonspecific nature of the symptoms means that diagnosis may be delayed during the prodromal phase, which can be anything from 2 weeks to 2 months.

The clinical staging developed by the British Medical Research Council based on status at time of diagnosis is helpful (128). In early (Stage I) disease, there is no disturbance of consciousness or focal neurological signs. In medium severity (Stage II), consciousness is disturbed but without coma or delirium; focal neurological signs and cranial nerve palsies may be present. In advanced (Stage III) disease, patients are comatose or stuperose, with or without focal neurological signs.

The meningeal process is accompanied by a low-grade fever, with some neck stiffness in adults, and irritability or drowsiness in children with neck retraction, and in infants, tense fontanelles. Papilloedema is not uncommon and may not be accompanied by raised intracranial pressure and is usually without reduced visual acuity. Choroidal tubercles may occasionally be seen. Cranial nerve palsies occur in a significant proportion of patients. Third- and sixth-nerve palsies are commoner than seventh- or eight-nerve palsies. Less common but of more serious import are lateral gaze palsies or internuclear ophthalmoplegias, which carry a poorer prognosis because of involvement of vital structures in the brain stem (129). A variety of other neurological signs can develop, including cerebellar signs and extrapyramidal movements including choreoathetosis, monparesis, and hemiparesis. Involvement of the spinal meninges can lead to reduced or absent deep tendon reflexes; occasionally this is a dominant feature (129), with a paraplegia with urinary and anal sphincter involvement. Epilepsy can occur at any stage (I–III) but is commoner in children.

13.2. Diagnosis

Diagnosis depends substantially on CSF examination, blood tests giving nonspecific abnormalities, and the tuberculin test being maybe negative. If there is associated miliary tuberculosis (Table 9.2), or changes of pulmonary tuberculosis, which in some series (130) is present in up to 50% cases, the diagnosis is much more easily suggested and made. The CSF pressure is usually raised, but lumbar puncture is safe in this form of chronic meningitis and can be carried out even if papilloedema is present. The leukocyte count is raised but seldom above 500/ mm^3 (131). In the early stages, the leukocytosis is of polymorph predominance (131) but changes to a lymphocyte-predominant pattern as the disease proceeds. The CSF is usually clear, but the CSF protein is almost invariably raised (131); the CSF glucose is usually reduced (132) but may be within the normal range. A completely normal CSF result in protein, cell count, and glucose effectively excludes tuberculous meningitis. In the early stages, particularly if a polymorph leukocytosis is present, the differential diagnosis includes a partially treated bacterial meningitis if antibiotics have been given. It may be necessary to perform serial lumbar punctures to establish a diagnosis if no acid-fast bacilli are seen on microscopy. The CSF changes of a partially treated bacterial meningitis improve over weeks, whereas those of TB meningitis do not. The identification of acid-fast bacilli on microscopy or on culture clinches the diagnosis. A thorough search should be made on microscopy, with success being greater if a larger (up to 10 ml) volume of CSF is analyzed. The first sample taken is most likely to give a positive result (133), but all samples taken should be sent for culture, the majority being negative on microscopy. If the clinical diagnosis is felt to be tuberculous meningitis, consideration should be given to sending sputum, urine, or gastric washings for culture, but the initiation of treatment should not be delayed. Because culture can take several weeks to give a positive result, early confirmation of diagnosis by detecting the presence of mycobacterial constituents (e.g., tuberculostearic acid or mycobacterial DNA) would be most useful. Of the techniques tried, the polymerase chain reaction (PCR) (134) is the most promising, but the technical resources and expense may largely limit its use to developed countries.

13.3. Medical Treatment

Antituberculosis drug treatment should be commenced as soon as possible, which may well be before the diagnosis is proven, unless the CSF is microscopy positive for acid-fast bacilli when the diagnosis is suspected. The penetration of antituberculosis drugs into the CSF depends partly on their serum protein binding and on whether the blood-brain barrier is intact.

Isoniazid penetrates very well (135,136), reaching many times the required minimal inhibitory concentration (MIC), even when the blood-brain barrier is intact. Streptomycin penetrates adequately only when the blood-brain barrier is defective (135). Rifampicin penetrates poorly (137–139) and its penetration may be related partly to inflammation (140). Pyrazinamide penetrates well (141,142) and reaches the MIC required (143,144) independent of stage or activity of disease.

The recommendations for the treatment of tuberculous meningitis are based on clinical experience and series, not on prospective clinical trials. In countries with low rates of isoniazid resistance, treatment with rifampicin and isoniazid for 12 months supplemented by 2 months of pyrazinamide is recommended (34). In areas where there is a higher rate of isoniazid resistance, a fourth drug should be used initially. In Hong Kong, streptomycin is used for the initial 2 months when CSF penetration is better. Alternative fourth drugs for the initial phase of treatment are ethambutol and ethionamide. Ethambutol penetrates the CSF poorly except when there is inflammation (145–147). Ethionamide has good CSF penetration not dependent on inflammation (148,149), and in South Africa, it is preferred to streptomycin (150).

Streptomycin should be given intramuscularly, intrathecal use is not required, but should be avoided in pregnancy. Ethionamide should also be avoided in pregnancy as it maybe teratogenic. In comatose patients, drugs should be administered on an empty stomach via a nasogastric tube (34).

The optimum duration of treatment is unknown. Twelve months' treatment is probably adequate, although some authors give up to 18 months for Stage III disease and between 18 and 24 months for tuberculomata (151). Satisfactory results, however, have been obtained from 6 months' (152,153) and 9 months' treatment (154). When to use corticosteroids in tuberculous meningitis can be controversial. Their use in meningitis was investigated (155) and shown to improve survival in Stage II and III disease and is supported by earlier literature reviews (156). Their use is also definitely indicated for tuberculous encephalopathy (157). Their use in spinal arachnoiditis is more controversial but are usually given although no definite evidence that outcome is altered is available. Their use in Stage II and III disease is, therefore, recommended (34,151).

13.4. CT Scanning and Surgery

Computer tomographic scanning in CNS tuberculosis is useful (158–161) and should be performed if available at diagnosis and if there is any clinical deterio-

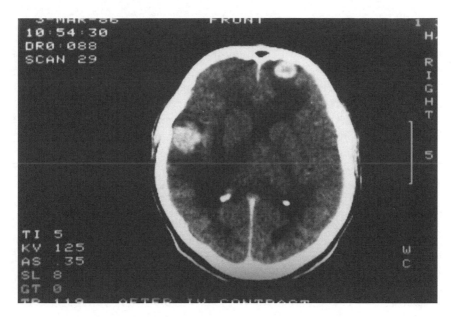

Figure 9.7. CT scan of 40-year-old taken 6 weeks after commencement of treatment of Stage I meningitis when initial CT scan normal. Multiple tuberculomas present. Resolved with continuation of drug therapy and additional corticosteroids.

ration thereafter (Fig. 9.7). The main values are in detecting tuberculomata, giving evidence of associated infarction, and detecting hydrocephalus. The demonstration of the development of tuberculomata on treatment is now well demonstrated (162,163), and paradoxical expansion of those present at diagnosis on treatment is also recorded (164). Surgery is rarely needed for tuberculomata, being required in only 6% (165), but may be necessary if a vital structure (e.g., optic chiasm) is compromised (166).

Computer tomography may show hydrocephalus that ventricular shunting is required. Early drainage of hydrocephalus is required (159,167,168), those drained doing better than those not drained (169). If an intracranial tuberculous abscess forms, neurosurgical drainage may be needed (170).

13.5. Prognosis

The main determinants of outcome with more modern treatment are patient age and stage of disease at presentation (171), with children under 3 years having a worse outcome, independent of stage. A study in Chinese children (131) again showed age and stage at presentation as the only significant factors on multivariate analysis, demonstrating the importance of early diagnosis and prompt treatment. This same study (131) showed recovery in 96%, 78%, and 21% respectively with Stage I, II, and III disease. These results are clearly better than those achieved at the beginning of chemotherapy with streptomycin alone, when mortalities of 46%, 66%, and 81% were recorded for Stages I, II, and III, respectively. Mortality has also been shown to be related to the severity of the hydrocephalus (168,172). HIV coinfection does not appear to alter presentation, symptoms, or prognosis, except in patients with CD4 counts of under 200 μL, when survival is reduced (8).

14. Tuberculous Pericarditis

Pericardial tuberculosis is uncommon in developed countries, making up approximately 1% of cases in the United Kingdom (3) and between 1% and 2% in the United States (173). Conversely, in developed countries, tuberculosis accounts for between 4% and 7% of cases of acute pericarditis, tamponade, or constrictive pericarditis (174,175). In developing countries, it is more important, and in some areas (e.g., southern Africa), it is an important cause of congestive heart failure (176).

The infection usually reaches the pericardium by direct extension from adjacent mediastinal glands, but occasional hematogenous spread can occur with pericardial involvement in miliary disease. An acute pericarditis can be seen, which is thought to represent an allergic response to tuberculoprotein. In chronic pericardial effusion and pericardial constriction, the pathology is granulomatous, which can proceed to fibrosis and calcification at a later date. Although the pericardium is the major site of cardiac involvement, postmortem studies (177) show lesser degrees of involvement of the myocardium and endocardium in some cases.

14.1. Clinical

The onset, as with some other forms of tuberculosis, is insidious with fever, malaise, sweats, weight loss, cough, retrosternal discomfort, and tachycardia. The peak occurrence is in the third to fifth decades (178). If the effusion is sufficient to cause tamponade, dyspnea may be the major symptom. With an effusion, the major signs are low blood pressure, narrow pulse pressure, pulsus paradoxus, edema, and raised venous pressure, with the latter two being more pronounced if tamponade is present. The electrocardiogram is usually of low voltage and shows widespread T-wave changes. The chest x-ray shows an enlarged cardiac outline in over 80% of cases (175), with associated pleural effusions in over 50%.

Pericarditis can progress to constriction any time from a few weeks to several years after the onset of pericarditis. Tuberculosis is the commonest cause of constrictive pericarditis in Africa and Asia, responsible for over 60% of cases in one Indian series (179). The symptoms of constriction are dyspnea, abdominal distension, and edema. The heart sounds are quiet, with pulsus paradoxus, and a raised venous pressure which rises in inspiration (Kussmauhls sign). Constrictive pericarditis causes apparent cardiac enlargement in over 50% of cases (179).

14.2. Diagnosis

The diagnosis should be suspected in patients with a combination of fever and pericardial effusion or signs of tamponade, particularly if from a high incidence ethnic group. The chest x-ray, in addition to an enlarged heart shadow, may show pericardial calcification in constrictive pericarditis and associated active pulmonary tuberculosis in up to 30% (180,181). Sputum smear and culture should be performed if the chest x-ray shows any evidence of pulmonary tuberculosis. In common with other serous membrane tuberculosis, the tuberculin test is positive

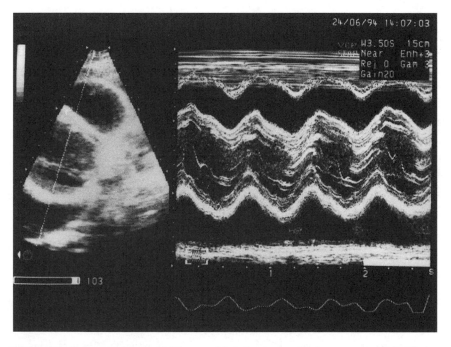

Figure 9.8. Echocardiogram of 24-year-old Asian man with large pericardial effusion showing substantial pericardial fluid and debris. Fluid lymphocyte-rich exudate with positive culture for *M. tuberculosis*. Rapid response to corticosteroids and drug therapy.

in between 80% and 100% of cases (175). Echocardiography is the best way of confirming effusion (Fig. 9.8). In addition to fluid, this may also show pericardial thickening or amorphous material within the pericardial space. Computerized tomography and nuclear magnetic resonance have also been used successfully to confirm pericardial effusion and/or thickening (182). Pericardiocentesis, done either from the subxiphisternal approach or the apical approach, preferably under echocardiographic guidance particularly with the latter approach, gives fluid which is an exudate (protein > 35 g/L) and usually lymphocyte predominant on cytology. The cell count can be of polymorph preponderance in the initial stages. The definitive diagnosis depends on the isolation of acid-fast bacilli from the fluid or obtaining histological confirmation. The fluid can be positive on microscopy for acid-fast bacilli with rates of up to 40% described. Positive culture rates of 59% and positive histology rates of 70% were found in the Transkei study (180). Pericardial biopsy usually requires thoracotomy, but a nonsurgical biopsy technique under x-ray control has been described (183) with encouraging initial results.

14.3. Treatment

MEDICAL

Studies from the Transkei (180,185) have shown that rifampicin/isoniazid for 6 months supplemented by pyrazinamide and streptomycin for the initial 3 months is effective in both effusion and constriction. These studies also assessed the usefulness of prednisolone. When given to patients with constriction, for 11 weeks tapering from an initial dose of 30–60 mg/day, the active group improved more rapidly, needed pericardectomy less, and had a lower death rate (4% versus 11%). In patients with effusion, the need for open drainage and repeated pericardiocentesis was lessened, and the death rate was also lessened (4% versus 14%). Corticosteroids should, therefore, be given in tapering dose over the first 2–3 months of treatment (34).

SURGERY

There are conflicting views on when this is necessary. An active policy of pericardial window procedure with pericardectomy if thickening is present has been advocated (185). A conservative approach is suggested by others (186), who found no constrictive pericarditis in 14 patients treated medically on follow-up. Since the evidence from the Transkei (180,185) of the beneficial effects of steroids, it is probably best if surgical intervention is reserved for those who present late with constriction and calcification, or who fail to respond to the initial 6–8 weeks of medical treatment and still have a raised venous pressure, or those with life-threatening tamponade at any stage (175,176).

15. Tuberculosis of the Skin

Skin involvement is uncommon but can occur in a number of ways. There are forms of skin disease, which involve infection of the skin with direct inoculation or by blood-borne spread, and the tuberculides, which are thought to be cutaneous immunological reactions to tuberculous infections elsewhere in the body.

Primary infection of the skin can be seen particularly in children where inoculation occurs due to minor skin trauma, often from a sputum smear-positive family member. A primary infection develops at the site of inoculation, usually a limb which may ulcerate and proceed to regional lymphadenopathy. This type of infection has also been recently described in needle-stick injury from a HIV-positive patient (187). Other forms of cutaneous tuberculosis are verrucosa cutis, a warty form, and verruca necrogenita, which is more painful and acute. Post-mortem workers, and those who may come into contact with *M. bovis* (e.g., veterinary surgeons, abattoir workers, and butchers) are at risk.

Skin involvement in acute hematogenous tuberculosis, miliary disease, is described with multiple, usually extensive, small papular lesions from which acid-fast bacilli can sometimes be cultured. This form may be becoming more common with the advent of HIV disease, being described in such patients (188) and with coinfection with *M. avium-intracellulare* (189).

The commonest form of skin tuberculosis is lupus vulgaris. This slowly progressive form, usually in older patients, can progress insidiously for months or even years. The extremeties, head, and face are the usual sites, with dull reddish or violaceous lesions, sometimes with a plaquelike psoriaform edge of active disease and residual scarring of a tissue paper type where past infection has occurred. A mutilating form can involve deeper tissue (e.g., cartilage) in ears and nose leading to deformity, and occasionally squamous carcinoma can complicate lupus vulgaris.

Of the forms of tuberculosis without direct skin involvement, the tuberculides, a number of patterns are described. Erythema nodosum associated with tuberculosis is seen usually 3–8 weeks after initial infection and is associated with tuberculin conversion so the skin test is strongly positive (39). Erythema induratum (Bazin's disease), papular and papulonecrotic tuberculides, and other forms of panniculitis with perivascular inflammation of arterioles and venules but also of fat and subcutaneous tissues are also described. The tuberculin test is strongly positive, there may be evidence of tuberculosis elsewhere, and the lesions respond to antituberculosis drug treatment.

16. Miscellaneous

16.1. Adrenal

Tuberculosis is now an uncommon cause of hypoadrenalism; in developing countries autoimmune adrenalitis is more common. Adrenal tuberculosis is seldom an isolated occurrence and is usually part of disseminated tuberculosis (190). Rifampicin, which is included in all short-course regimens, can unmask subclinical adrenal involvement, with adrenal crisis 2–4 weeks after the commencement of treatment being described (191). Being a potent inducer of hepatic microsomal drug metabolism, rifampicin significantly reduces the plasma half-life of corticosteroids (192) and may, thus, precipitate adrenal crisis in those whose adrenals are just producing enough cortisol under maximum stress to maintain minimum serum levels. This phenomenon has been postulated as a factor in the unexplained deaths soon after the commencement of antituberculosis drug therapy (193).

16.2. Liver

The liver is usually involved by hematogenous spread, particularly in miliary disease (Table 9.2), with a diffuse infiltration (194) which is usually diagnosed on needle liver biopsy. A nodular form of hepatic tuberculosis is also described which can mimic carcinoma or cirrhosis (195). This form can present with bleeding gastric varices (196) or portal hypertension (197). The miliary involvement of liver can produce a "bright" liver on ultrasound (198), which is nonspecific. Diagnosis depends on finding granulomas and acid-fast bacilli on biopsy, so liver tissue should routinely be cultured if tuberculosis is suspected. Localized abscesses can develop (82) that cannot be distinguished from amoebic or pyogenic abscesses on ultrasound or CT scan appearances (199,200). Occasionally, multiple microabscesses occur or pseudotumor masses (201).

Conclusion

Nonpulmonary forms of tuberculosis are challenging to clinicians because of the wide diversity of types and presentations. Clinical awareness needs to be maintained so that the diagnosis is suspected, the appropriate investigations are performed, ensuring there is no undue delay in reaching a diagnosis. The management of the drug treatment of nonpulmonary forms should be by physicians experienced in tuberculosis treatment (34), with any surgical aspect dealt with in conjunction with surgical colleagues in a team approach. Because of the often paucibacillary nature of samples of fluid (e.g., ascites, pus or tissue in extrapulmonary forms), the development of reliable molecular biological methods of rapid diagnosis has considerable potential over the next few years in the earlier confirmation of the diagnosis.

References

1. Innes JA (1981) Non-respiratory tuberculosis. J Roy Coll Phys (Lond) 15:227–231.

2. Medical Research Council Cardiothoracic Epidemiology Group (1992) National survey of notifications of tuberculosis in England and Wales in 1988. Thorax 47:770–775.

3. Medical Research Council Tuberculosis and Chest Diseases Unit (1987) National survey of tuberculosis notifications in England and Wales in 1983: characteristics of disease. Tubercle 68:19–32.

4. Weir MR, Thornton GF (1985) Extrapulmonary tuberculosis. Am J Med 79:467–478.

5. Pitchenik AE, Fertel D, Bloch AB (1988) Pulmonary effects of AIDS; mycobacterial disease—epidemiology, diagnosis, treatment and prevention. Clin Chest Med 9:425–441.

6. Chest Service, Medical and Health Department (1990) Annual report. 1990. Hong Kong Government: Hong Kong.

7. Pitchenik AE, Cole C, Russell BW, Fischl MA, Spira JJ, Snider DE (1984) Tuberculosis, atypical mycobacteriosis and acquired immunodeficiency syndrome among Haitian and non-Haitian patients in South Florida. Ann Intern Med 101:641–645.

8. Berenguer J, Santiago M, Laguna F, Morena S, Vincente T, Adrados M, Ortega A et al. (1992) Tuberculous meningitis in patients infected with the human immunodeficiency virus. N Engl Med J 326:668–672.

9. Coleblunders RL, Ryder RV, Nzilambi N, Dikulu K, Willame J-C, Kaboto M et al. (1989) HIV infection in patients with tuberculosis in Kinshasa: Zaire. Am Rev Respir Dis 139:1082–1085.

10. Elliot AM, Luo N, Tembo G (1990) Impact of HIV on tuberculosis in Zambia: a cross-sectional study. Br Med J 301:412–415.

11. Harries AD (1990) Tuberculosis and HIV infection in developing countries. Lancet 335:387–390.

12. Modilevsky T, Sattler FR, Barnes PF (1989) Mycobacterial disease in patients with HIV infection. Arch Intern Med 149:2201–2205.

13. Soriano E, Mallolas J, Gatell JM, Lattore X, Miro JM, Pecchiar M et al. (1988) Characteristics of tuberculosis in HIV-infected patients: a case control study. AIDS 2:429–432.

14. Campbell IA, Dyson AJ (1977) Lymph node tuberculosis: a comparison of various methods of treatment. Tubercle 58:171–179.

15. Mehta JB, Dutt A, Harvill L, Matthews KM (1991) Epidemiology of extrapulmonary tuberculosis. Chest 99:1134–1138.

16. Alvarez S, McCabe WR (1984) Extrapulmonary tuberculosis revisited: a review of experience at Boston City and other hospitals. Medicine 63:25–53.

17. Lau SK, Kwan S, Lee J, Wei WI (1991) Source of tubercle bacilli in cervical lymph nodes: a prospective study. J Laryngol Otol 105:558–561.

18. British Thoracic Society Research Committee (1985) Short course chemotherapy for tuberculosis of lymph nodes: a controlled trial. Br Med J 290:1106–1108.

19. Finch PJ, Millard FJC, Maxwell JD (1991) Risk of tuberculosis in immigrant Asians: culturally acquired immunodeficiency? Thorax 46:1–5.

20. Huhti E, Brander E, Ploheimo S, Sutinen S (1975) Tuberculosis of cervical lymph nodes; a clinical, pathological and bacteriological study. Tubercle 56:27–36.

21. British Thoracic Society Research Committee (1992) Six-months versus nine-months chemotherapy for tuberculosis of lymph nodes: preliminary results. Respir Med 86:15–19.

22. Lau SK, Wei WI, Hau C, Engzell UC (1990) Efficacy of fine needle aspiration cytology in the diagnosis of tuberculosis cervical lymphadenopathy. J. Laryngol Otol, Jan 104(1):24–7.

23. Lau SK, Wei WI, Kwan S, Engzell UC (1988) Fine needle aspiration biopsy of tuberculous cervical lymphadenopathy. Aust NZ J Surg 58:947–950.

24. Lau SK, Wei WI, Kwan S, Yew WW (1991) Combined use of fine-needle aspiration cytologic examination and tuberculin skin test in the diagnosis of cervical tuberculous lymphadenitis: a prospective study. Arch Otolaryngol Head Neck Surg 117:87–90.

25. Shaha A, Webber C, Marti J (1986) Fine needle aspiration in the diagnosis of cervical lymphadenopathy. Am J Surg 152:420–423.

26. Campbell IA, Dyson AJ (1977) Lymph node tuberculosis: a comparison of various methods of treatment. Tubercle 58:171–179.

27. British Thoracic Society Research Committee (1985) Short course chemotherapy for tuberculosis of lymph nodes: a controlled trial. Br Med J 290:1106–1108.

28. British Thoracic Society Research Committee (1988) Short course chemotherapy for lymph node tuberculosis: final report at 5 years. Br J Dis Chest 82:282–284.

29. Jindani A, Aber VR, Edwards EA, Mitchison DA (1980) The early bactericidal activity of drugs in patients with pulmonary tuberculosis. Am Rev Respir Dis 121:139–148.

30. Ormerod LP (1988) A retrospective comparison of two drug regimens RHZ2/RH10 and RHE2/RH10 in the treatment of tuberculous mediastinal lymphadenopathy. Br J Dis Chest 82:274–278.

31. British Thoracic Society Research Committee (1992) Six-months versus nine-months chemotherapy for tuberculosis of lymph nodes: preliminary results. Respir Med 83:15–19.

32. British Thoracic Society Research Committee (1993) Six-months versus nine-months chemotherapy for tuberculosis of lymph nodes: final results. Respir Med 87:621–623.

33. McCarthy OR, Rudd RM (1989) Six-months chemotherapy for lymph node tuberculosis. Respir Med 89:425–427.

34. Ormerod LP for the Joint Tuberculosis Committee (1990) Chemotherapy and management of tuberculosis in the United Kingdom; recommendations of the Joint Tuberculosis Committee of the British Thoracic Society. Thorax 45:403–408.

35. Jawahar MS, Sivasubramanian S, Vijayan VK, Ramakrishnan CV, Paramasivan CN, Selvakumar et al. (1990) Short course chemotherapy for tuberculous lymphadenitis in children. Br Med J 301:359–361.

36. Campbell IA, Dyson AJ (1979) Lymph node tuberculosis; a comparison of treatments 18 months after completion of chemotherapy. Tubercle 60:95–98.

37. Prissick FH, Masson AM (1956) Cervical lymphadenitis in children caused by chromogenic mycobacteria. Can Med Assoc J 75:798–803.

38. McKellar A (1976) Diagnosis and management of atypical mycobacterial lymphadenitis in children. J Paediatr Surg 11:85–89.

39. Wallgren A (1948) The timetable of tuberculosis. Tubercle 29:245–251.

40. Davies PDO, Humphries MJ, Byfield SP, Nunn AJ, Darbyshire JH, Citron KM et al. (1984) Bone and joint tuberculosis in a national survey of notifications in England and Wales in 1978/9. J Bone Joint Surg 66B:326–330.

41. Hodgson SP, Ormerod LP (1990) Ten-year experience of bone and joint tuberculosis in Blackburn 1978–87. J Roy Coll Surg (Edin) 35:259–262.

42. Humpries MJ, Sister Gabriel M, Lee YK (1986) Spinal tuberculosis presenting as abdominal symptoms—a report of two cases. Tubercle 67:303–307.

43. Walker GF (1968) Failure of early recognition of skeletal tuberculosis. Br Med J i:682.

44. Gorse GJ, Pais JM, Kurske JA, Cesario TC, Ip M, Chen NK, So SY, Chiu SW, Lam WK (1983) Tuberculous spondylitis: a report of six cases and a review of the literature. Medicine 62:178–193.

45. Ip M, Chen NK, So SY, et al. (1989) Unusual rib destruction in pleuropulmonary tuberculosis. Chest 95:242–244.

46. Lin-Greenberg A, Cholankeril J (1990) Vertebral arch destruction in tuberculosis: CT features. J Comput Assist Tomogr 14(2):300–302.

47. Bell GR, Stearns KL, Bonutti PM, Boumphrey FR (1990) MRI diagnosis of tuberculous vertebral osteomyelitis. Spine 15(6):462–465.

48. Smith DF, Smith FW, Douglas JG (1989) Tuberculous radiculopathy: the value of magnetic resonance imaging of the neck. Tubercle 70:213–216.

49. Angtuaco EGC, McConnell JR, Chadock WM, Flannigan S (1987) Magnetic resonance imaging of spinal epidural sepsis. Am J Roentgen 42:1249–1253.

50. Medical Research Council Working Party on Tuberculosis of the Spine (1973) A controlled trial of ambulant outpatient treatment and inpatient rest in bed in the management of tuberculosis of the spine in young Korean patients on standard chemotherapy. A study in Masan, Korea. J Bone Joint Surg 55B:678–697.

51. Medical Research Council Working Party on Tuberculosis of the Spine (1973) A controlled trial of plaster-of-Paris jackets in the management of ambulant outpatient treatment of tuberculosis of the spine in children on standard chemotherapy: a study in Pusan, Korea. Tubercle 54:261–282.

52. Medical Research Council Working Party of Tuberculosis of the Spine (1974) A controlled trial of anterior spinal fusion and debridement in the surgical management of tuberculosis of the spine in patients on standard chemotherapy: a study in Hong Kong. Br J Surg 61:853–866.

53. Medical Research Council Working Party on Tuberculosis of the Spine (1974) A controlled trial of debridement and ambulatory treatment in the management of tuberculosis of the spine in patients on standard chemotherapy. A study in Bulawayo, Rhodesia. J Trop Med Hyg 77:72–92.

54. Medical Research Council Working Party on Tuberculosis of the Spine (1982) A 10-year assessment of a controlled trial comparing debridement and anterior spinal fusion in the management of tuberculosis of the spine in patients on standard chemotherapy in Hong Kong. J Bone Joint Surg 64B:393–398.

55. Medical Research Council Working Party on Tuberculosis of the Spine (1985) A ten-year assessment of controlled trials of inpatient and outpatient treatment and of plaster-of-Paris jackets for tuberculosis of the spine in children on standard chemotherapy: studies in Masan and Pusan. J Bone Joint Surg 67B:103–110.

56. Medical Research Council Working Party on Tuberculosis of the Spine (1986) A controlled trial of six-month and nine-month regimens of chemotherapy in patients undergoing radical surgery for tuberculosis of the Spine in Hong Kong. Tubercle 67:243–259.

57. Parkinson RW, Hodgson SP, Noble J (1990) Tuberculosis of the elbow: a report of 5 cases. J Bone Joint Surg 72B:523–524.

58. Valdazo JP, Perez-Ruiz F, Albarracin A, Sanchez-Nievas G, Perez-Benegas J, Gonzales-Lanza M et al. Tuberculous arthritis: Report of a case with multiple joint involvement and periarticular tuberculous abscesses. J Rheumatol 17(3):399–401.

59. Ormerod LP, Grundy M, Rahman MA (1989) Multiple tuberculous bone lesions simulating metastatic disease. Tubercle 70:305–307.

60. Ormerod LP (1993) Why does genitourinary tuberculosis occur less often than expected in ISC ethnic patients. J Infect 27:27–32.

61. Stoller JK (1985) Late recurrence of *Mycobacterium bovis* genitourinary tuberculosis: case report and review of the literature. J Urol 134:565–566.

62. Yaqoob M, Goldsmith HJ, Ahmad R (1990) Bovine genitourinary tuberculosis revisited. Quart J Med 74:105–109.

63. Lichtenstein IH, MacGregor RR (1983) Mycobacterial infections in renal transplant recipients: report of five cases and a review of the literature. Rev Infect Dis 5(2):216–226.

64. Ferrie BG, Rundle JSH (1985) Genitourinary tuberculosis in patients under twenty five years of age. Urology XXV(6):576–578.

65. Datta SK (1987) Renal tuberculosis presenting as hypertension. J Assoc Physicians India 35(11):798–799.

66. Benn JJ, Scoble JE, Thomas AC, Eastwood JB (1988) Cryptogenic tuberculosis presenting as a preventable cause of end-stage renal failure. Am J Nephrol 8(4):306–308.

67. Morgan SH, Eastwood JB, Baker LRI (1990) Tuberculous interstitial nephritis—the tip of the iceberg? Tubercle 71:5–6.

68. al-Sulaiman MH, Dhar JM, al-Hasani MK, Haleem A, al-Khader A (1990) Tuberculous interstitial nephritis after kidney transplantation. Transplantation 50(1):162–164.

69. Gorse GJ, Belshe RB (1985) Male genital tuberculosis: a review of the literature with instructive case reports. Rev Infect Dis 7:511–524.

70. Sutherland AM (1985) Gynaecological tuberculosis: analysis of a personal series of 710 cases. Aust NZ J Obstet Gynaecol 25:203–207.

71. Dolev E, Bass A, Nossinowitz N (1985) Frequent occurrence of renal calculi in tuberculous kidneys in Israel. Urology 26:544–545.

72. Okazawa N, Sekiya T, Tada S (1985) Computed tomographic features of renal tuberculosis. Radiat Med 3:209–213.

73. Goldman SM, Fishman EK, Hartman DS (1985) Computed tomography of renal tuberculosis and its pathological correlates. J Comput Assist Tomogr 9:771–776.

74. Gow JG, Barbosa S (1984) Genitourinary tuberculosis. A study of 1117 cases over a period of 34 years. Br J Urol 56:449–455.

75. Morgan SH, Eastwood JB, Baker LRI (1990) Tuberculous interstitial nephritis—the tip of the iceberg? Tubercle 71:5–6.

76. Psihramis KE, Donahoe PK (1986) Primary genitourinary tuberculosis: rapid progression and tissue destruction during treatment. J Urol 135:1033–1036.

77. Osborn DE, Rao NJ, Blacklock NJ (1986) Tuberculous stricture of ureter. A new method of intubated ureterotomy. Br J Urol 58:103–104.

78. Nwokolo C (1961) Ascites in Africa. Br Med J 1:33.

79. Menzies RI, Alsen H, Fitzgerald JM, Mohapeola RG (1986) Tuberculous peritonitis in Lesotho. Tubercle 67:47–54.

80. Lucas SB, De Cock KM, Hounnou A, Peacock C, Diomande M, Honde M, et al. (1994) Contribution of tuberculosis in slim disease in Africa. Br Med J 308:1531–1533.

81. Mukerjee P, Singal AK (1979) Intestinal tuberculosis: 500 operated cases. Proc Assoc Surg East Africa 2:70–75.

82. Klimach OE, Ormerod LP (1985) Gastrointestinal tuberculosis: a retrospective review of 109 cases in a district general hospital. Quart J Med 56:569–578.

83. deMas R, Lombeck G, Rieman JF (1986) Tuberculosis of the oesophagus masquerading as an ulcerated tumour. Endoscopy 18:153–155.

84. Gupta SP, Arora A, Bhargava DK (1992) An unusual presentation of oesophageal tuberculosis. Tuberc Lung Dis 73:174–176.

85. Addison NV (1983) Abdominal tuberculosis—a disease revived. Ann Roy Coll Surg Eng 65:105–111.

86. Lambrianides AL, Ackroyd N, Shorey B (1980) Abdominal tuberculosis. Br J Surg 67:887–889.

87. Sherman S, Rohwedder JJ, Ravikrishnan KP, Weg JG (1980) Tuberculous enteritis and peritonitis—report of 36 general hospital cases. Arch Intern Med 140:506–508.

88. Bastani B, Shariatzadeh MR, Dehdashti F (1985) Tuberculous peritonitis—report of 30 cases and a review of the literature. Quart J Med 56:549–557.

89. Shukla HS, Hughes LE (1978) Abdominal tuberculosis in the 1970s: a continuing problem. Br J Surg 65:403–405.

90. Gilinsky NH, Marks IN, Kottler RE (1983) Abdominal tuberculosis. A 10 year review. S Afr Med J 64:849–857.

91. Das Pritam, Shukla HS (1976) Clinical diagnosis of abdominal tuberculosis. Br J Surg 63:941–946.

92. Pozniak AL, Dalton-Clarke HJ (1985) Colonic tuberculosis presenting as massive rectal bleeding. Tubercle 66:295–299.

93. Rodriguez de Lope C, San Miguel Joglar G, Pons Romero F (1982) Laparoscopic diagnosis of tuberculous ascites. Endoscopy 14:178–179.

94. Sochocky S (1967) Tuberculous peritonitis. A review of 100 cases. Am Rev Respir Dis 95:398–401.

95. Hanson RD, Hunter TB (1985) Tuberculous peritonitis: CT appearance. Am J Radiol 144:931–932.

96. Hulnick DH, Megibow AJ, Naidich DP, Hilton S, Cho KC, Balthazar EJ (1985) Abdominal tuberculosis—a CT-evaluation. Radiology 157:199–204.

97. Denath FM (1990) Abdominal tuberculosis in children: CT findings. Gastrointest Radiol 15:303–306.

98. Epstein BM, Mann JH (1982) CT of abdominal tuberculosis. Am J Radiol 139:861–866.

99. Levine H (1967) Needle biopsy of the peritoneum in exudative ascites. Arch Intern Med 120:542–545.

100. Singh MM, Bhargava AN, Jain KP (1969) Tuberculous peritonitis: an evaluation of pathogenic mechanisms, diagnostic procedures and therapeutic measures. N Engl J Med 281:1091–1096.

101. Shukla HS, Naitrani YP, Bhatia S, Das P, Gupta SC (1982) Peritoneal biopsy for diagnosis of abdominal tuberculosis. Postgrad Med J 58:226–228.

102. Jorge AD (1984) Peritoneal tuberculosis. Endoscopy 16:10–12.

103. Manohar A, Simjee AE, Haffejee AA (1990) Symptoms and investigative findings in 145 patients with tuberculous peritonitis diagnosed by peritoneoscopy and biopsy over a five year period. Gut 31:1130–1132.

104. Kalvaria I, Kottler RE, Marks IN (1988) The role of colonoscopy in the diagnosis of tuberculosis. J Clin Gastroenterol 10:516–523.

105. Kochhar RJ, Rajwanshi A, Goenka MK, Nijwahan R, Sood A, Nagi S et al. (1991) Colonoscopic fine needle aspiration cytology in the diagnosis of ileo-caecal tuberculosis. Am J Gastroenterol 86:102–104.

106. Tandon HD, Prakash A (1972) Pathology of intestinal tuberculosis and its distinction from Crohns disease. Gut 13:260–269.

107. Abrams JS, Holden WD (1964) Tuberculosis of the gastrointestinal tract. Arch Surg 89:282–293.

108. Dineen P, Homan WP, Grafe WR (1976) Tuberculous peritonitis: 43 years experience in diagnosis and treatment. Ann Surg 84:717–723.

109. McMillen MA, Arnold SD (1979) Tuberculous peritonitis associated with alcoholic liver disease. NY State J Med 79:922–924.

110. Byrom HB, Mann CV (1969) Clinical features and surgical management of ileocaecal tuberculosis. Proc Roy Soc Med 62:1230–1233.

111. Bobrowitz ID (1982) Active tuberculosis undiagnosed until autopsy. Am J Med 72:650–658.

112. Monie RDH, Hunter AM, Rocchiccioli KMS, White JP, Campbell IA, Kilpatrick GS (1983) Retrospective survey of the management of miliary tuberculosis in South and West Wales 1976–78. Thorax 38:369–373.

113. Sahn SA, Neff TA (1968) Miliary tuberculosis. Am J Med 56:495–505.

114. Price M (1968) Lymphangitis reticularis tuberculosa. Tubercle 49:377–384.

115. Proudfoot AT, Akhtar AJ, Douglas AC, Horne NW (1969) Miliary tuberculosis in adults. Br Med J ii:273–277.

116. Proudfoot AT (1971) Cryptic disseminated tuberculosis. Br J Hosp Med 5:773–780.

117. Grieco MH, Chmel H (1974) Acute disseminated tuberculosis as a diagnostic problem. Am Rev Respir Dis 109:554–560.

118. Medd WE, Hayhoe FGJ (1955) Tuberculous miliary necrosis with pancytopenia. Quart J Med 24:351–364.

119. Cooper W (1959) Pancytopenia associated with disseminated tuberculosis. Ann Intern Med 50:1497–1501.

120. Hughes JT, Johnstone RM, Scott AC, Stewart PD (1959) Leukaemoid reactions in disseminated tuberculosis. J Clin Pathol 12:307–311.

121. Twomey JJ, Leavell BS (1965) Leukaemoid reactions to tuberculosis. Arch Intern Med 116:21–28.

122. Oswald NC (1963) Acute tuberculosis and granulocytic disorders. Br Med J ii:1489–1496.

123. So SY, Yu D (1981) The adult respiratory distress syndrome associated with miliary tuberculosis. Tubercle 62:49–53.

124. Heap MJ, Bion JF, Hunter KR (1989) Miliary tuberculosis and the adult respiratory distress syndrome. Respir Med 83:153–156.

125. Dyer RA, Chappell WA, Potgeiter PD (1985) Adult respiratory distress syndrome associated with miliary tuberculosis. Crit Care Med 13:12–15.

126. Bishburg E, Sunderam G, Reichman LB, Kapila R (1986) Central nervous system tuberculosis with the acquired immunodeficiancy syndrome and its related complex. Ann Intern Med 105:201–213.

127. Rich AR, McCordock HA (1933) The pathogenesis of tuberculous meningitis. Bull Johns Hopkins Hosp 52:5–37.

128. Medical Research Council Streptomycin in Tuberculosis Trials Committee (1948) Streptomycin treatment of tuberculous meningitis. Lancet i:582–597.

129. Teoh R, Humphries MJ, Chan JCN, Ng HK, O'Mahoney G (1989) Internuclear ophthalmoplegia in tuberculous meningitis. Tubercle 70:61–64.

130. Wadia NH, Dastur DK (1969) Spinal meningitis with radiculomyelopathy. Part 1: clinical and radiological features. J Neurol Sci 8:239–260.

131. Humphries MJ, Teoh R, Lau J, Gabriel M (1990) Factors of prognostic significance in Chinese children with tuberculous meningitis. Tubercle 71:161–168.

132. Jeren T, Beus I (1982) Characteristics of cerebrospinal fluid in tuberculous meningitis. Acta Cytol 26:678–680.

133. Kennedy DH, Fallon RJ (1979) Tuberculous meningitis. J Am Med Assoc 241:264–268.

134. Shankar P, Manjunath N, Mohan KK, Prasad K, Behari M, Shrinivas et al. (1991) Rapid diagnosis of tuberculous meningitis by polymerase chain reaction. Lancet 2:5–7.

135. Ellard GA, Humphries MJ, Allen BW (1993) Penetration of isoniazid, rifampicin and streptomycin into the cerebrospinal fluid and the treatment of tuberculous meningitis. Am Rev Respir Dis 148:650–655.

136. Fletcher AP (1953) CSF isoniazid levels in tuberculous meningitis. Lancet ii:694–697.

137. D'Olivera JJ (1972) Cerebrospinal fluid concentrations of rifampicin in meningeal tuberculosis. Am Rev Respir Dis 106:432–437.

138. Forgan-Smith R, Ellard GA, Newton D, Mitchison DA (1973) Pyrazinamide and other drugs in tuberculous meningitis. Lancet ii:374.

139. Kaojarern S, Supmonchai K, Phuapradit P, Mokkhavesa C, Krittiyanunt S (1991) Effect of steroids on cerebrospinal fluid penetration of antituberculosis drugs in tuberculous meningitis. Clin Pharmacol Therapeut 49:6–12.

140. Woo J, Humphries MJ, Chan K, O'Mahoney G, Teoh R (1987) Cerebrospinal fluid and serum levels of pyrazinamide and rifampicin in patients with tuberculous meningitis. Curr Therap Res 42:235–242.

141. Ellard GA, Humphries MJ, Gabriel M, Teoh R (1987) The penetration of pyrazinamide into the cerebrospinal fluid in patients with tuberculous meningitis. Br Med J 294:284–285.

142. Donald PR, Seifart H (1988) Cerebrospinal fluid pyrazinamide concentrations in children with tuberculous meningitis. Paediatr Infect Dis J 7:469–471.

143. Stottmeier KD, Beam RE, Kubica GP (1967) Determination of drug suspectibility of mycobacteria to pyrazinamide in 7H10 agar. Am Rev Respir Dis 96:1072–1075.

144. Carlone NA, Acocella G, Cuffini AN, Forno-Pizzoglio M (1985) Killing of macrophage-ingested mycobacteria by rifampicin, pyrazinamide, pyrazinoic acid alone and in combination. Am Rev Respir Dis 132:1274–1277.

145. Gundert-Remy U, Lett M, Weber E (1964) Concentration of ethambutol in cerebrospinal fluid in man as a function of non-protein-bound fraction in serum. J Clin Pharmacol 6:133–136.

146. Borrowitz ID (1972) Ethambutol in tuberculous meningitis. Chest 61:629–632.

147. Place VA, Pyle MM, de la Huerga J (1969) Ethambutol in tuberculous meningitis. Am Rev Respir Dis 99:783–785.

148. Hughes IE, Smith HV, Kane PO (1962) Ethionamide and its passage into the cerebrospinal fluid fluid in man. Lancet i:616–617.

149. Donald PR, Seifart HI (1989) Cerebrospinal fluid concentrations of ethionamide in children with tuberculous meningitis. J Pediatr 115:383–386.

150. Donald PR, Schoeman JF, O'Kennedy A (1987) Hepatic toxicity during chemotherapy for severe tuberculous meningitis. Am J Dis Child 141:741–743.

151. Humphries MJ (1992) The management of tuberculous meningitis. Thorax 47:577–581.

152. Phuapradit P, Vejjajiva A (1987) Treatment of tuberculous meningitis: the role of short course chemotherapy. Quart J Med 239:249–258.

153. Jacobs RF, Sunakorn P, Chotitayasunonah T, Pope S, Kelleher K (1992) Intensive short course chemotherapy for tuberculous meningitis. Pediatr Infect Dis 11:194–198.

154. Acharya VN, Kudva BT, Retnam VJ, Mehta PJ (1985) Adult tuberculous meningitis: comparative study of different chemotherapeutic regimens. J Assoc Physicians India 33:583–585.

155. Shaw PP, Wang SM, Tung SG, Niu QW, Lu TS, Yu XC, et al. (1984) Clinical analysis of 445 adult cases of tuberculous meningitis. Chin J Tuberc Respir Dis 3:131–132.

156. Horne NW (1966) A critical evaluation of corticosteroids in tuberculosis. Adv Tuberc Res 15:1–54.

157. Udani PM, Dastur PK (1971) Tuberculous encephalopathy with and without meningitis: clinical features and pathological correlations. J Neurol Sci 14:541–561.

158. Bhargava S, Gupta AK, Tandon PN (1982) Tuberculous meningitis—a CT study. Br J Radiol 55:189–196.

159. Bullock MR, Welchman JM (1982) Diagnostic and prognostic features of tuberculous meningitis on CT scanning. J Neurol Neurosurg Psychol 45:1098–1101.

160. Teoh R, Humphries MJ, Hoare RD, O'Mahoney G (1989) Clinical correlation of CT changes in 64 Chinese patients with tuberculous meningitis. J Neurol 236:48–51.

161. Kingsley DPE, Hendrickse WA, Kendall BE, Swash M, Singh V (1987) Tuberculous

meningitis: role of CT scan in management and prognosis. J Neurol Neurosurg Psychol 50:30–36.

162. Teoh IR, Humphries MJ, O'Mahoney G (1987) Symptomatic intracranial tuberculoma developing during treatment of tuberculosis; a report of 10 cases and review of the literature. Quart J Med 241:449–460.

163. Lees AJ, Macleod AF, Marshall J (1980) Cerebral tuberculomas developing during treatment of tuberculous meningitis. Lancet i:1208–1211.

164. Chambers T, Hendrickse WA, Record C, Rudge P, Smith H (1984) Paradoxical expansion of intracranial tuberculomas during treatment. Lancet ii:181–184.

165. Tandon PN, Bhargava S (1985) Effect of medical treatment on intracranial tuberculoma—a CT study. Tubercle 66:85–97.

166. Teoh R, Poon W, Humphries MJ, O'Mahoney G (1988) Suprasellar tuberculoma developing during treatment of tuberculous meningitis requiring urgent surgical decompression. J Neurol 235:321–322.

167. Roy TK, Sircar PK, Chandar V (1979) Peritoneal-ventricular shunt in the management of tuberculous meningitis. Indian J Paediatr 16:1023–1027.

168. Palur R, Rajshekar V, Chandy MJ, Joseph T, Abraham J (1991) Shunt surgery for hydrocephalus in tuberculous meningitis. A long-term follow-up study. J Neurosurg 74:64–69.

169. Peacock WJ, Deeny JE (1984) Improving the outcome of tuberculous meningitis in childhood. S Afr Med J 66:597–598.

170. Tang ESC, Chau A, Fong D, Humphries MJ (1991) The treatment of multiple intracranial abscesses: a case report. J Neurol 238:183–185.

171. Ogawa SK, Smith MA, Brennessel DJ, Lowy FD (1987) Tuberculous meningitis in an urban medical centre. Medicine 66:317–326.

172. Arens LJ, Deeny JE, Molteno CD, Kibel MA (1977) Tuberculous meningitis in children in the western Cape: neurological sequelae. Paediatr Rev Commun 1:257–275.

173. Larrieu AJ, Tyers GF, Williams EH, Derrick JR (1980) Recent experience with tuberculous pericarditis. Ann Thorac Surg 29:464–468.

174. Lorell BH, Braunwald E (1988) Pericardial disease: tuberculous pericarditis. In: Braunwald E, ed. Heart Disease—A Textbook of Cardiovascular Medicine, 3rd Ed., pp. 1509–1511. Philadelphia: WB Saunders.

175. Fowler NO (1991) Tuberculous pericarditis. J Am Med Assoc 266:99–103.

176. Strang JIG (1984) Tuberculous pericarditis in Transkei. Clin Cardiol 7:667–670.

177. Dave T, Narula JP, Chopra P (1990) Myocardial and endocardial involvement in tuberculous constrictive pericarditis. Int J Cardiol 28:245–251.

178. Rooney JJ, Crocco JA, Lyons HA (1970) Tuberculous pericarditis. Ann Intern Med 72:73–78.

179. Bashi VV, John S, Ravimukar, Rooney JJ, Crocco JA, Lyons HA (1988) Early and

late results of pericardectomy in 118 cases of constrictive pericarditis. Thorax 43:637–641.

180. Sagrosta-Sauleda J, Permanyer-Miralda G, Soler-Soler J (1988) Tuberculous pericarditis: ten year experience with a prospective protocol for diagnosis and treatment. J Am Coll Cardiol 11:724–728.

181. Strang JIG, Kakaza HHS, Gibson DG, Girling DJ, Nunn AJ, Fox W (1988) Controlled trial of complete open drainage and prednisolone in the treatment of tuberculous pericardial effusion in Transkei. Lancet ii:759–763.

182. Pohost GM, O'Rourke RA, eds. (1991) Principles and Practice of Cardiovascular Imaging, p. 457. Boston: Little Brown.

183. Endrys J, Simo M, Shafie MZ, et al. (1988) New non-surgical technique for multiple pericardial biopsies. Cathet Cardivasc Diagn 15:92–94.

184. Strang JIG, Kakaza HHS, Gibson DG, Girling DJ, Nunn AJ, Fox W (1987) Controlled trial of prednisolone as adjuvant in treatment of tuberculous constrictive pericarditis in Transkei. Lancet ii:1418–1422.

185. Quale JM, Lipshick GY, Heurich AE (1987) Management of tuberculous pericarditis. Ann Thorac Surg 43:653–655.

186. Long R, Younes M, Patton N, Hershfield E (1989) Tuberculous pericarditis: long term outcome in patients who received medical treatment alone. Am Heart J 117:1133–1139.

187. Kramer F, Sasse SA, Simms JC, Leedom JM (1993) Primary cutaneous tuberculosis after a needle stick injury from a patient with AIDS and undiagnosed tuberculosis. Ann Intern Med 119:594–595.

188. Stack RJ, Bickley LK, Coppel IG (1990) Miliary tuberculosis presenting as skin lesions in a patient with the acquired immunodeficiency syndrome. J Am Acad Dermatol 23:1031–1035.

189. Lombardo PC, Weitzman I (1990) Isolation of *Mycobacterium tuberculosis* and *M. avium* complex from the same skin lesions in AIDS. N Engl J Med 323:916–917.

190. Van Kralingen KW, Slee PH (1987) A patient with miliary tuberculosis and acute adrenal failure. Neth J Med 30:235–241.

191. Wilkins EGL, Hnizdo E, Cope A (1989) Addisonian crisis induced by treatment with rifampicin. Tubercle 70:69–73.

192. McAllister WAC, Thompson FJ, Al-Habet S, Rodgers S, Nagai H, Shimizu S, Kawamoto H, Yamanove M, Tsuchiya T, Yamamoto M (1982) Adverse effects of rifampicin on prednisolone deposition. Thorax 37:792.

193. Ellis ME, Webb AK (1983) Cause of death in patients admitted to hospital with pulmonary tuberculosis. Lancet i:665–667.

194. Essop AR, Posen JA, Hodgkinson JH, Segal I (1984) Tuberculous hepatitis: a clinical review of 96 cases. Quart J Med 212:465–477.

195. Nagai H, Shimizu S, Kawamoto H, et al. (1989) A case of solitary tuberculosis of the liver. Japan J Med 28:251–255.

196. Sheen-Chen SM, Chou FF, Tai DI, Eng HL (1990) Hepatic tuberculosis; a rare cause of bleeding gastric varices. Tubercle 71:225–227.

197. Gibson JA (1973) Granulomatous liver disease and portal hypertension. Proc Roy Soc Med 66:502–503.

198. Andrew WK, Thomas RG, Gollach BL (1982) Miliary tuberculosis of the liver—another cause of 'bright liver' on ultrasound examination. S Afr Med J 62:808–809.

199. Spiegel CT, Tuazon CU (1984) Tuberculous liver abscess. Tubercle 65:127–131.

200. Epstein BM, Leibowitz CB (1987) Ultrasonographic and computed tomographic appearance of focal tuberculosis of the liver. A case report. S Afr Med J 71:461–462.

201. Denath FM (1990) Abdominal tuberculosis in children: CT findings. Gastrointest Radiol 15:303–306.

10

Chemotherapy of Nontuberculous Mycobacterial Diseases

I.A. Campbell, P.A. Jenkins, and Richard J. Wallace, Jr.

Nontuberculous mycobacteria can be found throughout the environment and readily gain access to clinical specimens. The significance of an isolate can, therefore, be doubtful and it is necessary to establish criteria which determine significance and which indicate whether or not treatment is required. The most important considerations are the type of specimen from which the organism is isolated, the number of isolates, the degree of growth, and the identity of the organism. Predisposing factors on the part of the patient and the clinical presentation can also help to determine the significance.

Nontuberculous mycobacteria can cause both pulmonary and extrapulmonary disease. It is convenient to consider these separately as they present quite different problems.

1. Pulmonary Disease in the HIV Negative Host (1)

Pulmonary disease is by far the commonest manifestation of infection by nontuberculous mycobacteria. A single isolate recovered from multiple pulmonary specimens is rarely of significance. However, isolation of the same mycobacterium from specimens taken several days apart, together with appearances on the chest radiograph suggesting mycobacterial infection, indicate that chemotherapy is necessary, especially if symptoms suggestive of pulmonary infection are present. Although there are geographical variations in incidence, the species most commonly causing disease are *Mycobacterium kansasii, M. avium, M. intracellulare, M. malmoense, M. xenopi* and *M. abscessus* (2,3). Small numbers of patients have disease due to *M. fortuitum* (2,3), *M. szulgai* (4), or *M. simiae* (5). Very occasionally, species usually taken to be saprophytes can cause infection

279

and disease (e.g., *M. gastri*, *M. gordonae* (6), *M. terrae* complex (including *M. nonchromogenicum*), and *M. triviale* (7).

1.1. Susceptibility Testing

It is usual to determine either the proportion of the bacterial population that is resistant to a critical concentration of the drug on solid media or in BACTEC 12B broth, or to determine the resistance ratio by titrating the test strain and comparing its end point with that of a standard strain of *M. tuberculosis* such as H37Rv or a panel of naturally occurring "wild" strains (8). For *M. tuberculosis* results obtained by either of these means have correlated well with the outcome of treatment in controlled clinical trials. These have shown, for example, that a resistance ratio of at least eight times that of the standard strain means that the drug will not generally be effective in treatment.

Extending this principal to nontuberculous mycobacteria provides patterns of resistance/susceptibility as shown in Table 10.1. The susceptibility of *M. kansasii* to rifampicin and ethambutol correlates with the clinical response to treatment with these drugs (9,10). Susceptibility of *M. avium* (11,12), *M. intracellulare* (13), *M. chelonae* (14), and *M. abscessus* (14), to clarithromycin correlates with clinical response to therapy with either clarithromycin or azithromycin-containing regimens, and resistant isolates are associated with treatment failures and have been shown to have a point mutation in the 23S rDNA macrolide binding site (12–14). For the other species and drugs, the situation is less clear. Retrospective studies demonstrated that clinical response to treatment does not correlate with single-drug *in vitro* susceptibility results for the antimycobacterial drugs (15–17,17a). It has been demonstrated that synergy occurs between rifampicin and ethambutol for many species of nontuberculous mycobacteria, for example, when strains of the *M. avium* complex, *M. malmoense,* and *M. xenopi* were tested against rifampicin and ethambutol individually in one study, most were resistant. When the two drugs were combined, however, 100% of *M. malmoense* strains, 86% of *M. xenopi* strains, and 31% of *M. avium* complex strains were susceptible (18). The meaning of *in vitro* susceptibility results to the fluoroquinolones, clofazimine, and the rifamycins (especially rifabutin) is subject to the same qualifications and reservations. Some species appear susceptible, others resistant, to selected critical concentrations. What this means in terms of response in the patient with disease is unknown. The clinical efficacy of clofazimine and amikacin (except the latter in the setting of rapidly growing mycobacterial disease) is unproven.

1.2. Treatment

Clinical features and radiographic appearance of pulmonary infection are similar, irrespective of the species of mycobacterium involved. The majority of pa-

Table 10.1 Susceptibility of opportunistic mycobacteria to individual drugs *in vitro*

Species (Test concentration)	Streptomycin (2 μg/ml)	Isoniazid (1.0 μg/ml)	Rifampicin (1.0 μg/ml)	Ethambutol (5 μg/ml)	Clarithromycin (8 μg/ml)	Ethionamide (5 μg/ml)	Capreomycin (10 μg/ml)	Ciprofloxacin (2 μg/ml)
M. kansasii	B[a]	V	S	S	S	S	B	S
M. avium complex	R	R	R	R	S	S	R	V
M. xenopi	S	R	V	R	S	S	S	S
M. malmoense	R	R	V	V	S	S	R	S
M. abscessus	R	R	R	R	S	R	R	R
M. chelonae	R	R	R	R	S	R	R	R
M. fortuitum	R	R	R	R	V	R	V	S

[a]S = sensitive; R = resistant; B = borderline; V = variable.

Note: Pyrazinamide is tested by a different technique but all opportunistic mycobacteria are resistant.

tients have preexisting lung disease and some have conditions associated with reduced immune responsiveness. Heavy smoking with chronic obstructive lung disease, occupational dust exposure, and alcohol abuse are risk factors for *M. kansasii* and the *M. avium* complex. Peptic ulcer and previous gastroduodenal surgery are common (9,15–17), although whether or not these latter disorders are true associations has not been determined. Bronchiectasis is frequently found in elderly nonsmoking women with the nodular bronchiectasis form of disease due to *M. avium* complex. Patients present with a combination of one or more symptoms of fatigue, cough, sputum production, hemoptysis, weight loss, malaise, and increasing breathlessness. Radiological appearances generally do not allow the clinician to differentiate among the various nontuberculous mycobacterial species or from infection by *M. tuberculosis* (19). The recently described pattern of mid-lung field multifocal bronchiectasis and multiple small nodules seen on high-resolution computed tomography (CT) is relatively specific for the *M. avium* complex (20).

1.3. Treatment of M. kansasii Infection

The availability and use of rifampicin has resulted in marked improvement in the treatment of this disease. In one retrospective series, rifampicin was included in the treatment of all 30 of the patients and ethambutol in 26 (87%). The duration of treatment ranged from 3 to 24 months and there was 100% cure with no relapses in the mean follow-up period of 5 years. The authors suggested that therapy should usually be with rifampicin and ethambutol for 15 months, perhaps including ethionamide or prothionamide until the susceptibility of the strain to rifampicin and ethambutol was confirmed (9). More recently, the British Thoracic Society (BTS) has published the results of a prospective study of rifampicin and ethambutol given for 9 months. Most patients received a third or fourth drug for the first 2 or 3 months, treatment having been started before it had been ascertained that the infection was due to *M. kansasii* rather than *M. tuberculosis*. Of the 173 patients, only one had sputum still positive on culture during 2 of the last 3 months of treatment: this man admitted noncompliance. On average, patients gained 2.2 kg by the end of treatment. Two out of three consistently showed satisfactory clinical progress during and after treatment and, of the remainder, in only 20% was unsatisfactory progress attributed to *M. kansasii*. A quarter of the patients died during the period of the study, but none because of *M. kansasii* infection. Radiological "healing" occurred in 80% within 3 years of completing chemotherapy. During the 51-month period of follow-up after the end of chemotherapy, 10% of the patients produced two or more sputum specimens which were positive on culture. Factors contributing to this could be identified in half these (e.g., noncompliance with treatment, severe malnourishment, severe bronchiectasis, and steroid treatment). Another 25% of the radiological appearances suggested rein-

Table 10.2. Suggested daily regimens for the three common nontuberculous mycobacterial species causing chronic lung disease in the HIV uninfected host

Pathogen	Treatment	Duration
M. avium complex	Clarithromycin 500 mg BID	Culture (−)
	Ethambutol 25 mg/kg × 2 months, then 15 mg/kg	1 Year
	Rifampicin 600 mg	
	or	
	Rifabutin 300 mg	
M. kansasii	Regimen A: Rifampicin 600 mg	9–15 Months
	(UK) Ethambutol 15 mg/kg	
	or	
	Regimen B: Rifampicin 600 mg	18 Months or
	(U.S.A.) Ethambutol 25 mg/kg × 2	culture (−) 1 year
	months, then 15 mg/kg	
	Isoniazid 300 mg	
M. abscessus	Clarithromycin 500 mg BID	3 Months[a]
	Cefoxitin 3 g IV 96 h	2–4 Weeks[a]

[a] Designed to produce clinical improvement; curative therapy (probably 4–6 months) not possible.

fection rather than relapse. In the remaining quarter, no reason was apparent and they were considered as genuine relapses. All of the relapses/reinfections responded satisfactorily to further treatment with rifampicin and ethambutol (21).

Therefore, options for chemotherapy are either treatment with (1) ethambutol and rifampicin for perhaps 9 to 15 months or (2) isoniazid, rifampin, and ethambutol for 18 months or until culture negative 12 months (American Thoracic Society recommendations) (See Table 10.2) (22). Clarithromycin should be considered in cases of ethambutol or rifampicin resistance or intolerance. A successful regimen has been used for rifampicin-resistant *M. kansasii* (10) that utilizes a sulfonamide, but it antedates the availability of the newer quinolones and the newer macrolides. (The success of clarithromycin with *M. kansasii* is unproven, but is highly likely given the low MICs and its success with other nontuberculous mycobacteria.)

1.4. Treatment of Infection with the M. avium Complex

Outcome of treatment in the premacrolide era was not always satisfactory. Yeager and Raleigh used five or six drugs with only a 43% response rate, 11% dying from their *M. avium* complex infection, and 20% relapsing after the end of treatment (23). A similar relapse rate was found by Ahn et al. using a four-drug regimen (24). Even though a majority of patients showed an initial bacteriological response to a multiple-drug regimen, Dutt and Stead noted that only 46% were culture negative after 3–8 years (25). In another study by Etzkorn et al., no significant difference was noted in terms of sputum conversion rate among regimens containing three, four, or five drugs, sputum conversion sometimes taking up to 1 year (26). Hunter et al (15) found that although patients without symptoms could often be considered to have "benign" disease, some developed progressive disease, and those patients with symptoms who were not treated were likely to die. The same authors also found that when isoniazid, rifampicin, and streptomycin, or rifampicin, isoniazid, and ethambutol, were given for between 9 and 24 months, there was a satisfactory clinical, radiological, and bacteriological response in 84%. However, 14% relapsed within a year of the end of treatment and three out of four patients who did not respond to treatment died. When second- or third-line drugs or when four or more drugs were used, significant problems arose from toxicity, and noncompliance was common (15).

The introduction of the newer macrolides and rifabutin has markedly improved the prospects for successful therapy of *M. avium* complex lung disease. Both clarithromycin and azithromycin have been shown to have microbiologic *in vivo* activity (27,28,28a), and clarithromycin in multidrug regimens (29,30) has been shown to produce sputum conversion rates that generally exceed those reported in the premacrolide era. The duration of therapy in the two studies reported to date was until patients were sputum culture negative for 10–12 months. Essentially, no relapses were seen in patients who completed therapy with this length of sputum culture negativity with an average 2-year follow-up (29,30). The importance or advantage of rifabutin over rifampicin for *M. avium* complex lung disease is unanswered. With disseminated *M. avium* complex, rifabutin has been shown to be effective as a prophylactic agent (28a,31), in a short-term monotherapy trial (32), and in several combination studies (33,34,34a), activity considered unlikely or not present with rifampicin. Thus, it seems likely that rifabutin will contribute more to a multidrug regimen with a macrolide than rifampicin. This must be balanced, however, by the greater incidence of side effects (especially leukopenia, polymyalgia, and uveitis) seen with rifabutin compared to rifampicin. The recommendations of the recently revised American Thoracic Society Statement on nontuberculous mycobacteria (22) are that clarithromycin, ethambutol, and rifampicin or rifabutin be given as initial therapy for *M. avium* complex lung disease administered until the patient is culture negative 10–12 months on therapy (RJW, unpublished data) (see Table 10.2). The use of rifabutin and likely intermittent streptomycin are reasonable components to consider for patients who have failed previous therapy, especially if rifampicin was part of their regimen.

The role of surgical resection remains controversial. It should be considered in patients with localized disease who fail drug therapy, and can be curative in this setting although surgical complications are frequent.

1.5. Treatment of M. malmoense Infection

In two retrospective studies, it was found that patients treated for 18–24 months with regimens including ethambutol and rifampicin did better than those in whom other regimens with short durations of treatment were used (17,35). Again, it was noted that therapy with second- or third-line drugs was poorly tolerated and gave poor results. The potential benefit of ciprofloxacin or clarithromycin has not yet been adequately demonstrated *in vivo,* although results *in vitro* (35a) and results of therapy with other mycobacterial species (29,30,34) suggest that they will be helpful.

Resection of the affected lobe, as with patients with *M. avium* complex, remains an option in patients who do not respond satisfactorily to chemotherapy and who are fit enough for surgery (17). However, chemotherapy should be continued after surgery for at least 18 months. Over 10% of patients with *M. malmoense* pulmonary infection will die from causes other than mycobacterial infection and about 10% will die because of their *M. malmoense* infection.

1.6. Treatment of M. xenopi Infection

This condition is, like infection with the *M. avium* complex, more difficult to treat and cure than is infection with *M. kansasii* or *M. malmoense*. Constrini et al. reported only a 50% response rate using regimens which usually contained isoniazid and ethambutol (36), whereas Smith and Citron reported 70% cure in patients treated for between 8 and 24 months, follow-up extending for a minimum of 3 years (37). One of the 11 patients died despite treatment. Banks et al. reported that only 23% were cured after chemotherapy. In that survey, treatment with rifampicin and isoniazid, plus ethambutol and streptomycin, was preferable to regimens based on second- and third-line drugs. Despite chemotherapy, disease progressed in 10% of patients. Initial response followed by relapse was found in 26%, whereas 16% died from *M. xenopi* infection. When surgery was possible in those who showed poor response or relapse, cure resulted from a combination of surgery and continued chemotherapy. Disease progressed relentlessly in the four patients who were not given chemotherapy (16).

Clarithromycin has been reported to be active *in vitro* against *M. xenopi* (37a), and may improve on the relatively poor results reported with multidrug regimens with or without surgery in the past. One preliminary report by Dautzenberg reported sputum conversion in 9/11 (82%) of patients at 3 months on a clarithromycin-containing regimen, and 11/11 (100%) at 6 months. Long term follow-up is still pending, however (37a).

1.7. Treatment of Infection Due to Rapidly Growing Mycobacteria

Pulmonary disease due to rapidly growing mycobacteria shows considerable geographical variation in incidence and is usually third behind the *M. avium* complex and *M. kansasii* as a cause of nontuberculous lung disease in the United States (38,39). It is, however, currently the most difficult to treat (3). Approximately 80% of disease is due to *M. abscessus* and 15% to *M. fortuitum* (3). Pulmonary disease due to *M. abscessus* is currently not curable with drugs. Courses of clarithromycin or amikacin plus cefoxitin often result in clinical improvement but fail to eradicate the causative organism. With *M. abscessus* disease that follows prior granulomatous disease such as treated *M. tuberculosis*, the disease often stays localized to that area and is amenable to surgery (3). For the rare patient with underlying achalasia, esophageal repair often results in improvement or resolution of the lung disease. For the majority of patients with bilateral disease, surgical resection is not an option. For patients with lung disease due to *M. fortuitum,* multiple drugs are available for therapy such as clarithromycin (40), fluorinated quinolones (41), sulfonamides (42), amikacin (42), cefoxitin (42,43), and imipenem (43,44), although no clinical trials and few case reports (32,44–47) have been published due to the infrequent nature of this species as a lung pathogen.

1.8. Treatment of Pulmonary Infections Caused by Other Nontuberculous Mycobacteria

Pulmonary disease due to nontuberculous mycobacteria such as *M. szulgai, M. simiae,* and *M. gordonae* is very rare (2,4,5). As a result, it is difficult to offer any clinically derived advice on treatment. What reports there have been tend to be single case reports and it is not possible to say whether treatment did have any effect on the outcome. However, treatment with the drugs to which the organism is susceptible *in vitro* would seem to be logical in the current state of knowledge, as would the addition of clarithromycin, which is generally the most active drug *in vitro* against these species (47a).

1.9. Summary of the Treatment of Pulmonary Disease in the HIV (−) Host

In the current state of knowledge and including the *in vitro* work on drug synergy and the results of clinical studies described above, daily treatment with clarithromycin, ethambutol, and rifampicin or rifabutin until culture negative 10–12 months on treatment should be the first-line regimen for the *M. avium* complex (29,30). This is the recommended regimen of the recently revised and approved (1997) ATS Statement on the Nontuberculous Mycobacteria. For patients who fail to show any response in 6 months or remain culture positive at 12 months, the clinician should consider adding rifabutin in place of rifampicin (if the latter was being given) (30) and add one or more of streptomycin or amikacin, ciprofloxacin, or perhaps ethionamide. Surgical resection should be considered for patients with localized disease. For patients with *M. kansasii*, rifampicin plus ethambutol for 9–15 months, or isoniazid, rifampicin, and ethambutol for 18 months or until culture negative 12 months are the current best regimens, recognizing the exceptional antimycobacterial agent clarithromycin has yet to be assessed by clinical trial for this species. For other species of nontuberculous mycobacteria, the same regimen used for the *M. avium* complex may be a reasonable start, tempered by available drug susceptibilities. In patients who do not respond to chemotherapy but are fit enough to tolerate surgery, the resection of the affected lobe and continuation of chemotherapy offers a good chance of cure.

For the rapidly growing nontuberculous mycobacteria (especially *M. abscessus*), the newer drugs such as the quinolones and the newer macrolides would appear to be the most useful. However, with disease due to *M. abscessus,* microbiologic cure is not attainable with available drug therapy.

1.10. Treatment of Disseminated M. avium *Complex in Patients with HIV Infection*

Infection with nontuberculous mycobacteria in the setting of HIV disease is not usually confined to the lung and is frequently associated with bacteremia. Diagnosis is usually made by blood cultures (48). Treatment of disseminated *M. avium* complex has improved dramatically in the past few years thanks to a series of controlled or comparative clinical treatment trial. Optimal treatment regimens, however, are still evolving. Short-term monotherapy trials have shown clarithromycin (49,50), azithromycin (51), ethambutol (52), and rifabutin (32) to decrease levels of *M. avium* bacteremia, and most clinicians agree that a macrolide plus ethambutol are an essential component of the usual treatment regimen. The optimal regimen for clarithromycin is 500 mg BID (higher doses are associated with a higher mortality) (50), whereas the optimal dose of azithromycin has not been determined. A three-drug regimen of clarithromycin, ethambutol, and rifabutin was recently shown to be much more effective in clearing bacteremia and prolonging survival than the four-drug regimen of rifampicin, ethambutol, ciprofloxacin, and clofazimine (34). This would likely be the preferred regimen were it not for the complex relationship between clarithromycin, rifabutin, and the recently introduced antiretroviral protease inhibitors retonavir, saquinavir, and indinavir related to their differing influences on the cytochrome P-450 system that metabolizes all three of the drugs. In general, rifabutin induces the metabolism of clarithromycin and the protease inhibitors, whereas the latter two inhibit the metabolism of rifabutin. This leaves the choice of rifabutin as the third agent somewhat in doubt. The place of ciprofloxacin, clofazimine, and an aminoglycoside, all of which have been used, remains uncertain. Their greatest use will likely be in patients intolerant of the first-line agents or with a macrolide-resistant organism. Treatment should last for the life of the patient, until more is learned as patients survive longer with better drugs.

1.11. Prophylaxis of Disseminated M. avium Complex Disease in Patients with AIDS

Long-term studies have shown that up to 40% of patients with AIDS will develop disseminated *M. avium* complex infection. Major risk factors include the degree of fall in the CD4 lymphocyte count and the presence of previous episodes of opportunistic infections. Recently completed clinical trials have shown that daily rifabutin (300 mg) (31), twice-daily clarithromycin (500 mg twice daily) (53), once-weekly azithromycin (1200 mg) (54), or the combination of once-weekly azithromycin (1200 mg) plus daily rifabutin (300 mg) (54), are all effective regimens in reducing the incidence of bacteremia. In patients who fail prophylaxis, up to 40% of those on clarithromycin (53) but 0% of those on rifabutin (31) will have developed resistance to the treatment drug. Most clinicians and treatment groups now recommend prophylaxis with one of these agents in patients with CD4 counts of < 75 cells, especially with a history of a prior opportunistic infection.

2. Extrapulmonary Disease

Nontuberculous mycobacteria can infect a wide variety of extrapulmonary sites. It is convenient to consider superficial lymph node infections separately, as they form a reasonably distinct entity.

2.1. Superficial Lymph Node Infections

These are caused primarily by strains of the *M. avium* complex or *M. malmoense* and occur predominantly in the cervical lymph nodes of children. The nodes are usually unilateral and may be "hot" or "cold." There is little systemic upset and the chest radiograph is clear. Routine hematological tests are unhelpful. Histologically, the appearances are identical with disease due to *M. tuberculosis* and this can lead to unnecessary treatment with toxic antituberculosis drugs. Skin test antigens have been prepared from the strains most often implicated, and a differential Mantoux test using these antigens can assist the diagnosis (these antigens are not currently commercially available in the United States) (55).

The treatment of choice is total excision of the affected nodes; antimycobacterial chemotherapy is not indicated. Incision and drainage alone is not recommended, as it can leave a discharging sinus which can eventually cause ugly scarring (56). Clarithromycin with or without rifabutin should be considered in children with disease recurrence or if surgical excision is incomplete (55a,55b) (e.g., when the disease involves the facial nerve). For children able to undergo visual testing, ethambutol may be preferred as the second drug (55c).

2.2. Other Extrapulmonary Disease

These are primarily caused by the rapidly growing mycobacteria. For example, *M. chelonae* and *M. abscessus* cause a wide variety of extrapulmonary disease, including cellulitis, tenosynovitis, synovitis, osteomyelitis, corneal infections, surgical wound infections, catheter-related sepsis, and disseminated cutaneous disease (57–59). The recent introduction of the new macrolides has made almost all skin and soft tissue disease due to the rapidly growing mycobacteria readily treatable, in contrast to the relatively bleak picture with *M. abscessus* lung disease. More specifically, clarithromycin is the first oral agent with activity against all initial isolates of *M. chelonae* and *M. abscessus* (40). In a monotherapy trial of clarithromycin 500 mg twice daily for 6 months in patients with cutaneous disease due to *M. chelonae,* 14/14 patients responded clinically with all being culture negative within 4 weeks. With a minimum follow-up after therapy of 6 months, all patients had either died of unrelated causes or remained culture negative (60). Although only case reports are available for treatment of *M. abscessus* cutaneous disease with clarithromycin, the results have generally been excellent (61,62). Clarithromycin monotherapy will not provide all the answers for *M. chelonae* and *M. abscessus,* as acquired resistance appears to occur in approximately 10% of patients treated for disseminated *M. chelonae* disease (60) and an estimated similar number of patients with *M. abscessus* lung disease (14). This resistance relates to a point mutation involving adenine 2058 or 2059 in the peptidyltransferase region of the 23S ribosomal RNA gene, as has been described for *M. intracellulare* and *M. avium* (12,13). Thus, combination therapy will be needed to prevent this resistance. For *M. chelonae* the best combination of drugs are the injectables tobramycin and imipenem (42–44,60), whereas for *M. abscessus,* it is amikacin and cefoxitin or imipenem (42,43).

For treatment of skin and soft tissue infections due to *M. fortuitum, in vitro* studies and case reports have shown a number of drugs to be effective. This included up to 100% of isolates for the injectable agents amikacin and imipenem, and for the oral agents of fluorinated quinolones (ciprofloxacin and ofloxacin) and the sulfonamides. Approximately 80% of isolates are inhibited by 4 μg/ml or less of clarithromycin (40), 50% by the tetracycline analogs doxycycline and minocycline (42), and 70% by 32 μg/ml of cefoxitin (42,43,63). For serious disease, patients usually receive an initial course of amikacin plus cefoxitin or imipenem for several weeks, combined with an oral agent which is continued for approximately 6 months (64,65). For lesser disease, completely oral therapy is given for 3–6 months. Follow-up cultures are essential to monitor outcome, and initial *in vitro* drug susceptibilities are needed to guide therapy as susceptibility among the taxonomic groups of *M. fortuitum* to the three commonly used drugs (doxycycline, clarithromycin, and cefoxitin) is variable (40,42,43). Recent studies with tetracycline (66) have suggested that this resistance may relate to tetracycline-resistance genes (*tet K/L, Otr A, B, C*) previously known in *Streptomyces* sp. and gram-positive bacteria. Acquired mutational resistance on therapy seems to be greatest with ciprofloxacin and amikacin (67,68), so monotherapy with these agents should be avoided when possible. Acquired resistance to the remaining agents (sulfonamides, imipenem, clarithromycin, doxycycline, cefoxitin) has not been reported following monotherapy [although resistance to clarithromycin with other species has been reported (12,13) as already noted].

Surgery remains an important part of the treatment of extrapulmonary disease due to rapidly growing mycobacteria (69), but its role is diminishing as the number and quality of drugs for therapy increases. This is especially true for the macrolides, as they offer potential therapy for *M. chelonae* and *M. abscessus* where surgery plus amikacin were for many years the mainstay of therapy (64,67). When infection occurs in the presence of a foreign body such as a breast implant, peritoneal or intravenous catheter, and nasolacrimal duct stents, the foreign body should be removed. With soft tissue or bone necrosis, extensive sternal osteomyelitis following cardiac surgery and/or abscess formation, surgical debridement is important to the clinical response. For patients with posttraumatic infections of the cornea, excision of the cornea is often required to cure the disease. These concepts are not dissimilar to those for bacterial disease involving the same tissues.

3. General Treatment Summary

In the current state of knowledge, it is quite acceptable to treat infections caused by the *M. avium* complex, *M. malmoense,* or *M. xenopi* with clarithromycin, rifampicin or rifabutin, and ethambutol until culture negative for 12 months. The ongoing British Thoracic Society multicenter, prospective, clinical trial comparing ethambutol plus rifampicin with ethambutol, rifampicin, and isoniazid, given for 24 months, in the treatment of pulmonary infection with the above three species should, in due course, indicate whether there is any point in including isoniazid in the regimen. A further study of pulmonary infection with the *M. avium* complex, *M. malmoense,* or *M. xenopi* by the British Thoracic Society incorporating immunotherapy with *M. vaccae* and also incorporating a comparison of ciprofloxacin and clarithromycin as third drugs added to the basic rifampicin and ethambutol regimen should help resolve some of the current dilemmas in therapy. Treatment for pulmonary infections caused by *M. kansasii* with rifampicin and ethambutol given for 9–15 months should suffice, although the American Thoracic Society recommends a regimen of isoniazid, rifampicin, and ethambutol for 18 months or until culture negative for 12 months. If resistance to rifampicin or drug intolerance to any of the agents develops, clarithromycin should be substituted for the problem drug. For those with AIDS and disseminated *M. avium* complex, clarithromycin or azithromycin, ethambutol, and, if possible, rifabutin is the current optimum regimen administered for the lifetime of the patient. Prophylaxis to prevent disseminated *M. avium* complex is now the standard of care in high-risk patients with AIDS, with drug choice being related to cost and individual treatment factors. Rifampicin should be avoided in patients on protease inhibitors, with rifabutin or clarithromycin offering reasonable alternatives.

References

1. Campbell IA and Jenkins PA (1995) Opportunist mycobacterial infections. In: Brewis RAL, Corrin B, Geddes DM, Gibson GJ, eds. Respiratory Medicine. London: WB Saunders.

2. Wallace RJ Jr, Swenson JM, Silcox VA, Good RC, Tischen JA, Stone MS (1983) Spectrum of disease due to rapidly growing mycobacteria. Rev Infect Dis 5:657.

3. Griffith DE, Girard WM, Wallace RJ, Jr (1993) Clinical features of pulmonary disease caused by rapidly growing mycobacteria. Am Rev Respir Dis 147:1271.

4. Schaefer WB, Wolinsky E, Jenkins PA, Marks J (1973) *Mycobacterium szulgai,* a new pathogen. Am Rev Respir Dis 108:1320.

5. Rose HD, Dorff GJ, Louwasser M, Sheth NK (1982) Pulmonary and disseminated *Mycobacterium simiae* infection in humans. Am Rev Respir Dis 126:1110.

6. Clague H, Hopkins CA, Roberts C, Jenkins PA (1985) Pulmonary infection with *Mycobacterium gordonae* in the presence of bronchial carcinoma. Tubercle 66:61.

7. Tsukamura M, Kia N, Otsuka W, Shimoide H (1983) A study of the taxonomy of the *Mycobacterium nonchromogenicum* complex and report of six cases of lung infection due to *Mycobacterium nonchromogenicum.* Microbiol Immunol 27:219.

8. Marks J (1961) The design of sensitivity tests on tubercle bacilli. Tubercle 42:314.

9. Banks J, Hunter AM, Campbell IA, Jenkins PA, Smith AP (1983) Pulmonary infection with *Mycobacterium kansasii* in Wales, 1970–9: review of treatment and response. Thorax 38:271.

10. Wallace RJ Jr., Dunbar D, Brown BA, Onyi G, Dunlap R, Ahn CH, Murphy DT (1994) Rifampin-resistant *Mycobacterium kansasii*. Clin Infect Dis 18:736.

11. Heifits L, Mor N, VanderKolk J (1993) *Mycobacterium avium* strains resistant to clarithromycin and azithromycin. Antimicrob Agents Chemother 37:2364.

12. Meier A, Heifets L, Wallace RJ Jr., Zhang Y, Brown BA, Sander P, Böttger EC (1996) Molecular mechanisms of clarithromycin resistance in *Mycobacterium avium:* observation of multiple 23S rDNA mutations in a clonal population. J Infect Dis 174:354.

13. Meier A, Kirschner P, Springer B, Steingrube VA, Brown BA, Wallace RJ Jr, Böttger EC (1994) Identification of mutations in 23S rRNA gene of clarithromycin-resistant *Mycobacterium intracellulare*. Antimicrob Agents Chemother 38:381.

14. Wallace RJ Jr, Meier A, Brown BA, Zhang Y, Sander P. Onyi GO, Böttger EC (1996)

Genetic basis for clarithromycin resistance among isolates of *Mycobacterium chelonae* and *Mycobacterium abscessus*. Antimicrob Agents Chemother 40:1676.

15. Hunter AM, Campbell IA, Jenkins PA (1981) Treatment of pulmonary infections caused by mycobacteria of the *Mycobacterium avium-intracellulare* complex. Thorax 36:326.

16. Banks J, Hunter AM, Campbell IA, Jenkins PA, Smith AP (1984) Pulmonary infection with *Mycobacterium xenopi:* review of treatment and response. Thorax 39:376.

17. Banks J, Jenkins PA, Smith AP (1985) Pulmonary infection with *Mycobacterium malmoense*—a review of treatment and response. Tubercle 66:197.

17a. Sison JP, Yao Y, Kemper CR, et al. (1996) Treatment of *Mycobacterium avium* complex infection: Do the results of *in vitro* susceptibility tests predict therapeutic outcome in humans? J Infect Dis 173:677.

18. Banks J, Jenkins PA (1987) The effect of combined versus single antituberculosis drugs on *in vitro* sensitivities patterns of non-tuberculous mycobacteria. Thorax 42:838.

19. Christensen EE, Dietz GW, Ahn CH, Chapman JS, Murry RC, Anderson J, Hurst GA (1981) Initial roentgenographic manifestations of pulmonary *Mycobacterium tuberculosis, M. kansasii*, and *M. intracellulare* infections. Chest 80:132.

20. Swensen SJ, Hartman TE, Williams DE (1994) Computed tomographic diagnosis of *Mycobacterium avium-intracellulare* complex in patients with bronchiectasis. Chest 105:49.

21. BTS Research Committee (1994) *Mycobacterium kansasii* pulmonary infection: a prospective study of the results of 9 months of treatment with rifampicin and ethambutol. Thorax 49:442.

22. Wallace RJ Jr, O'Brien R, Glassroth J, Raleigh J, Dutt A (1990) Diagnosis and treatment of disease caused by nontuberculous mycobacteria. NTM Statement, American Thoracic Society.

23. Yeager H Jr, Raleigh JW (1973) Pulmonary disease due to *Mycobacterium intracellulare*. Am Rev Respir Dis 108:547.

24. Ahn CH, Ahn SS, Anderson RA, Murphy DT, Mammo A (1986) A four drug regimen for initial treatment of cavitary disease caused by *Mycobacterium avium* complex. Am Rev Respir Dis 134:438.

25. Dutt AK, Stead WW (1979) Long-term results of medical treatment in *Mycobacterium intracellulare* infection. Am J Med 67:449.

26. Etzkorn ET, Aldarondo S, McAllister CK, Matthews J, Ognibene AJ (1986) Medical therapy of *Mycobacterium avium-intracellulare* pulmonary disease. Am Rev Respir Dis 134:442.

27. Wallace RJ Jr, Brown BA, Griffith DE, Girard WM, Murphy DT, Onyi GO, Steingrube VA, Mazurek GH (1994) Initial clarithromycin monotherapy for *Mycobacterium avium-intracellulare* complex lung disease. Am J Respir Crit Care Med 149:1335.

28. Griffith DE, Brown BA, Girard WM, Murphy DT, Wallace RJ Jr (1996) Azithromycin activity against *Mycobacterium avium* complex lung disease in HIV negative patients. Clin Infect Dis 23:983.

28a. Havlir DV, Dubé MP, Sattler FR, et al. (1996) Prophylaxis against disseminated *Mycobacterium avium* complex with weekly azithromycin, daily rifabutin, or both. N Engl J Med 335:392.

29. Dautzenberg B, Piperno D, Diot P, Truffot-Pernot C, Chauvin J-P, and the Clarithromycin Study Group of France (1995) Clarithromycin in the treatment of *Mycobacterium avium* lung infections in patients without AIDS. Chest 107:1035.

30. Wallace RJ Jr, Brown BA, Griffith DE, Girard WM, Murphy DT (1996) Clarithromycin regimens for pulmonary *Mycobacterium avium* complex. The first 50 patients. Am J Respir Crit Care Med 153:1766.

31. Nightingale SD, Cameron DW, Gordin FM, Sullam PM, Cohn DL, Chaisson RE, Eron LJ, Sparti PD, Bihari B, Kaufman DL, Stern JJ, Pearce DD, Weinberg WG, La Marca A, Siegal FP (1993) Two controlled trials of rifabutin prophylaxis against *Mycobacterium avium* complex infection in AIDS. N Engl J Med 329:828.

32. Dautzenberg B, Castellani P, Pellegrin J-L, Vittecoq D, Truffot-Pernot C, Pirotta N, Sassella D (1996) Early bactericidal activity of rifabutin versus that of placebo in treatment of disseminated *Mycobacterium avium* complex bacteremia in AIDS patients. Antimicrob Agents Chemother 40:1722.

33. Sullam PM, Gordin FM, Wynne BA, and the Rifabutin Treatment Group (1994) Efficacy of rifabutin in the treatment of disseminated infection due to *Mycobacterium avium* complex infection. Clin Infect Dis 19:84.

34. Shafrin SD, Singer J, Zarowny DP, et al. (1996) A comparison of two regimens for the treatment of *Mycobacterium avium* complex bacteremia in AIDS: rifabutin, ethambutol and clarithromycin versus rifampin, ethambutol, clofazimine and ciprofloxacin. N Engl J Med 335:377.

34a. Dautzenberg, B (1996) Rifabutin in the treatment of *Mycobacterium avium* complex infection: Experience in Europe. Clin Infect Dis 22(Suppl 1):S33.

35. France AJ, McLeod DT, Calder MA, Seaton A (1987) *Mycobacterium malmoense* infections in Scotland: an increasing problem. Thorax 42:593.

35a. Hoffner SE, Hjelm U, Källenius (1993) Susceptibility of *Mycobacterium malmoense* to antibacterial drugs and drug combinations. Antimicrob Agents Chemother 37:1285.

36. Costrini AM, Mahler DA, Gross WM, et al. (1981) Clinical and roentographic features of nosocomial pulmonary disease due to *Mycobacterium xenopi*. Am Rev Respir Dis 123:104.

37. Smith MJ, Citron KM (1983) Clinical review of pulmonary disease caused by *Mycobacterium xenopi*. Thorax 38:373.

37a. Dautzenberg B, Papillon F, Lepitre M, Truffot-Pernod CH, Chauvin JP (1993) *Mycobacterium xenopi* infections treated with clarithromycin-containing regimens. Abstract no. 1125 Thirty-third Interscience Conference of Antimicrob Agents and Chemother, New Orleans, LA, USA.

38. O'Brien RJ, Geiter LJ, Snider DE Jr (1987) The epidemiology of non-tuberculous mycobacterial diseases in the United States. Am Rev Respir Dis 135:1007.

39. Kennedy TP, Weber DJ (1994) Non-tuberculous mycobacteria: an underappreciated cause of geriatric lung disease. Am J Respir Crit Care Med 149:1654.

40. Brown BA, Wallace RJ Jr, Onyi GO, DeRosas V, Wallace RJ III (1992) Activities of four macrolides, including clarithromycin, against *Mycobacterium fortuitum, Mycobacterium chelonae* and *M. chelonae*-like organisms. Antimicrob Agents Chemother 36:180.

41. Wallace RJ Jr, Bedsole G, Sumter G, Sanders CV, Steel LC, Brown BA, Smith J, Graham DR (1990) Activities of ciprofloxacin and ofloxacin against rapidly growing mycobacteria with demonstration of acquired resistance following single-drug therapy. Antimicrob Agents Chemother 34:65.

42. Swenson JM, Wallace RJ Jr, Silcox VA, Thornsberry C (1985) Antimicrobial susceptibility of five subgroups of *Mycobacterium fortuitum* and *Mycobacterium chelonae*. Antimicrob Agents Chemother 28:807.

43. Wallace RJ Jr., Brown BA, Onyi GO (1991) Susceptibilities of *Mycobacterium fortuitum* biovar. *fortuitum* and the two subgroups of *Mycobacterium chelonae* to imipenem, cefmetazole, cefoxitin, and amoxicillin-clavulanic acid. Antimicrob Agents Chemother 35:773.

44. Yew WW, Kwan SYL, Wong PC, Lee J (1990) Ofloxacin and imipenem in the treatment of *Mycobacterium fortuitum* and *Mycobacterium chelonae* lung infections. Tubercle 71:131.

45. Burns DN, Rohatgi PK, Rosenthal R, Seiler M, Gordin FM (1990) Disseminated *Mycobacterium fortuitum* successfully treated with combination therapy including ciprofloxacin. Am Rev Respir Dis 142:468.

46. Pacht ER (1990) *Mycobacterium fortuitum* lung abscess: resolution with prolonged trimethoprim/sulfamethoxazole therapy. Am Rev Respir Dis 141:1599.

47. Ichiyama S, Tsukamura M (1987) Ofloxacin and the treatment of pulmonary disease due to *Mycobacterium fortuitum*. Chest 92:1110.

47a. Brown BA, Wallace RJ Jr, Onyi GO (1992) Activities of clarithromycin against eight slowly growing species of nontuberculous mycobacteria, determined by using a broth microdilution MIC system. Antimicrob Agents Chemother 36:1987–1990.

48. Hawkins CC, Gold JWM, Whimbey E, Kiehn TE, Brannon P, Cammarata R, Brown AE, Armstrong D (1986) *Mycobacterium avium* complex infections in patients with the acquired immunodeficiency syndrome. Ann Intern Med 105:184.

49. Dautzenberg B, Truffot C, Legris S, et al. (1991) Activity of clarithromycin against *Mycobacterium avium* infection in patients with the acquired immune deficiency syndrome. Am Rev Respir Dis 144:564.

50. Chaisson RE, Benson CA, Dube MP, et al. and the Aids Clinical Trial Group Protocol 157 Study Team (1994) Clarithromycin therapy for bacteremic *Mycobacterium avium* complex disease. A randomized, double-blind, dose ranging study in patients with AIDS. Ann Intern Med 121:905.

51. Young LS, Wiviott L, Wu M, Kolonski P, Bolan R, Inderlied CB (1991) Azithromycin for treatment of *Mycobacterium avium-intracellulare* complex infection in patients with AIDS. Lancet 338:1107.

52. Kemper CA, Havlir D, Haghighat D, et al. (1994) The individual microbiologic effect

of three antimycobacterial agents, clofazimine, ethambutol, and rifampin, on *Mycobacterium avium* complex bacteremia in patients with AIDS. J Infect Dis 170:157.

53. Pierce M, Crampton S, Henry D, et al. (1996) A randomized trial of clarithromycin as prophylaxis against disseminated *Mycobacterium avium* complex infection in patients with advanced acquired immunodeficiency syndrome. N Engl J Med 335:384.

54. Havlir DV, Dubé MP, Sattler FR, et al. and the California Collaborative Treatment Group (1996) Prophylaxis against disseminated *Mycobacterium avium* complex with weekly azithromycin, daily rifabutin, or both. N Engl J Med 335:392.

55. Stanford JL. Newer tuberculins, their reactivity profile and impact on treatment in developing countries. Indian Paediatrics (1991). Ed. V. Seth.

55a. Jadavji T, Wong A (1996) Atypical mycobacteria cervical adenitis in normal children—Is clarithromycin effective? In: Program and Abstracts of The Third International Conference on the Macrolides Azalides and Streptogramins, Lisbon, Portugal, January 24–26, Abstract 7.23, p. 67.

55b. Berger CH, Pfyffer GE, Nadal D (1996) Drug therapy of lymphadenitis with non-tuberculous mycobacteria (NTM). In: Program and Abstracts of The Third International Conference on the Macrolides Azalides and Streptogramins, Lisbon, Portugal, January 24–26, Abstract 7.22, p. 67.

55c. Green PA, von Reyn CF, Smith RP, Jr (1993) *Mycobacterium avium* complex parotid lymphadenitis: Successful therapy with clarithromycin and ethambutol. Ped Infect Dis J 12:615–616.

56. White MP, Bangash H, Goel KM, Jenkins PA (1986) Non-tuberculous mycobacterial lymphadenitis. Arch Dis Child 61:368.

57. Wallace RJ Jr, Swenson JM, Silcox VA, Good RC, Tschen JA, Stone MS (1983) Spectrum of disease due to rapidly growing mycobacteria. Rev Infect Dis 5:657.

58. Wallace RJ Jr, Musser JM, Hull SI, Silcox VA, Steele LC, Forrester GD, Labidi A, Selander SK (1989) Diversity and sources of rapidly growing mycobacteria associated with infections following cardiac surgery. J Infect Dis 159:708.

59. Wallace RJ Jr, Brown BA, Onyi GO (1992) Skin soft tissue, and bone infections due to *Mycobacterium chelonae:* importance of prior corticosteroid therapy, frequency of disseminated infections, and resistance to oral antimicrobials other than clarithromycin. J Infect Dis 166:405.

60. Wallace RJ Jr., Tanner D, Brennan PJ, Brown BA (1993) Clinical trials of clarithromycin for cutaneous (disseminated) infection due to *Mycobacterium chelonae.* Ann Intern Med 119:482.

61. Mushat DM, Witzig RS (1995) Successful treatment of *Mycobacterium abscessus* infections with multidrug regimens containing clarithromycin. Clin Infect Dis 20:1441.

62. Maxson S, Schutze GE, Jacobs RF (1994) *Mycobacterium abscessus* osteomyelitis: treatment with clarithromycin. Infect Dis Clin Prac 3:203.

63. Cynamon MH, Palmer GS (1982) *In vitro* susceptibility of *Mycobacterium fortuitum*

to *N*-formimidoyl thienamycin and several cephamycins. Antimicrob Agents Chemother 22:1079.

64. Wallace RJ Jr, Swenson JM, Silcox VA, Bullen MG (1985) Treatment of nonpulmonary infections due to *Mycobacterium fortuitum* and *Mycobacterium chelonei* on the basis of *in vitro* susceptibilities. J Infect Dis 152:500.

65. Yew WW, Kwan SYL, Ma WK, Mok CK (1989) Combination of ofloxacin and amikacin in the treatment of sternotomy wound infection. Chest 95:1051.

66. Pang Y, Brown BA, Steingrube VA, Wallace RJ Jr., Roberts MC (1994) Tetracycline resistance determinants in *Mycobacterium* and *Streptomyces* species. Antimicrob Agents Chemother 38:1408.

67. Wallace RJ Jr., Hull SI, Bobey DG, Price KE, Swenson JM, Steele LC, Christenson L (1985) Mutational resistance as the mechanism of acquired drug resistance to aminoglycosides and antibacterial agents in *Mycobacterium fortuitum* and *Mycobacterium chelonei*. Am Rev Respir Dis 132:409.

68. Franklin DJ, Starke JR, Brady MT, Brown BA, Wallace RJ Jr (1994) Chronic otitis media after tympanotomy tube placement caused by *Mycobacterium abscessus:* a new clinical entity? Am J Otol 15:313.

69. Dalovisio JR, Pankey GA, Wallace RJ Jr, Jones DB (1981) Clinical usefulness of amikacin and doxycycline in the treatment of infection due to *Mycobacterium fortuitum* and *Mycobacterium chelonei*. Rev Infect Dis 3:1068.

11

Preventive Therapy for Tuberculosis

Lawrence J. Geiter

1. Introduction

Most individuals infected by *Mycobacterium tuberculosis* do not immediately develop disease, and the interval between infection and illness can range from months to decades. The development of infectious, pulmonary tuberculosis, epidemiologically the most important form of disease because it is the source transmission and new infections, is usually the result of the reactivation of such a latent infection. This period of latency provides an opportunity for secondary prevention of tuberculosis. This strategy takes on increasing importance as the risk of infection with *M. tuberculosis* in a community decreases and the proportion of incident cases resulting from reactivation of old infections increases. Secondary prevention also takes on an increasing clinical importance as evidence increases that the development of active tuberculosis in persons also infected with HIV increases the speed of progression of the HIV infection. The concept of preventive therapy for tuberculin-positive individuals developed along with isoniazid and preventive therapy with isoniazid has become a central part of the national tuberculosis control program in the United States.

2. Clinical Trials of Isoniazid Preventive Therapy

Isoniazid was considered as a preventive therapy agent very shortly after its discovery. Lincoln first noted that children receiving isoniazid did not develop complications of a primary tuberculosis infection (1). Clinical trials of isoniazid to prevent the development of tuberculosis in previously infected individuals were then undertaken in the late 1950s in Italy (2), France (3), and the United States

(4). Since then, placebo-controlled trials of preventive therapy have been conducted in a wide variety of places and situations, including large cohorts of institutionalized individuals and household contacts of infectious cases and smaller studies of individuals infected in outbreak situations, where the time of infection and degree of exposure was better understood. The best of these studies are summarized in an excellent review of preventive therapy for tuberculosis (5).

In these trials, the average reduction in tuberculosis was about 60%, with a range of 25–92%. These results are based on a comparison of the groups to which the participants were randomized in an analysis by intent to treat. The variation in effectiveness in these studies is almost entirely attributable to the degree of adherence of the subjects with the isoniazid regimen. In studies where drug-taking could be very closely controlled, the reduction in the number of tuberculosis cases among those assigned to take isoniazid, compared to a placebo group, was 90% or more. Information from some of these studies indicates that the protection against tuberculosis conferred by isoniazid appears to last at least 19 years and may be lifelong (6).

The most recent, large controlled trial of preventive therapy with isoniazid was a study in Eastern European countries of individuals with fibrotic lesions but no signs of active tuberculosis (7). The trial was designed to evaluate different lengths of preventive therapy and demonstrated that a regimen of 3 months of isoniazid was not likely to be effective. Comparing the isoniazid regimens to a placebo group, those assigned to a 3-month isoniazid regimen had a 21% reduction in tuberculosis, the 6-month regimen a 65% reduction, and the 12-month regimen a 75% reduction. When the results were stratified by adherence with the regimen (adherence defined as taking 80% or more of the assigned medication), the protection conferred on adherent individuals by the 3- and 6-month regimens was similar to that for the groups as a whole, 31% and 69%, respectively, but the protection afforded by the 12-month regimen increased to 93%.

3. Preventive Therapy for Children

Soon after isoniazid came into use and its chemoprophylactic role in adults has been established, Hsu (8) conducted extensive studies of prophylactic chemotherapy in children, with two objectives: (a) Would isoniazid therapy of tuberculin reactions present early manifestations of the disease in childhood. (b) Would isoniazid therapy of tuberculin reactors prevent later manifestations of the disease in adolescence and adult. In her reports on a 20 year follow up, dealing with 1881 children, she reported a morbidity of only 3.2/1000. Among the 591 positive cases, none reactivated. These findings were confirmed in a later report (9) dealing with 30 year follow up. Of the 1882 children who were followed up, only a morbidity of 4.2/1000 was noted. Among the 1566 children with a subclinical infection, only 8 (4.5/1000) had experienced the overt disease.

4. Preventive Therapy for HIV-Infected Individuals

Human immunodeficiency virus (HIV) infection has been identified as the most potent risk factor for the development of tuberculosis, given infection with *M. tuberculosis.* One estimate places the risk of TB at over 100 times more likely in the HIV-infected person, compared with a seronegative individual (10). Selwyn estimated that 7–10% of tuberculin-positive IV drug users develop tuberculosis each year (11), but the risk is even higher for HIV-infected individuals recently infected with *M. tuberculosis.* In two tuberculosis outbreaks involving HIV-infected individuals, one reported from Italy (12) and one in the United States (13), 40% of those subsequently infected with *M. tuberculosis* developed tuberculosis in the first 2–4 months following infection.

The efficacy of preventive therapy with isoniazid in HIV-infected individuals was implied in the study by Selwyn. In that study, none of 13 tuberculin-positive individuals who had reported receiving preventive therapy with isoniazid developed tuberculosis during the observation period, compared with 7 of 13 individuals who reported not receiving preventive therapy. Selwyn reported further evidence of the efficacy in a follow-up study, where no cases occurred among 27 HIV-infected individuals who received 12 months of isoniazid preventive therapy, whereas 4 cases occurred among 27 patients who failed to complete therapy or who received no therapy (14). The short-term efficacy of preventive therapy in HIV-infected individuals has been more clearly demonstrated in a number of clinical trials. However, only one of these, by Pape et al., has been presented in a peer-reviewed study (15). Preliminary results from trials conducted in Zambia (16) and Haiti (17) have been presented at international meetings in abstract form. The efficacy of these for preventive therapy demonstrated in these studies has been similar to that observed in non-HIV-infected persons, 60–75% protective efficacy for regimens of isoniazid preventive therapy lasting from 6 to 12 months. However, it should also be noted that in most studies of isoniazid preventive therapy in countries with high rates of tuberculosis and a high risk of infection with *M. tuberculosis,* the survival of HIV-infected individuals receiving isoniazid preventive therapy was not significantly better than those who received placebo or no treatment.

Although a key argument for isoniazid preventive therapy for non-HIV-infected individuals is the public health importance of preventing future cases of infectious tuberculosis, particularly in those in occupations or living situations where they are more likely to transmit the disease to others, there may be little clinical benefit for those who are likely to receive prompt diagnoses and treatment for tuberculosis, when and if it develops. However, there may be a clear clinical benefit for those with HIV infection. First, as noted above, tuberculosis in HIV-infected individuals is more likely to progress rapidly. Second, even though there was no survival benefit in many studies, the Pape study noted a delayed progression to an autoimmune deficiency syndrome (AIDS)-defining illness. This finding is supported by a study by Whalen et al. (18), which showed that HIV-infected individuals who develop tuberculosis suffer an increase in the number rate of AIDS-defining opportunistic infections and had significantly reduced survival.

5. Toxicity of Isoniazid Preventive Therapy

Isoniazid is generally well tolerated and is one of the least toxic of the antituberculosis drugs. Most of the side effects are mild, and a study by Pitts estimated that approximately 5.4% of all patients receiving isoniazid experience an adverse reaction (19). The reactions included rash (2%), fever (1.2%), peripheral neuropathy (0.2%), and jaundice (0.6%). Peripheral neuropathy is most likely caused by interference with pyridoxine metabolism, and although it is uncommon at doses of 5 mg/kg or less, it is most likely to occur in persons with diabetes, uremia, alcoholism, and malnutrition and can be prevented by administering pyridoxine with the isoniazid.

The occurrence of isoniazid-induced hepatitis is of greatest concern, as severe and fatal hepatitis may occur. Elevation of serum aminotransferase levels (indicative of hepatic damage) may occur in 10–20% of patients receiving isoniazid, but these elevations are often transient, the patients remain asymptomatic, and aminotransferase levels return to normal without discontinuation of the drug (20). Isoniazid-induced hepatitis is rare under the age of 20, but the risk increases with age to a peak of 2–3% in the 50–64-year-old age group (21). Because of these findings, recommendations limited the use of preventive therapy to high-risk reactors (e.g., HIV-infected individuals and close contacts of infectious cases) and only those at moderate risk if under 35 years of age. Also, monthly monitoring is advised for those at higher risk of hepatotoxicity. Following these guidelines has been shown to reduce the risk of serious isoniazid hepatitis (22).

The risk of isoniazid-associated hepatitis mortality and the decision to recommend isoniazid preventive therapy has been a subject of continued debate. Conflicting decision analysis and cost–benefit analysis have been published supporting (23,24) and opposing (25,26) the use of isoniazid preventive therapy. However, it is important to bear in mind that the controversy has only concerned low-risk, adult, tuberculin reactors; few question that the benefits of preventive therapy with isoniazid far outweigh the risk of hepatitis for tuberculin reactors at a high risk of tuberculosis, such as those infected with HIV, household contacts, and recent skin test converters.

Some of the controversy over the use of preventive therapy has resulted from the estimate of the risk of isoniazid hepatitis mortality. A 57/100,000 risk of isoniazid hepatitis mortality was estimated in a United States Public Health Service (USPHS) surveillance study (21). Whereas this mortality risk was based on eight cases of fatal hepatitis that occurred during the study and were attributed to isoniazid, seven of the deaths occurred in one study site, Baltimore, Maryland. A subsequent study of deaths in Maryland revealed a significant increase in deaths due to liver disease in the general population during the study period. The risk of mortality at the other study sites was estimated as 9.4/100,000. Using program-related data in the United States, the isoniazid-associated hepatitis mortality risk in the United States was estimated to be about 14/100,000 persons starting preventive therapy or 23/100,000 persons completing preventive therapy (27). These estimates are much lower than the USPHS study derived figure most often used in the various cost–benefit and decision analyses published on preventive therapy. The authors warn that the estimates are very crude and suggest that the truly important question is which groups are actually at highest risk for toxicity and death. Data from their paper and other sources suggest that women, particularly black and Hispanic women in the postpartum period, may be at an increased risk of hepatotoxicity.

6. Current Recommendations for Preventive Therapy in the United States

6.1. Candidates for Preventive Therapy

There are a variety of guidelines on the application of preventive therapy with isoniazid, but, in the United States, the most authoritative guidelines are probably those jointly issued by the American Thoracic Society (ATS) and the Centers for Disease Control and Prevention (CDC), most recently updated in 1994 (28). In an attempt to balance the risks and benefits of preventive therapy, several categories for a positive tuberculin skin test would make individuals eligible for preventive therapy. A tuberculin skin test producing induration of ≥5 mm is defined as positive for (1) persons with HIV infection or unknown HIV status and risk factors for HIV infection, (2) persons who have had recent close contact with an infectious case of tuberculosis, and (3) persons with fibrotic lesions on chest radiograph that is consistent with tuberculosis. Others are considered to have positive tests if the induration is ≥10 mm and the person (1) is an injecting drug user known to be HIV negative, (2) has conditions known to increase the risk of developing tuberculosis, including diabetes mellitus, prolonged therapy high-dose therapy with adrenocorticosteroids and other immunosuppressive therapy, leukemia, lymphoma, and other malignancies, end-stage renal disease, weight loss ≥10% below ideal body weight, and gastrectomy, jejunoileal bypass, and other clinical conditions associated with rapid weight loss, (3) is a resident or an employee of high-risk congregate settings, (4) is born outside the United States and arrived within the last 5 years from countries with a high prevalence or incidence of tuberculosis, (5) is a member of a medically underserved, low-income population, including migrant farm workers and homeless persons, (6) is a member of a locally defined high-risk racial or ethnic minorities, or (7) is a child <4 years of age or an adolescent exposed to adults in high-risk categories. Finally, anyone with a tuberculin skin test result ≥15 mm would be considered positive. These classification and other recommendations on screening were updated by the Centers for Disease Control and Prevention in 1995 (29).

When considering preventive therapy for a person, first and foremost, the presence of active tuberculosis, pulmonary and extrapulmonary must be ruled out to avoid the possibility of monotherapy with isoniazid and the development of drug resistance. Candidates for preventive therapy should be excluded if they have already been adequately treated for tuberculosis or preventive therapy. Finally, individuals should be excluded from preventive therapy if there is a history of adverse reactions to isoniazid and contraindications to the use of isoniazid (especially a high risk for hepatic injury or the presence of peripheral neuropathy). Pregnant women should generally be excluded from preventive therapy, unless there is some special circumstance indicating a high risk for tuberculosis, as pregnancy does not seem to increase the risk for tuberculosis (note that isoniazid is safely administered during pregnancy for the treatment of tuberculosis). Once these individuals are excluded, recommendations for preventive therapy for individuals with positive tuberculin skin tests are then based on the age of the individual. The following individuals with positive tests should be considered for preventive therapy, regardless of age; (a) those who are HIV infected or have risk factors for HIV infection and are of unknown HIV-infection status, (b) close contacts of persons with infectious tuberculosis, (c) recent skin test converters, and (d) those who have any of the conditions described in category 2 in the previous paragraph. Individuals under the age of 35 should also be considered for preventive therapy if the are in categories 3–7 described above. Individuals under the age of 35 with a positive test but no risk factors (i.e., those with an induration of ≥15 mm) should be considered individually and the risks and benefits assessed.

Anergy to tuberculin has been defined as a problem when testing individuals with HIV infection or other immune-compromising conditions. Guidelines have been published that suggest testing with purified protein derivative (PPD) tuberculin and two companion delayed-type hypersensitivity (DTH) antigens to assess anergy (30). Any induration to a DTH antigen at 48–72 h is evidence of DTH responsiveness, and a failure to respond to all three antigens is evidence of anergy. These guidelines further suggest that anergic individuals with a personal risk of infection with *M. tuberculosis* of ≥10% should be considered for preventive therapy. However, the lack of standardization of many of the DTH antigens and the uncertainty about the implications of anergy needs to be considered in the decision to start preventive therapy. One decision analysis, weighing the risks and benefits of preventive therapy, concluded that preventive therapy would be justified for HIV-positive individuals, except black women and men (because of a possible increased risk for isoniazid-induced hepatitis) in the absence of a tuberculin test, even when the risk of infection is a low as 3–8% (31).

6.2. Recommended Regimens

Preventive therapy with isoniazid is usually prescribed at a dose of 10–15 mg/ kg body weight daily for children and adolescents, up to a maximum adult dose of 300 mg daily. For individuals requiring directly observed therapy to ensure adherence, a twice-weekly dose of 15 mg/kg to a maximum of 900 mg should be considered when directly observed daily therapy is not possible. Most clinical trials studied 12-month regimens of isoniazid preventive therapy, but there is evidence that, a nine month regimen may be equally effective (5). On an intention to treat basis, that a 6-month regimen approaches the effectiveness of the 12 month regimen? Regimens of less than 6-months are clearly less effective. Regimens of greater than 12 months have not been shown to confer any additional benefit in people without immune compromise.

All individuals prescribed preventive therapy for tuberculosis should receive a minimum of 6 months and up to 12 months of continuous preventive therapy. A cost-effectiveness study showed that the 6 month regimen was more cost-effective than either the 3- or 12-month regimen (32). However, the cost-effectiveness estimate uses data from the intention to treat analysis (65% protection for the 6-month regimen versus 75% protection for the 12-month regimen), but there is ≥90% protection provided to patients adherent with the 12-month regimen. Therefore, every effort should be made to ensure that all individuals prescribed preventive therapy complete at least a 6-month regimen and the option should be provided for the completion of a 12-month course of preventive therapy. The American Academy of Pediatrics recommends a regimen of 9 months of isoniazid preventive therapy for children (33).

As described earlier, preventive therapy for HIV-infected individuals is of particular importance offering both clinical and public health benefits. A 12-month regimen of isoniazid preventive therapy is recommended in the ATS/CDC guidelines for all HIV-infected individuals with positive tuberculin skin tests. This regimen is also recommended by the USPHS/Infectious Disease Society of America Working Group on the Prevention of Opportunistic Infections (34). These same guidelines also recommend a 12-month regimen of isoniazid preventive therapy for anergic, HIV-infected individuals, living in or coming from areas with a high prevalence (≥10%) of positive tuberculin tests, recognizing that efficacy has not yet been demonstrated. In the United Kingdom, the Tuberculosis Committee of the British Thoracic Society has recommended that HIV-positive individuals who have not received Bacille–Calmette–Guérin (BCG) and are tuberculin positive should receive either a 12-month regimen of isoniazid preventive therapy or a 6-month regimen of preventive therapy with both isoniazid and rifampin.

Data from environments with high rates of tuberculosis indicate that HIV-infected individuals may be particularly susceptible to relapse after completion of treatment and data from the Zambian (Wadhawan) and Haitian (Halsey) studies cited earlier indicates that the protection conferred by preventive therapy may not be long-lived. This may be because the therapy never kills all of the bacilli in a patient with disease and a well-functioning immune system is necessary for long-term protection. This could also be due to a greater susceptibility for reinfection of HIV-infected people. In either case, longer periods of preventive therapy, perhaps lifelong, may be desirable.

A minimum of 12 months of preventive therapy had been the standard recommendation for preventive therapy with isoniazid for individuals with positive tuberculin skin tests and either silicosis or a chest radiograph with fibrotic lesions consistent with old-healed tuberculosis. Although the 12-month regimen is still an acceptable regimen, a 4-month regimen of both isoniazid and rifampin may be a more attractive alternative. This recommendation is based on studies showing this regimen to be suitable for the treatment of individuals with active pulmonary tuberculosis but negative sputum smears and cultures (35,36). In theory, this regimen should also be suitable for individuals who are at high risk for pulmonary tuberculosis but have not yet developed any symptoms.

6.3. Monitoring Preventive Therapy

Individuals receiving isoniazid should be seen monthly and questioned carefully for the signs and symptoms of adverse reactions to isoniazid. No more than a single month's supply of drugs should be given at any time, to help ensure regular contact between the patient and provider. Because the risk of isoniazid-associated hepatitis increases with age, individuals 35 years of age and older should have a baseline transaminase measurement, prior to starting therapy and then monthly during the course of treatment. Others at high risk for isoniazid-associate hepatitis should also be considered for more careful monitoring, including those who use alcohol daily, injection drug users, and those with chronic liver disease. Postpubertal black and Hispanic women may also be at increase risk. As guidelines point out, liver function tests are not a substitute for careful clinical evaluation or the prompt assessment of signs and symptoms of adverse reactions.

7.0. Recommendations for Low-Income Countries

Both the World Health Organization and the International Union Against Tuberculosis and Lung Disease have made recommendations about preventive therapy with isoniazid in their TB control guidelines (37,38). Preventive therapy is never recommended as a core part of a national tuberculosis control program for a low-income country, where the priority has to be on the diagnosis, treatment, and cure of smear-positive pulmonary tuberculosis patients, in order to break the chain of infection. However, both sets of guidelines have advised that tuberculin-positive children under the age of 5, residing with smear-positive, infectious tuberculosis patients who are under treatment, be considered for a 6-month course of preventive therapy with isoniazid while the case is also under treatment. This should only be considered as an addition to an already well-functioning program, and, in practice, the implementation of such programs will be rare.

The World Health Organization and the International Union Against Tuberculosis and Lung Disease have also recommended that isoniazid preventive therapy be considered as an individual health measure for tuberculin-positive individuals who are HIV infected (39). However, this statement cautions that this should not be an activity incorporated into national tuberculosis control programs. Due to issues of cost and feasibility, it is unlikely that isoniazid preventive therapy programs will be widely implemented in low-income countries. Unresolved issues for these programs include the appropriate length of therapy (e.g., fixed duration versus lifetime), the cost–benefit of these regimens, the role of skin testing in these programs, and an efficient means of screening for tuberculosis.

8.0. Adherence with Isoniazid Preventive Therapy

Tuberculosis preventive therapy is core component of the U.S. effort to control and eliminate tuberculosis. But, there is a major problem in its application; namely the problem of maintaining patient and provider adherence with a preventive regimen that lasts 6–12 months and is associated with some hepatotoxicity. State and local health departments are asked to report on the results of tuberculosis program activities to the Division of Tuberculosis Elimination in the Centers for Disease Control and Prevention. About 90 program areas voluntarily participate in this system each year. According to these program management reports (CDC, unpublished data), approximately nine contacts are identified for each tuberculosis (TB) case reported, about 90% of the contacts are examined and about 20% of them found to be infected. Whereas program guidelines would call for virtually all of the infected contacts under age 35 to be started on preventive therapy, about 90% were actually started in 1993. Of those started on therapy in 1992, only 68% of this group completed what is considered to be an adequate course of therapy. Therefore, only about 60% of the contacts identified as candidates for preventive therapy received an adequate course of preventive therapy. These figures are improvements on the experience in the United States in the 1980s and indicate the difficulty in delivering preventive therapy.

To improve provider's adherence with preventive therapy, additional efforts are required to assist providers in identifying high-priority candidates for preventive therapy. Much of the resistance to preventive therapy may be a reluctance to expose relatively low-risk individuals to the potential hazards of isoniazid. Promoting patient adherence will be very challenging, as they are asymptomatic and may have difficulty accepting the need for a 6–12-month regimen of preventive therapy. Some TB control programs in the United States have begun to use directly observed therapy (DOT) combined with incentives and enablers, a strategy very successful in the treatment of tuberculosis, to help ensure adherence among patients at very high risk of developing tuberculosis or at high risk of infecting others if they develop active disease.

Even greater problems have been experienced in organizing isoniazid preventive therapy programs in low-income countries. In one published study reporting on a feasibility study of isoniazid preventive therapy in Uganda, only 24% of HIV-infected persons who agreed to be skin tested (which was 56% of the persons who returned for skin readings and found to be positive) completed a 6-month course of therapy, at a cost of about $18–19 per person completing treatment (40). The study acknowledged a number of problems in the organization of the program that created barriers to patient adherence and the cost-effectiveness of the program could very likely be improved. However, in countries which lack the funding to adequately diagnose and treat existing infectious, smear-positive cases with pulmonary tuberculosis, preventive therapy programs like these will not be feasible. In some settings, particularly if incorporated within comprehensive HIV/AIDS clinical care, isoniazid preventive therapy could be feasible, even in low-income countries. This will require excellent coordination between the national tuberculosis and HIV/AIDS programs.

9.0. Alternatives to Isoniazid Preventive Therapy

Alternatives to isoniazid preventive therapy are needed both to address infection with drug-resistant organisms and to increase adherence. Preventive therapy with rifampin for infected contacts of isoniazid-resistant cases is recommended in the ATS/CDC guidelines. The rifampin preventive therapy is recommended for the same duration isoniazid would have be given to the person if the source case was not resistant. The rifampin dosage should be the same as used in the treatment of tuberculosis. A shorter course of rifampin may actually be possible. This was indicated by a published study of tuberculosis preventive therapy for silicotic patients in Hong Kong, which demonstrated a low toxicity for a rifampin-only regimen and provides an estimate of relative efficacy (41). The study was placebo controlled and regimens of 6 months of isoniazid, 3 months of isoniazid and rifampin, and 3 months of rifampin alone were studied. None of the regimens proved acceptable as preventive therapy for this high-risk population, reducing the risk of disease by only 41–63% (thus, the recommendations for either 12 months of isoniazid or 4 months of a multidrug regimen). However, all of the regimens were more effective than placebo and, although not statistically significantly different from the other two regimens, the 3-month rifampin regimen was the most effective. Examining toxicity, more patients on the isoniazid-containing regimens had significant increases in serum ALT, compared with the rifampin-only regimen and the mean serum level for the ALT barely changed from baseline for the rifampin-only group.

Outbreaks and an increasing incidence of tuberculosis resistant to both isoniazid and rifampin have made regimens that rely on neither isoniazid nor rifampin a necessity, for contacts of these multidrug resistant tuberculosis (MDR-TB) cases. Although there are no controlled studies of preventive therapy regimens with other drugs, the Centers for Disease Control and Prevention has published guidelines for the management contacts of MDR-TB (42) recommending that those at the highest risk of infection with MDR-TB bacilli and at very high risk of developing tuberculosis following infection with *M. tuberculosis* (e.g. those with HIV infection) be placed on preventive therapy with two drugs to which the presumed source case bacilli are shown to be sensitive. Possibly effective regimens suggested in the document are ethambutol–pyrazinamide and pyrazinamide plus one of the licensed quinolones shown to be active against *M. tuberculosis* (ciprofloxacin, ofloxacin, or temafloxacin in the United States). A minimum of 6 months of preventive therapy is recommended for immune-competent individuals and 12 months of preventive therapy are recommended for HIV-infected and other immune-compromised individuals.

Because of the limitations placed on preventive therapy by the length of therapy and the concerns about toxicity, short-course regimens of preventive therapy that avoid the use of isoniazid have been investigated. A meeting of experts, convened to discuss alternatives to isoniazid preventive therapy, concluded that the most promising alternatives were a short-course rifampin regimen or a rifampin–pyrazinamide combination (43). Support for rifampin and rifampin–pyrazinamide regimens later came from the results of studies of preventive therapy in a mouse model (44). Four regimens were tested, isoniazid alone for 6 months, rifampin alone for 4 months, rifampin and pyrazinamide for 2 months, and isoniazid, rifampin, and pyrazinamide for 2 months. The most effective of the four regimens was the rifampin–pyrazinamide regimen. Interestingly, it was even more effective than the same two drugs in combination with pyrazinamide. The rifampin-only regimen was also very promising.

Preliminary results have been presented from studies of the safety and acceptability of short-course preventive therapy, comparing a 2-month rifampin–pyrazinamide and a 4-month rifampin-only regimen with a control regimen of 6 months of isoniazid preventive therapy. These regimens were tested in tuberculin reactors with special risk factors for reactivation in the United States and Canada

(45), Germany (46), and Poland (47). The study population at the North American centers was limited to adults over 18 years of age. In Germany, the patients were enrolled at a pediatric clinic in Berlin and only children were enrolled. The study population in Poland was mixed, with about equal numbers of children and adults.

All three regimens were extremely well tolerated by the children. Adherence with the regimens approached 100% for the children studied in both Germany and Poland. However, in the adult populations, there was much more toxicity and lower rates of adherence than expected. Approximately 10% of the patients on the 2-month rifampin–pyrazinamide had medication withdrawn due to adverse reactions, many of which were hepatotoxic reactions. However, there was very little evidence of toxicity for the rifampin-only regimen. Adherence with the regimens was best for the group assigned to the 6-month isoniazid regimen and the adherence rate was significantly lower for the rifampin-only group. Possible explanations for the lower adherence with the therapy was the description of the short-course regimens as experimental, warnings about adverse reactions in the informed consent statement, and a dislike of the orange coloration of body fluids caused by rifampin.

A number of studies in the United States and in low-income countries are already underway evaluating regimens with rifampin and rifampin–pyrazinamide for short-course preventive therapy and results from some of these studies should soon be available. Given the adverse reaction profile of the pyrazinamide-containing regimens, it is unlikely that they will be acceptable for implementation. Also, in low-income countries, the inability to easily rule out tuberculosis, and the generally accepted policy to only use rifampin in the context of directly observed therapy, makes it unlikely that these regimens will be useful.

10. Conclusions

Isoniazid preventive therapy has been a core part of the national tuberculosis program in the United States and is gaining acceptance in other industrialized countries, particularly for use in pediatric and foreign-born populations. The potential to prevent the development of active tuberculosis in an individual with a latent infection is extremely attractive. However, preventive therapy has its limitations and is an intervention in need of improvement.

Shorter and more widely spaced intermittent regimens would facilitate the use of directly observed therapy. One of the long-acting rifamycins may be useful in this regard and a 3–4-month regimen of once-weekly preventive therapy is not impossible with these drugs. A reduction in toxicity with a rifamycin-based regimen may also help increase the acceptability and the cost–benefit ratio. However, these regimens are only likely to be of use in industrialized countries.

There are also a number of research questions related to the provision of preventive therapy for HIV-infected individuals. One question is operational in nature, and that is how feasible these programs can be and how to integrate them with the national HIV/AIDS programs. The second question has to do with the length of therapy. It is quite possible that a lifelong preventive therapy regimen may be required.

Other methods of prevention may be more attractive than preventive therapy, particularly an effective vaccine. However, unless a vaccine can provide protection against an already existing infection, some form of preventive therapy will be required and is likely to be a common feature of tuberculosis programs in low-incidence countries.

References

1. Lincoln EM (1954) The effect of antimicrobial therapy on the prognosis of primary tuberculosis in children. Am Rev Tuberc 69(5):682–689.

2. Zorini AO (1958) Antituberculous chemoprophylaxis with isoniazid: preliminary note. Dis Chest 32(1):1–17.

3. Debre R, Perdrizet S, Lotte A, Naveau M, Lert F (1973) Isoniazid chemoprophylaxis of latent primary tuberculosis: in five trial centres in France from 1959 to 1969. Int J Epidem 2:153–160.

4. Ferebee SH, Mount FW (1962) Tuberculosis morbidity in a controlled trial of the prophylactic use of isoniazid among household contacts. Am Rev Respir Dis 85(4):490–521.

5. Ferebee SH (1969) Controlled chemoprophylaxis trials in tuberculosis. A general review. Adv Tuberc Res 17:28–106.

6. Comstock GW, Baum C, Snider DE (1979) Isoniazid prophylaxis among Alaskan eskimos: A final report of the Bethel isoniazid studies. Am Rev Respir Dis 119(5):827–830.

7. International Union Against Tuberculosis Committee on Prophylaxis (1982) Efficacy of various durations of isoniazid preventive therapy for tuberculosis: Five years of follow-up in the IUAT trial. Bull WHO 60:555–564.

8. Hsu KHK (1974) Isoniazid in the prevention and treatment of tuberculosis. A 20-year study of the effectiveness in children. J. Amer. Med. Assoc. 229:528–533.

9. Hsu KHK (1984) Thirty years after isoniazid. Its impact on tuberculosis in children and adolescents. J. Amer. Med. Assoc. 251:1283–1285.

10. Rieder HL, Cauthen GM, Comstock GW, Snider DE Jr (1989) Epidemiology of tuberculosis in the United States Epidemiol Rev 11:79–98.

11. Selwyn PA, Hartel D, Lewis VA, Schoenbaum EE, Vermund SH, Klein R, Walker AT, Friendland GH (1989) A prospective study of the risk of tuberculosis among intravenous drug users with human immunodeficiency virus infection. N Engl J Med 320:545–550.

12. DiPerri G, Danzi MC, DeChecchi G, Pizzighella S, Solbiati M, Cruciani M, Luzzati R, Malena M, Mazzi R, Concia E, Bassetti D (1989) Lancet 8678:1502–1504.

13. Centers for Disease Control and Prevention (CDC) (1991) Tuberculosis outbreak among persons in a residential facility for HIV infected persons—San Francisco. 40:649–652.

14. Selwyn PA, Sckell BM, Alcabes P, Friedland GH, Klein RS, Schoenbaum EE (1992) High risk of active tuberculosis in HIV-infected drug users with cutaneous anergy. J Am Med Assoc 268:504–509.

15. Pape JW, Jean SS, Ho JL, Hafner A, Johnson WD Jr (1993) Effect of isoniazid prophylaxis on incidence of active tuberculosis and progress of HIV infection. Lancet 342:268–272.

16. Halsey N, Coberly J, Atkinson J, Boulos R (1994) Twice weekly INH for TB prophylaxis. In: X International Conference on AIDS. Yokohama, abstract No. PB0681.

17. Wadhawan D, Hira S, Mwansa N, Sunkutu R, Adera T, Perine P (1993) Preventive tuberculosis chemotherapy with isoniazid (INH) among patients infected with HIV-1. In: IX International Conference on AIDS, abstract No. PO-B07-1114.

18. Whalen C, Horsburgh CR, Hom D, Lahart C, Simberkoff M, Ellner J (1995) Accelerated course of human immunodeficiency virus infection after tuberculosis. Am J Respir Crit Care Med 151:129–135.

19. Pitts FW (1977) Tuberculosis prevention and therapy. In: Hook EW, Mandel GL, Gwatney JM Jr, eds. Current Concepts in Infectious Diseases, pp. 181–194. New York: John Wiley and Sons.

20. Mitchell JR, Zimmerman HJ, Ishak KG, Thorgeirsson UP, Timbrell JA, Snodgrass WR, Nelson SD (1976) Isoniazid liver injury: clinical spectrum, pathology and probable pathogenesis. Ann Intern Med 84:181–192.

21. Kopanoff DE, Snider DE Jr, Caras GJ (1978) Isoniazid-related hepatitis: a U.S. Public Health Service cooperative surveillance study. Am Rev Respir Dis 117(6):991–1001.

22. Dash LA, Comstock GW, Flynn JPG (1980) Isoniazid preventive therapy retrospect and prospect. Am Rev Respir Dis 121:1039–1044.

23. Colice GL, Decision analysis, public health policy, and isoniazid chemoprophylaxis for young adult tuberculin skin reactors. Arch Intern Med 150:2517–2522.

24. Rose DN, Schecter CB, Fahs MC, Silver AL (1988) Tuberculosis prevention: Cost-effectiveness analysis of isoniazid chemoprophylaxis. Am J Prevent Med 4:102–109.

25. Taylor WC, Aronson MD, Delbanco TL (1981) Should young adults with a positive tuberculin test take isoniazid? Ann Intern Med 94:808–813.

26. Tsevat J, Taylor WC, Wong JB, Pauker SG (1988) Isoniazid for the tuberculin reactor: Take it or leave it. Am Rev Respir Dis 137:215–220.

27. Snider DE Jr, Caras GJ (1992) Isoniazid-associated hepatitis deaths: A review of available information. Am Rev Respir Dis 145:494–497.

28. American Thoracic Society and the Centers for Disease Control and Prevention (1994)

Treatment of tuberculosis and tuberculosis infection in adults and children. Am J Crit Care Med 149:1359–1374.

29. Centers for Disease Control and Prevention (1995) Screening for tuberculosis and tuberculosis infection in high-risk populations. Recommendations of the Advisory Council for the Elimination of Tuberculosis. Morbid Mortal Wkly Rep 44(RR-11):19–34.

30. Centers for Disease Control (1991) Purified protein derivative (PPD)-tuberculin anergy and HIV infection: Guidelines for anergy testing and management of anergic persons at risk of tuberculosis. Morbid Mortal Wkly Rep 40(RR-5):27–33.

31. Jordan TJ, Lewit EM, Montgomery RL, Reichman LB (1991) Isoniazid as preventive therapy in HIV-infected intravenous drug abusers. J Am Med Assoc 265(22):2987–2991.

32. Snider DE Jr, Caras GJ, Koplan JP (1986) Preventive therapy with isoniazid: Cost-effective of different durations of therapy. J Am Med Assoc 255:1579–1583.

33. American Academy of Pediatrics (1991) Report of the Committee on Infectious Diseases, 22nd ed., pp. 487–508. Elk Grove, IL: American Academy of Pediatrics.

34. Kaplan JE, Masur H, Holmes KK, McNeil MM, Schonberger LB, Navin TR, Hanson DL, Gross PA, Jaffe HW, and the USPHS/IDSA Prevention of Opportunistic Infections Working Group (1995) USPHS/IDSA guidelines for the prevention of opportunistic infections in persons infected with human immunodeficiency virus. Disease specific recommendations. Clin Infect Dis 21(Suppl 1):S32–S43.

35. Hong Kong Chest Service/Tuberculosis Research Centre, Madras/British Medical Research Council (1989) A controlled trial of 3-month, 4-month, and 6-month regimens of chemotherapy for sputum-smear negative pulmonary tuberculosis. Results at 5 years. Am Rev Respir Dis 139:871–876.

36. Dutt AK, Moers D, Stead WW (1989) Smear- and culture-negative pulmonary tuberculosis: four-month short course chemotherapy. Am Rev Respir Dis 139:867–870.

37. International Union Against Tuberculosis and Lung Disease (1994) Tuberculosis Guide for Low Income Countries, 3rd ed. Paris: IUATLD.

38. World Health Organization (1993) Treatment of Tuberculosis: Guidelines for National Programmes. Geneva: World Health Organization.

39. World Health Organization and the International Union Against Tuberculosis and Lung Disease (1994) Tuberculosis preventive therapy in HIV-infected individuals. WHO Wkly Epidemiol Rec 68:361–364.

40. Aisu T, Raviglione MC, van Praag E, Eriki E, Narain J, Barugahare L, Tembo G, McFarland D, Adatu Engwau F (1995) Preventive chemotherapy for HIV-associated tuberculosis in Uganda; an operational assessment at a voluntary counseling and testing centre. AIDS 9:267–273.

41. Hong Kong Chest Service/Tuberculosis Research Centre, Madras/British Medical Research Council (1992) A double-blind placebo-controlled clinical trial of three antituberculosis chemoprophylaxis regimens in patients with silicosis in Hong Kong. Am Rev Respir Dis 145:36–41.

42. Centers for Disease Control (1992) Recommendations for the management of persons exposed to multidrug-resistant tuberculosis. Morbid Mortal Wkly Rep 41(RR-11):59.

43. Geiter LJ, O'Brien RJ (1987) Conference on new approaches for tuberculosis preventive therapy. Journal of Infectious Diseases (JID) 156(3):536–537.

44. Lecoeur HF, Truffot-Pernot C, Grosset JH (1989) Experimental short-course preventive therapy of tuberculosis with rifampin and pyrazinamide. Am Rev Respir Dis 140:1189–1193.

45. Geiter LJ, O'Brien RJ, Kopanoff DE (1990) Short-course preventive therapy. Am Rev Respir Dis 141(4, part 2):A437.

46. Magdorf K, Arizzi-Rusche AF, Geiter LJ, O'Brien RJ, Wahn U (1991) Short-course preventive therapy for tuberculosis: a pilot study of rifampin and rifampin-pyrazinamide regimens in children (abstract). Am Rev Respir Dis 143(4, part 2):A119.

47. Grazcyk J, O'Brien RJ, Bek E, Nimerowska H, Geiter LJ (1991) Assessment of rifampin containing regimens for tuberculosis preventive therapy: preliminary results of a pilot study in Poland (abstract). Am Rev Respir Dis 143(4, part 2):A119.

12

Failures in Tuberculosis Chemotherapy

Pierre Chaulet and Noureddine Zidouni

1. Introduction

The failure of chemotherapy is characterized by the persistence of active tuberculosis (i.e., bacilli in the sputum or in extrapulmonary lesions) in a patient who has received specific chemotherapy. This broad definition includes both failures observed during or at the end of chemotherapy and relapses observed in the years following the end of treatment.

The progress made in tuberculosis chemotherapy in the last 40 years has considerably reduced the risk of treatment failure. The short-course chemotherapy regimens currently recommended by the World Health Organization (WHO) (1) and applied worldwide have a cumulative failure and relapse rate of below 5% in patients who have received a complete course of chemotherapy.

However, the two principal factors of chemotherapy failure, recognized for some time (2), are still a problem today:

1. Treatment that is unsatisfactory in composition (number of drugs, dosage) or in duration. This is more often a result of prescription error on the part of the doctor or errors in distribution by health care personnel than patient negligence (3,4). Inadequate treatment leads to the reactivation of the disease, which increases in likelihood depending on the relative shortness of the treatment. This is the main cause of chemotherapy failure in national programs.
2. Resistance of *Mycobacterium tuberculosis* to drugs. The emergence of resistance is the consequence of inappropriate chemotherapy following deliberate (prescription error, selection of a single drug by the nurse or the patient, or nonavailability of the necessary drugs) or accidental (severe unknown resistance) monotherapy. It is rare today for drug resistance to be the main cause

of chemotherapy failure if appropriate short-course chemotherapy is applied correctly. Although it is more rare than before, chemotherapy failure is, in certain cases, more serious when it is a result of the excretion of multiresistant bacilli (i.e., resistant to at least isoniazid and rifampicin).

One of the characteristics of modern tuberculosis chemotherapy is that the failure rate observed in programs in the community is very close to that observed in clinical trials. But as the definition of failure and its detection are not the same in the two situations, they will be presented as follows in this chapter:

• Failure of chemotherapy in clinical trials
• Failure of chemotherapy in treatment programs
• Methods of correcting and preventing failure

2. Failure of Chemotherapy in Clinical Trials

Clinical trials allow the precise measurement of the effectiveness of different chemotherapy regimens. Trials take place in optimal conditions and prospectively according to predefined protocols. Patients admitted to the trials are selected; the diagnosis of pulmonary tuberculosis is proven by positive bacteriological tests before treatment, and cases of tuberculosis with concomitant disease are generally excluded. For the final analysis of the trial, only those patients are retained who have completed their course of treatment in keeping with the initially defined protocol and whose bacteriological status at the end of treatment and during a follow-up period of 2–5 years is known. Patients who have not completed their treatment for reasons other than failure or who have had their treatment modified (temporarily or totally) because of side effects are analyzed separately, as are patients whose bacteriological status was not evaluated at the end of treatment.

For the groups of analyzable cases to be sufficiently large (so that the results are significant), clinical trials are based on cases of initially smear or culture-positive pulmonary tuberculosis. Sensitivity tests are performed on all positive cultures isolated before treatment, or during and after treatment.

In these conditions, failures and relapses are defined by rigorous bacteriological criteria which indicate whether a new course of chemotherapy should be prescribed for the patients.

2.1. Definition of Failure

Failure of chemotherapy after a complete course of treatment is defined by the following:

- Either the persistence of bacilli in the patient's sputum in all monthly checks up to the fifth month, or at least at the end of the second, third, and fifth months of treatment (1).
- Or the reappearance of bacilli in the patient's sputum from the fifth month of treatment, or later, in the period between the fifth month and the end of treatment, after temporary bacteriological conversion of tests before the fifth month. This reappearance of bacilli must be confirmed twice in sputum samples collected at intervals of at least 2 weeks; this avoids classifying as a failure the single, isolated excretion of a few "dead" bacilli, which are stainable but do not grow in culture, or an isolated positive culture (with a limited number of colonies—fewer than 20—in a single test tube) without deterioration of radiological lesions (5).

2.2. Definition of Relapse

Relapse after a complete course of chemotherapy is defined by the reappearance of bacilli in the sputum of a patient considered as cured at the end of treatment. In the majority of cases (at least 90%), relapse occurs in the year following the end of treatment; for the rest, it occurs during the 2 following years, and later only exceptionally. The presence of bacilli must be confirmed in two sputum samples collected at intervals of at least 2 weeks; this avoids classifying as a relapse the isolated excretion of bacilli without radiological deterioration (5). However, if there is aggravation of radiological lesions with or without accompanying clinical signs, a single positive bacteriological result can suffice to confirm relapse.

2.3. Failure and Relapse in Cases of Smear-Positive Pulmonary Tuberculosis with Initially Sensitive Bacilli

Several clinical trials have been performed worldwide on this type of patient. All the results are similar: Chemotherapy regimens combining isoniazid and rifampicin for 6 months, with a supplement of pyrazinamide during the initial phase, are very effective. The cumulative rate of failure and relapse is always between 1% and 4% (Table 12.1).

These regimens can be given either daily for the full 6 or 8 months, or daily for the first month and then 3 times per week during the continuation phase, or three times per week throughout treatment. In patients with bacilli sensitive to antibiotics before treatment who are receiving a 6-month regimen, the addition of a fourth drug (streptomycin or ethambutol) during the initial phase of treatment does not significantly reduce the rate of relapse and failure. The length of administration of rifampicin can be reduced to 4 months or even 2 months without reducing the overall effectiveness of the treatment, on condition that the total duration of treatment is 8 months.

Table 12.1. Highly effective regimens of chemotherapy in pulmonary tuberculosis (smear-positive cases with sensitive strains pretreatment)

Study (Ref.)	Regimens[a]	Duration (months)	Follow-up (months)	Patients assessed	Failures and relapses	
					n	%
2nd BTA (6)	2SRHZ/4RH	6	30	125	1	0.8
	2ERHZ/4RH	6	30	132	3	2
Singapore (7)	2SRHZ/4RH	6	24	80	2	2
5th East Africa (8)	2SRHZ/4RH	6	24	160	4	2
Algerian Sahara (9)	2ERHZ/4RH	6	30	126	4	3
Algeria (10,11)	2SRHZ/4RH	6	24	593	18	3
Singapore (12)	$2SRHZ/4R_3H_3$	6	60	96	2	2
	$2RHZ/4R_3H_3$	6	60	105	3	3
Hong Kong (13)	$4S_2R_3H_3Z_3/2R_3H_3$	6	24	142	2	1
	$6R_3H_3Z_3$	6	24	135	6	4
Algiers (14)	3RHZ/3RH	6	24	121	3	2
U.S. trial 21 (15)	2RHZ/4RH	6	24	273	10	4
Algiers (16)	2RHZ/4RH	6	30	159	1	0.6
Algeria (10,11)	2SRHZ/2RH/4H	8	24	578	20	3
3rd East Africa (17)	2SRHZ/6TH	8	24	81	0	0
5th East Africa (18)	2SRHZ/6H	8	12	119	4	3

[a]The drugs utilized in these regimens are conventionally represented by the following letters: H = isoniazid; R = rifampicin; S = streptomycin; Z = pyrazinamide; T = thiacetazone; E = ethambutol. The number preceding a group of letters indicates the duration in months of the initial phase and continuation phase of treatment. The number in the subscript to a letter represents the number of weekly doses if the regimen is intermittent. Without a subscript the regimen is daily

One last and important note: Bacillary strains isolated after failure or relapse in these patients are usually sensitive to isoniazid and rifampicin whether the treatment is 6 or 8 months long.

2.4. Failure and Relapse in Cases of Smear-Positive Pulmonary Tuberculosis with Primary Resistance

In patients with bacilli initially resistant to isoniazid and/or streptomycin, clinical trials demonstrate the great effectiveness of 6-month chemotherapy regimens combining isoniazid and rifampicin with a supplement of pyrazinamide and of a fourth drug (streptomycin or ethambutol) during the first months. The cumulative total of failures and relapses is, on average, 8% in reported trials (Table 12.2). In other series using the same regimens, this rate varies from 5% to 13% (12,19).

However, in patients with primary resistance to isoniazid at least, the emergence of acquired resistance to rifampicin can be observed when rifampicin has been given for 6 months and a fourth drug (ethambutol or streptomycin) has not been added during the initial phase (16).

Eight-month chemotherapy regimens containing rifampicin for only 2 or 4 months have a lower effectiveness than 6-month regimens in patients with primary bacterial resistance. The cumulative rate of failures and relapses is, on average, 15% in reported trials (Table 12.2). However, in patients with primary resistance to isoniazid at least, these failures are not accompanied by acquired resistance to rifampicin (20) because this drug has usually only been given in combination with two other drugs and uniquely during the initial phase.

There is a close correlation between the failure rate of short-course chemotherapy in patients with primary bacterial resistance and the total duration of treatment with rifampicin (Table 12.3). Contrary to the results observed in patients with sensitive bacilli before treatment, in those patients with primary resistance, treatment failure is more frequent than relapse.

Primary resistance to streptomycin has no impact on the results of short-course chemotherapy regimens. However, primary resistance to isoniazid alone or combined with streptomycin leads to 8% of failures/relapses after the application of a 6-month regimen and 15% of failures/relapses after an 8-month regimen. These rates are considerably lower than the 50% and 60% failure rates observed after

Table 12.2. Highly effective regimens of chemotherapy in pulmonary tuberculosis (smear-positive cases with primary resistance)

Study	Regimens[a]	Duration (months)	Follow-up (months)	Patients assessed	Failures and relapses	
					n	%
Singapore (7)	2SRHZ/4RH	6	24	2	0	0
5th East Africa (18)	2SRHZ/4RH	6	24	10	2	20
Algerian Sahara (9)	2ERHZ/4RH	6	30	15	1	7
Algeria (10,11)	2SRHZ/4RH	6	24	23	1	4
Singapore (12)	2SRHZ/4R$_3$H$_3$	6	60	9	0	0
	2RHZ/4R$_3$H$_3$	6	60	11	1	9
Algiers (14)	3RHZ/3RH	6	24	10	0	0
Algiers (16)	2RHZ/4RH	6	30	9	2	22
Algeria (10,11)	2SRHZ/2RH/4H	8	24	28	5	18
3rd East Africa (17)	2SRHZ/6TH	8	24	9	1	11
5th East Africa (18)	2SRHZ/6H	8	12	6	1	16

[a]See footnote for Table 12.1.

Table 12.3. Results of short-course chemotherapy regimens in patients with primary resistance

Regimen[a]	Total duration (months)	Duration of rifampicin (months)	Patients assessed	Failures and relapses	
				n	%
6RHZ (+ S, E, or SE)	6	6	110	1	1
4SRHZ/$S_2H_2Z_2$	6–8	4	37	2	5
3SRHZ/$S_2H_2Z_2$ or TH	5–7	3	52	8	15
2SRHZ/$S_2H_2Z_2$ or TH	6–8	2	89	11	12
1SRHZ/$S_2H_2Z_2$ or TH	6–8	1	33	10	30

[a]See footnote for Table 12.1.
Source: Ref. 20.

the application of conventional chemotherapy using regimens of 12–18 months, without rifampicin, 20 years ago (21,22).

2.5. Influence of Primary Resistance on the Overall Results of Modern Chemotherapy

The great improvement in the results of modern chemotherapy, thanks to the application of short-course regimens, is due to the reduction of the overall number of failure and relapses in patients with sensitive bacilli before treatment, as well as in those with primary resistance. These failures and relapses, after the application of 6- or 8-month regimens in cases of smear-positive pulmonary tuberculosis, are patients who after treatment still excrete mainly isoniazid- and rifampicin-sensitive bacilli. In fact, failures (and at times relapses) due to severe primary resistance to isoniazid alone or associated with streptomycin represent only a fraction of the total number of failures in most situations (Table 12.4). This limited influence of primary resistance to isoniazid (alone or associated with streptomycin) was observed in a community clinical trial undertaken in a region of 4 million inhabitants in Algeria, where the rate of primary resistance was, on average, 4% (10,11).

Although the 8-month regimen (containing 4 months of rifampicin) is a little less effective than the 6-month regimen for cases with primary resistance to isoniazid at least, it is noted that for all cases, the majority of whom had sensitive bacilli, there is no statistically significant difference in the cumulative rate of failures and relapses of the two regimens, with a follow-up period of 24 months after the end of treatment (Table 12.5).

Table 12.4. Failure[a] due to primary resistance to isoniazid (at least) among new cases of pulmonary tuberculosis treated by short-course chemotherapy regimens in different epidemiological situations

Primary resistance rate (%)	Patients with sensitive bacilli		Patients with primary resistance		Failures of chemotherapy	
	Total	Failures	Total	Failures	Total (100%)	Due to primary resistance (%)
(Six-month regimens)						
0	1000	30	0	0	30	0
5	950	28	50	4	32	12.5
10	900	27	100	8	35	23
15	850	25	150	12	37	32
(Eight-month regimens)						
0	1000	30	0	0	30	0
5	950	28	50	7	35	20
10	900	27	100	15	42	36
15	850	25	150	22	47	47

[a]Failure rates: 3% in patients with sensitive bacilli pretreatment treated with 6- or 8-month regimens; 8% in patients with primary resistance treated with 6-month regimens (with rifampicin throughout); 15% in patients with primary resistance treated with 8-month regimens (with 2 months of rifampicin).

Table 12.5. Comparison of treatment outcomes of two short-course chemotherapy regimens (6 months and 8 months) 2 years after completion of chemotherapy

Patients with initially	6-month regimen 2SRHZ/4RH			8-month regimen 2SRHZ/2RH/4H		
	Patients assessed	Failure/relapse		Patients assessed	Failure/relapse	
		N	%		n	%
Sensitive bacilli	593	18	3	578	20	3.4
Bacilli resistant to isoniazid (at least)	23	1	4	28	5	18
Total	616	19	3	606	25	4

Source: Ref. 11.

2.6. Failures and Relapses in Cases of Smear-Negative Pulmonary Tuberculosis

Factor analysis of short-course chemotherapy failures in cases of smear-positive pulmonary tuberculosis has shown that the different factors connected to the high number of bacilli in initial pulmonary lesions are closely linked to the frequency of failures and relapses (23,24). These factors are the extent of initial radiological lesions, the number of colonies in the sputum culture collected before treatment, the rate of positivity and number of colonies in cultures performed after 2 and 3 months of treatment, and the length and quality of the chemotherapy applied.

Following this analysis, it was suggested that failures and relapses could be more uncommon in groups of patients whose bacterial population is lower, and that a shorter course of treatment could be applied to cases of smear-negative pulmonary tuberculosis identified either by positive culture or by radiological deterioration and excretion of bacilli in the absence of treatment if the initial culture is negative and the patients are strictly monitored.

In a multicentered study performed in Hong Kong and Madras (25), short-course chemotherapy regimens associating isoniazid, streptomycin, rifampicin, and pyrazinamide, given every day or three times per week for 3, 4, or 6 months, were applied to patients with smear-negative pulmonary tuberculosis. The cumulative rate of failures and relapses (after an observation period of 5 years after the end of treatment) is 2% for cases with sensitive bacilli pretreatment and 8% for cases with primary resistance to isoniazid who received daily or intermittent regimens of 4 months: There is no significant difference between the rates of failures and relapses observed in patients receiving 6-month regimens. For initially culture-negative patients (more than 57% of whom become "positive" without treatment in a control group), the cumulative rate of failures and relapses is 2% after a 4-month course of treatment. In a former study (26), this rate was 11% after a very short chemotherapy course of 2 months, and 7% after 3 months of treatment.

All of these studies demonstrate the effectiveness of 4-month short-course chemotherapy regimens for cases of smear-negative pulmonary tuberculosis selected according to the rigorous criteria applied in the Hong Kong and Madras trials (25,26).

2.7. Failures and Relapses in Cases of Pulmonary Tuberculosis with Acquired Bacterial Resistance

- For patients who have already received chemotherapy and who excrete bacilli resistant to isoniazid, or to isoniazid and streptomycin, there are fewer clinical trials.

Table 12.6. Failures of retreatment regimens in patients with acquired resistance

	6-month regimen 2SREZ/4RE		8-month regimen 2SREZ/7RE	
	n	%	*n*	%
Patients assessed at the end of				
chemotherapy	113	100	113	100
Failures	1	1	1	1
Patients assessed 30 months after				
the end of chemotherapy	86	100	92	100
Relapses (total)	6	7	2	2
Strains resistant to isoniazid alone	72	72		
Relapses	3	2		
Strains resistant to isoniazid and				
streptomycin	14	20		
Relapses	3	0		

Former studies have shown the superiority of chemotherapy regimens associating rifampicin and ethambutol for 12 months given daily or twice weekly during the continuation phase, compared with retreatment regimens associating ethionamide and cycloserine for 12 or 18 months with an initial supplement of pyrazinamide for the first months (27–29). More recently (30), it was proven that 6- or 9-month chemotherapy regimens associating rifampicin and ethambutol daily with a supplement of streptomycin and pyrazinamide during the two first months was highly effective in cases of acquired bacterial resistance to isoniazid or isoniazid and streptomycin (Table 12.6).

- For patients who have already received one or more courses of chemotherapy and who excrete bacilli resistant to at least isoniazid and rifampicin (multiresistant strains), no comparative clinical trials currently give an analysis of new antituberculosis drugs or a definition of the optimal drug combination and duration of chemotherapy. These patients have been treated empirically and quite successfully with combinations of minor antituberculosis drugs and new drugs.

3. Failures of Chemotherapy in Programs

The definition of failure is less precise under national antituberculosis program conditions than in clinical trials.

3.1. A Broader Definition of Failure

The bacteriological criteria of failure and relapse, followed by the prescription of a new course of chemotherapy, are still a valid definition of failure in programs.

To these criteria should be added all those cases for whom a new course of chemotherapy has been prescribed before the end of treatment. This broader definition of failure also extends to the following cases:

- Cases of gradual reactivation of the disease who have been inadequately treated because of the premature interruption of chemotherapy for more than 1 or 2 months (consecutively or at different times) and who still have active tuberculosis on reexamination (bacilli in the sputum, caseating lesions in cases of extrapulmonary tuberculosis, sometimes simple deterioration of radiological pulmonary lesions, compatible with active tuberculosis).
- Cases of initially smear-negative pulmonary tuberculosis whose sputum show bacilli after more than a month of treatment.
- Sometimes also, cases of tuberculosis for whom the chemotherapy regimen originally prescribed was interrupted and completely changed because of major problems attributed to drug intolerance or toxicity. This last, very limited, group of cases is sometimes excluded or identified separately in a cohort analysis.

Patients who still have a positive bacteriological test when they interrupt their treatment too early are sometimes classified as treatment failures: In fact, these are not failures of chemotherapy but organizational failures of the program, as they have not received the appropriate treatment.

The prevalence of failure in treatment programmes can be measured in two ways:

1. In routine practice in programs using the notification system of tuberculosis cases and quarterly reports as recommended by the World Health Organization (WHO) and the International Union Against Tuberculosis and Lung Disease (IUATLD) (1,31): The failure rate is that which is recorded at the end of treatment or at the moment when treatment has been changed, and it is calculated as a percentage of all cases admitted to the cohort (some exclude death from the denominators in the report). The cohort must be analyzed separately for new cases and for retreatment cases.
2. During prospective studies undertaken in programs, the cumulative rate of failures and relapses can be measured in large groups of patients observed over a longer period (2, 3, or 5 years), as can the importance of "chronic" cases in the program in relation to all cases admitted to the prospective study.

3.2. Failures at the End of Treatment in Programs

The analysis of homogeneous cohorts of patients (new cases of smear-positive pulmonary tuberculosis or of old, already treated cases receiving a standardized retreatment regimen due to newly positive bacteriological results) allows the routine evaluation of a program's rate of successes and failures. Although the meaning of failure is different in each region or country depending on the quality of the local laboratories and on the criteria applied by different doctors for prescribing a new chemotherapy regimen, cohort analysis gives a reasonable estimate of the failure rate.

In routine practice, a doctor's prescription of a retreatment regimen can be decided upon following major intolerance (not mentioned or not reported) or following a "failure" that has not been bacteriologically proven but presumed due to the evolution of radiological lesions or severity of clinical symptoms; these cases are, however, in the minority.

The failure rate has been measured for several years in national programs where short-course chemotherapy regimens are routinely applied (32–34). In six countries, a regimen of 8 months, 2SRHZ/6TH, was applied to large groups of patients. The results published for the period 1983–1991 show that the failure rate calculated using the cohort analysis method recommended by WHO and the IUATLD is, on average, 2%; it is 2.2%–3.3% if one takes into account only those cases for whom the bacteriological results at the end of treatment are available (Table 12.7).

In the same countries, the standard retreatment regimen recommended by WHO and the IUATLD was applied to all cases of failure and gradual reactivation after a first course of chemotherapy. This retreatment regimen consists of an 8-month association of isoniazid, rifampicin, and ethambutol with a supplement of pyrazinamide for the first 3 months and streptomycin for the first 2 months. This regimen, which can be applied even in situations where sensitivity tests are not

Table 12.7. Failure of chemotherapy in program conditions among smear-positive pulmonary tuberculosis cases treated with an 8-month regimen (2SRHZ/6TH)

| Country | Patients assessed | Results at the end of treatment (%) | | | | | |
		Cured[a]	Completed treatment	Failed	Died	Lost	Transferred out
Tanzania (1983–91)	55,330	79[b]	—	2	7	8	4
Malawi (1984–90)	12,593	85[b]	—	1	8	2	4
Mozambique (1985–90)	6,015	70	7	2	2	11	8
Benin (1984–90)	5,112	80[b]	—	2	6	11	1
Nicaragua (1985–90)	2,475	60	18	2	3	12	5
Vietnam (1989–91)	8,953	90[b]	—	2	2	4	2

[a]Cured: smear negative after completion of chemotherapy; Treatment completed: without smear examination at the end of chemotherapy.
[b]Including cases with treatment completed and smear-negative results at the end of the second month.

Table 12.8. Failure of retreatment regimen under program conditions (8-month regimens: 2SERHZ/1ERHZ/5ERH or $5E_3R_3H_3$)

Country	Patients assessed	Cured[a]	Completed treatment	Failed	Died	Lost	Transferred out
			Results at the end of treatment (%)				
Tanzania (1983–91)	4,922	72	—	4	9	11	4
Malawi (1984–90)	1,410	89	—	3	4	4	—
Mozambique (1985–90)	3,124	72	—	3	3	14	8
Benin (1984–90)	612	66	—	5	8	18	3
Nicaragua (1985–90)	751	62	—	9	4	22	3
Vietnam (1989–91)	2,660	82	—	10	4	2	2

[a]Including cases who have completed treatment with smear-negative results at the end of the third month.

available (which is more often than not the case in countries with a high prevalence of tuberculosis), has a low failure rate; in fact, it leads to cure in already treated cases with bacilli still sensitive to isoniazid and rifampicin, as well as in cases with bacilli resistant to isoniazid but still sensitive to rifampicin. Under program conditions, the failure rate of this retreatment regimen is from 4% to 10%; it is from 3.2% to 12% if one takes into account only those cases for whom a bacteriological result at the end of treatment is available (Table 12.8).

3.3. Failures and Relapses Observed in Programs

In well-established antituberculosis programs, it has for several years been possible, over an observation period of 2–5 years after the end of treatment, to study all events that take place during and after retreatment. One can also measure the cumulative rate of failures and relapses, reclassify patients according to the results obtained after retreatment, and identify those cases who still had positive smears at the last test (35,36). In these studies, undertaken retrospectively or prospectively, all cases admitted to treatment during a given period and all available data and bacteriological results collected during the observation period are taken into account, whatever the history of the patients initially started on chemotherapy: Regimen applied as planned at the beginning or modified (in length, or due to toxicity, or following complications due to concomitant disease).

In a retrospective study based on the results of chemotherapy regimens (of 4, 5, or 6 months) in 10 Hong Kong outpatient clinics, the overall rate of relapses observed 5 years after chemotherapy was 6% for cases of smear-positive pulmonary tuberculosis and 1% of smear-negative but initially culture-positive pulmonary tuberculosis cases (37).

In a prospective study in Algeria (10), the overall results of two short-course chemotherapy regimens (6 and 8 months) were compared. In the case of failure or relapse, the retreatment regimen always consisted of 6 months of combined isoniazid and rifampicin. The rate of failure and relapse requiring a second course of chemotherapy was similar: 2.1% and 2.4%, whether the duration of chemotherapy originally prescribed was for 6 or 8 months. By the same token, the proportion of patients who were still culture positive at the last examination in the posttreatment observation period is also similar: 1.3% and 1.8%. Nine of these 35 patients were chronic cases (Table 12.9).

This study comparing the overall results of two short-course chemotherapy regimens applied under program conditions in 30 districts of Algeria demonstrates

Table 12.9. Failure and relapse of chemotherapy patients under program conditions; comparison of the global outcome of two short-course regimens 2 years after completion of allocated regimen

| | Regimens initially allocated | | | |
| | 6-month 2SRHZ/4RH | | 8-month 2SRHZ/2RH/4H | |
Final outcome	*n*	%	*n*	%
Cured	852 (19)	87.2	1068 (22)	86
12–36 months after treatment	748 (17)		914 (20)	
0–11 months after treatment	104 (2)		154 (2)	
Treatment completed (no bacteriology after treatment)	31	3.1	61 (1)	5
Culture positive when last seen	13 (2)	1.3	22 (7)	1.8
18 months after treatment	7 (2)		15 (7)	
0–17 months after treatment	6		7	
Defaulters during treatment	35	3.6	40	3.2
Died (all causes)	46	4.8	50	4
During treatment	39		33	
After treatment	7		17	
Total assessed	977 (21)	100	1241 (30)	100

Note: Numbers in parentheses refer to patients who received an additional course of chemotherapy due to failure, relapse, or development of a nonpulmonary lesion.

that the cumulative total of failures and relapses after short-course treatment is under 3%, whatever the duration—6 or 8 months—of the regimen initially prescribed, and that the rate of chronic cases after 2 years (true failure of a program) is under 1% (10).

4. Management and Prevention of Failure

4.1. *Management of failures*

BEFORE MANAGING FAILED CASES IT IS IMPORTANT TO DEFINE THEM CORRECTLY

- Bacteriologically confirmed failures of chemotherapy cannot usually be identified before the end of 5 months of treatment. Certain patients with smear-positive pulmonary tuberculosis receiving regular and appropriate treatment can continue to excrete bacilli at the end of the third or fourth month of treatment; this possibility, although rare, should be kept in mind.

  However, a patient excreting increasing numbers of bacilli at the second, third, and fourth monthly examinations could be classified as a failure before the fifth month if the case was smear negative before treatment or if the treatment was inappropriate. Before confirming the failure, it should be verified that the patient does, in fact, swallow the drugs prescribed.
- Certain failures can be linked to major digestive intolerance: Vomiting immediately after taking the drugs prevents them from being absorbed into the bloodstream. Likewise, severe side effects, such as generalized hypersensitivity, purpura, hepatitis, and jaundice, can oblige the patient to interrupt treatment for some time. These phenomena lead to delays in conversion and, more often, the prescription of a more adapted and better tolerated treatment. In clinical trials, these cases are analyzed separately; they are at times classified as failures in programs, as the reason for the change of treatment is not always indicated in routine reports.
- It is very unusual for a failure or relapse not to be proven by a positive bacteriological test (microscopy or culture). Isolated deterioration of radiological lesions with negative bacteriology can be an indication of the potential existence of a disease other than active tuberculosis: acute bacterial infection, bronchopulmonary mycosis, or cancer.

ANOTHER PRECAUTION IS NOT TO CONFOUND FAILURE WITH RISK OF FAILURE

In countries or health services where sensitivity tests are performed in all patients before treatment, it is not unusual for treating physicians to decide to modify the chemotherapy prescribed depending on the laboratory result, even when there is no unfavorable progress. This practice leads to the overestimation of the number of failures, to an increase in cases admitted to standard retreatment or atypical regimens, and thus to treatment chaos, which is not always of any benefit to the patients.

The optimal way of managing the problem of failures is the adoption of the treatment categories and standardized chemotherapy regimens recommended by WHO.

MANAGEMENT OF RISKS OF FAILURE AND FAILURES

- In patients who have never before been treated (new cases), it is unnecessary to request a sensitivity test to indicate the appropriate treatment. Short-course chemotherapy containing a combination of four drugs during the initial phase (and at least isoniazid, rifampicin, and pyrazinamide) cures more than 96% of cases with initially sensitive bacilli and almost 90% of cases with initially resistant bacilli.

   If primary resistance is recognized during treatment (from the laboratory results of pretreatment samples), it is preferable not to modify the chemotherapy prescribed, but simply to reinforce bacteriological surveillance of the patient (monthly or bimonthly checks), particularly if primary resistance involves isoniazid and/or rifampicin, rather than change the chemotherapy prematurely without proof of failure.

- In patients who have received treatment previously, a sensitivity test performed in satisfactory conditions, in a qualified and controlled laboratory, can help to lighten the standardized retreatment regimen. The administration of ethambutol can be stopped until the eighth month if sensitivity to isoniazid and rifampicin is conserved. Isoniazid can be interrupted if a high level of resistance to it is recognized (Table 12.10).

- It is only after the failure of two complete courses of short-course chemotherapy (the second at least, of which, should have been directly observed) (i.e., after more than 10 or 14 months in total of treatment with isoniazid and rifampicin) that a test of sensitivity to the drugs should be performed on the bacillary strain isolated from the patient, with a view to reorienting treatment toward a new course of chemotherapy. This new prescription must be adapted to the results of the sensitivity test and combine the drugs to which the strain is still sensitive, as well, if possible, as other second-line or new drugs if available. In such situations, more than half of patients are resistant to isoniazid and rifampicin at least, and the chances of successful treatment are limited (Table 12.11).

4.2. Prevention of Failure

Even if one uses the most effective antituberculosis drugs (i.e., isoniazid, rifampicin, and pyrazinamide combined during the first 2 months of treatment),

Table 12.10. Highly efficient short-course chemotherapy regimens recommended for tuberculosis programs

Patient category	Optimal regimens	Alternative regimens
New cases (smear positive)	2ERHZ/4RH or 4R₃H₃	2ERHZ/6TH or EH
Severe cases (pulmonary and extra pulmonary)	2SRHZ/4RH or 4R₃H₃	2SRHZ/6TH or 6EH
Failures, relapses (smear or culture positive) after one course of chemotherapy	2SERHZ/1ERHZ/5ERH or 5/E₃R₃H₃	

Table 12.11. Acquired bacterial resistance in 81 patients previously treated by one or more courses of chemotherapy containing isoniazid and rifampicin for 6 months; Algeria 1992

6-month courses of chemotherapy with isoniazid and rifampicin	Patients assessed	Strains resistant			
		to one drug at least		to isoniazid and rifampicin at least	
		n	%	n	%
One course	27	9	33	6	22
Two courses	22	14	64	12	55
Three courses or more	32	32	100	31	97
Total	81	55	68	49	60

Source: Ref. 38.

Table 12.12. Level of relapses in regimens containing isoniazid, pyrazinamide, and rifampicin in patients with smear-positive tuberculosis and fully drug-sensitive strains

Duration of chemotherapy (months)	Patients assessed	Bacteriological relapses (%)	95% Confidence limits
6	432	1	0.3–2
4.5–5	598	4	3–6
4	364	12	9–16
3	361	16	12–20

Source: Data from Refs. 19 and 39.

courses that are insufficiently long are the principal cause of failure (Table 12.12). To avoid this risk and thus prevent treatment failure, it is the responsibility of the health personnel situated at the different levels of the hierarchy to combine the prescription of the most efficient short-course chemotherapy regimens with organizational measures (40).

1. At the central level, the main responsibility of the program managers is to organize the regular supply of antituberculosis drugs and their distribution among the health services in order to avoid shortages (41). This involves selection of standardized chemotherapy regimens to be applied, quantification of needs depending on existing usable stocks and on the results of case notifications in recent years, purchase of quality-controlled drugs at the best prices on the international market, creation and maintenance of reserve stocks, organization and verification of stocks in central and provincial depots, and organization of distribution to districts.

If the whole of this plan of action has not been provided for, sudden budgetary restriction measures or devaluations in the local currency can seriously perturb the supply of antituberculosis drugs and compromise the application of treatment programs. Repeated shortages of drugs have been the most frequent cause of premature interruption of antituberculosis treatment in many low- or middle-income countries (in Latin America and Africa particularly) since the beginning of the nineties.

2. Regular prescription by physicians of the most effective standard regimens taking into account the country's available resources implies the continued support of medical training, both in medical schools and in postgraduate courses. Several surveys have shown that even when a tuberculosis program is well established and the standard treatments well codified, doctors often prescribe inappropriate and, at times, unacceptable treatment that is unnecessarily long or modified for no reason (37,42–44). Anarchic prescription of drugs contributes to the creation of drug shortages in countries with limited resources and to the undermining of patient confidence in the technical guidelines of the national tuberculosis program.

3. The distribution of drugs to patients is the responsibility of the health personnel.

- In public health services, the conditions under which the drugs are delivered to patients must be carefully planned, whether it be delivery of a supply of drugs for a given period or treatment that is directly observed by a health worker.

  To improve patient reception and reduce waiting times, it is a good idea to appoint special days or times for receiving the patients under treatment. This enables staff to establish personal contact, to transmit adapted health education, and to reinforce treatment compliance (45,46). In this way, it becomes possible to give special attention and more time to patients with particular problems—medical, socioeconomic, or psychological—and as a result if necessary to adapt the treatment or its method of administration accordingly (13).

The application of fully supervised treatment, daily during the initial phase or three times per week throughout treatment or only during the continuation phase, is a good way of checking a patient's compliance with treatment. This method is indispensable for treating certain categories of patient whose voluntary cooperation cannot be relied on (e.g., prisoners, drug users, alcoholics, and those with psychological disorders). It can also be useful in urban areas with easy access to health services and enough health personnel to put it into practice. Its effectiveness in reducing both the failure rate and the rate of bacterial resistance has recently been proven (47). However, it is not always possible to apply "directly observed treatment" in rural areas of poor countries with difficult access to health services and reduced health staffs. In these cases, solutions can be found in the form of fixed-dose combinations of drugs (isoniazid and rifampicin; isoniazid and thiacetazone; isoniazid and ethambutol), which facilitate self-administered treatment during the continuation phase while eliminating the risk of accidental monotherapy. The use of blister packs containing a week's worth of treatment is an alternative to consider if it is not too expensive for the program.

- In private sector health services (private clinics or medical practices, private pharmacists), which in some countries distribute most of the antituberculosis drugs used, it is difficult to obtain the same level of compliance from health personnel and patients, particularly as patients do not always have the financial resources to pay for each visit or for all of the drugs prescribed (44). In order to avoid increases in rates of bacterial resistance due to treatment chaos, drug-regulation measures have been imposed in certain low- or middle-income countries with a high prevalence of tuberculosis: free supplies of antituberculosis drugs delivered in health services only to those patients with proven, notified tuberculosis; prohibition of sales of separate antituberculosis drugs to the public; and sales authorized only for combined, fixed-dose preparations. But these measures often encounter opposition from physicians and chemists in the private sector or from other influential groups in society.

4. Visits made by the supervisor to the health centers where tuberculosis treatment takes place has a visible effect on the reduction of failure rates and of the number of cases lost to follow-up before the end of treatment. A national survey performed in Algeria in 1985 compared two similar groups in districts which had applied the same chemotherapy regimen for 5 years: One group received regular supervisory visits, the other did not. In those that were visited regularly, the rate of patients transferred out or lost to follow-up was half that of the other group (40).

5. With the creation of a permanent system of evaluation of treatment results, based on the analysis of treatment outcome in patient cohorts according to quarterly reports (48), the failure rates in chemotherapy regimens can be monitored

and the main deficiencies at province and district levels can be targeted. Corrective measures can thus be applied at central or local levels.

6. All of these organizational measures, which are the direct responsibility of the health personnel, gives the individual responsibility of patients in treatment failure its proper perspective. Patients' negligence and absence of motivation and, at times, their noncooperation with or refusal of treatment are often referred to by health personnel to disguise their own negligence. The identification of underlying causes such as psychological or social difficulties, encountered by some patients in following their treatment regularly, is still one of the responsibilities of the health workers who, with the help of dialoguing with the patients and their families, can find an appropriate solution.

5. Conclusion

Failures in modern chemotherapy are rare, both in clinical trials and in programs. However, as they can be more severe, because of the risks of the emergence of bacilli multiresistant to isoniazid and rifampicin, detection and prevention necessitate the adoption of short-course chemotherapy regimens for all patients and vigilance throughout treatment to prevent premature interruption.

The failure rate constitutes one of the main epidemiological indicators for monitoring tuberculosis chemotherapy programs in the community.

References

1. WHO (1993) Treatment of Tuberculosis, Guidelines for National Programmes. Geneva: World Health Organization.
2. Crofton J (1980) Failure in the treatment of pulmonary tuberculosis: potential causes and their avoidance. Bull Int Union Tuberc Lung Dis 55(3–4): 93–99.
3. Chaulet P (1987) Compliance with antituberculosis chemotherapy in developing countries. Tubercle 68(Suppl):19–24.
4. Gangadharam PRJ (1994) Chemotherapy of tuberculosis under program conditions— with special relevance to India. Tuberc Lung Dis 75:241–244.
5. IUATLD (1988) Antituberculosis regimens of chemotherapy. Recommendations from the Committee on Treatment of the IUATLD. Bull Int Union Tuberc Lung Dis 63:60–64.
6. British Thoracic Society (1984) A controlled trial of 6-month chemotherapy in pulmonary tuberculosis. Final report. Results during 36 months after the end of chemotherapy and beyond. Br J Dis Chest 78:330–336.
7. Singapore Tuberculosis Service/British Medical Research Council (1981) Clinical trial of six-month and four-month regimens in the treatment of pulmonary tuberculosis. The results up to 30 months. Tubercle 62:95–102.

8. Fifth East African/British Medical Research Council Study (1986) Final report: controlled clinical trial of four short course regimens of chemotherapy (three 6-month and one 8-month) for pulmonary tuberculosis. Tubercle 67:5–15.

9. Berkani M, Chaulet P, Darbyshire JH, Nunn A, Fox W (1986) Résultats d'un essai thérapeutique comparant un régime de 6 mois à un régime de 12 mois dans la tuberculose pulmonaire au Sahara Algérien. Rapport final: résultats trois ans après le début du traitement. Rev Mal Respir (Paris) 3:73–85.

10. Algerian Working Group/British Medical Research Council Co-operative Study (1991) Short course chemotherapy for pulmonary tuberculosis under routine programme conditions: a comparison of regimens of 28 and 36 weeks' duration in Algeria. Tubercle 72:88–100.

11. Zidouni N, Chaulet P (1991) Evaluation of the treatment results of pulmonary tuberculosis in a community survey in Algeria, including the comparison of two (6-month and 8-month) short course chemotherapy regimens. In: The Tuberculosis Surveillance Research Unit of the IUATLD. Progress Report Vol. 1, The Netherlands: KNCV pp. 29–44.

12. Singapore Tuberculosis Service/British Medical Research Council (1988) Five-year follow-up of a clinical trial of three six-month regimens of chemotherapy given intermittently in the continuation phase in the treatment of pulmonary tuberculosis. Am Rev Respir Dis 137:1147–1150.

13. Hong Kong Chest Service/British Medical Research Council (1991) Controlled trial of 2, 4 and 6 months of pyrazinamide in six-month three-times-weekly regimens for smear-positive pulmonary tuberculosis, including an assessment of a combined preparation of isoniazid, rifampicin and pyrazinamide. Results at 30 months. Am Rev Respir Dis 143:700–706.

14. Mazouni L, Tazir M, Boulahbal F, Chaulet P (1985) Enquête contrôlée comparant trois régimes de chimiothérapie quotidienne de six mois dans la tuberculose pulmonaire en pratique de routine à Alger. Résultats au 30e mois. Rev Mal Respir Paris (Paris) 2:209–214.

15. Combs DL, O'Brien R, Geiter L (1990) USPHS tuberculosis short course chemotherapy trial 21. Effectiveness, toxicity and acceptability. The report of final results. Ann Intern Med 112:397–406.

16. Chaulet P, Boulahbal F (1995) Essai clinique d'une combinaison en proportions fixes de trois médicaments dans le traitement de la tuberculose. Tubercle Lung Dis, 76:407–412.

17. Third East Africa/British Medical Research Council Study (1980) Controlled clinical trial of four short-course regimens of chemotherapy for two durations in the treatment of pulmonary tuberculosis. Second report. Tubercle 61:59–69.

18. East and Central Africa/British Medical Research Council (1983) Controlled trial of four short-course regimens of chemotherapy (three 6-month and one 8-month) for pulmonary tuberculosis. Fifth collaborative study. First report. Tubercle 64:153–166.

19. Fox W (1984) Short course chemotherapy for pulmonary tuberculosis and some

problems of its application with particular reference to India. Lung India 2(2):161–174.

20. Coates ARM, Mitchison DA (1983) The role of sensitivity tests in short-term chemotherapy. Bull Int Union Tuberc 58(2):111–114.

21. Hong Kong Tuberculosis Treatment Service/British Medical Research Council Investigation (1972) A study in Hong Kong to evaluate the role of pretreatment susceptibility tests in the selection of regimens of chemotherapy for pulmonary tuberculosis. Am Rev Respir Dis 106:1–22.

22. Abderrahim K, Chaulet P, Oussedik N, Amrane R, Si Hassen C, Mercer M (1976) Practical results of standard first-line treatment in pulmonary tuberculosis: influence of primary resistance. Bull Int Union Tuberc 51:359–366.

23. Aber VR, Nunn AJ (1978) Factors affecting relapse following short-course chemotherapy. Bull Int Union Tuberc 53:260–264.

24. Aber VR, Nunn AJ, Darbyshire JH (1984) Predicting a successful outcome in short course chemotherapy. Bull Int Union Tuberc 59:22–23.

25. Hong Kong Chest Service/Tuberculosis Research Centre Madras/British Medical Research Council (1989) A controlled trial of 3-month, 4-month and 6-month regimens for sputum smear-negative pulmonary tuberculosis. Results at 5 years. Am Rev Respir Dis 139:871–876.

26. Hong Kong Chest Service/Tuberculosis Research Centre Madras/British Medical Research Council (1984) A controlled trial of 2-month, 3-month and 12-month regimens of chemotherapy for sputum smear-negative pulmonary tuberculosis. Results at 60 months. Am Rev Respir Dis 130:23–28.

27. IUAT (1969) A comparison of regimens of ethionamide, pyrazinamide and cycloserine in retreatment of patients with pulmonary tuberculosis. Bull Int Union Tuberc 42:7–57.

28. Hong Kong Tuberculosis Treatment Service/Brompton Hospital/British Medical Research Council Investigation (1974) A controlled clinical trial of daily and intermittent regimens of rifampicin plus ethambutol in the retreatment of patients with pulmonary tuberculosis in Hong Kong. Tubercle 55:1–27.

29. Ait-Khaled N, Benadjila H, Loucif MS, Mounedji A, Chaulet P (1976) Controlled therapeutic trial on three second line chemotherapy regimens (one of short duration) in pulmonary tuberculosis. Bull Un Int Tuberc 51(1):95–102.

30. Swai OB, Aluoch JA, Githui WA, Thiongo'o R, Edwards EA, Darbyshire JH, Nunn AJ (1988) Controlled clinical trial of a regimen of two durations for the treatment of isoniazid-resistant pulmonary tuberculosis. Tubercle 69:5–14.

31. Enarson DA, Rieder HL, Arnadottir T (1994) Tuberculosis Guide for Low Income Countries, 3rd Ed., Paris: IUATLD.

32. Broekmans JF (1994) Control strategies and programme management. In: Porter JDH, McAdam KPWJ, eds. Tuberculosis, Back to the Future, pp. 171–179. Chichester: John Wiley and Sons.

33. Nyangulu DS (1991) The point of view of a high-prevalence country: Malawi. Bull Int Un Tub Lung Dis 66:173–174.

34. Salomao MA (1991) The national tuberculosis control program in Mozambique, 1985–1988. Bull Int Un Tuberc Lung Dis 66(4):175–178.

35. Chaulet P (1990) La lutte antituberculeuse dans le monde: stratégies et actions sur le terrain. Respiration 57:145–159.

36. Chaulet P, Zidouni N (1993) Evaluation of applied strategies of tuberculosis control in the developing world. In: Reichman LB, Hershfield ES, eds. Tuberculosis, a Comprehensive International Approach, pp. 601–627. New York: Marcel Dekker, Inc.

37. Chan SL, Wong PC, Tam CM (1994) 4-, 5- and 6-month regimens containing isoniazid, rifampicin, pyrazinamide and streptomycin for treatment of pulmonary tuberculosis under program conditions in Hong Kong. Tubercle Lung Dis 75:245–250.

38. Mazouni L, Zidouni N, Boulahbal F, Chaulet P (1992) Treatment of failure and relapse cases of pulmonary tuberculosis within a national programme based on short course chemotherapy (preliminary report). In: Tuberculosis Surveillance Research Unit of the IUATLD, Progress Report The Netherlands: KNCV Vol. 1, pp. 36–42.

39. Fox W (1981) Whither short course chemotherapy? Br J Dis Chest 75:331–357.

40. Chaulet P (1991) Compliance with chemotherapy for tuberculosis. Responsibilities of the health ministry and of physicians. Bull Int Union Tuberc Lung Dis 66(Suppl 90–91):33–36.

41. Chaulet P (1992) The supply of antituberculosis drugs and national drugs policies. Tubercle Lung Dis 73:295–304.

42. Fox W (1993) Compliance of patients and physicians: experience and lessons from tuberculosis I. Br Med J 287:33–35.

43. Fox W (1993) Compliance of patients and physicians: experience and lessons from tuberculosis II. Br Med J 287:101–105.

44. Uplekar MW, Rangan S (1993) Private doctors and tuberculosis control in India. Tubercle Lung Dis 74:332–337.

45. Snider DE (1982) Vue générale des problèmes d'adhérence aux prescriptions en matière de programmes de traitement de la tuberculose. Bull Union Int Tuberc 57(3–4):255–260.

46. Sbarbaro JA (1991) Patient compliance with chemotherapy operational considerations. Bull Int Un Tuberc Lung Dis 66(Suppl. 90–91):37–40.

47. Weis SE, Slocum PC, Blais FX, King B, Nunn M, Matney GB, Gomez E, Foresman BK (1994) The effect of directly observed therapy on the rates of drug resistance and relapse in tuberculosis. N Engl J Med 330:1179–1184.

48. WHO Tuberculosis Programme (1991) Tuberculosis surveillance and monitoring. Report of a WHO workshop. WHO/TB/91–163. Geneva: World Health Organization.

13

New Drugs and Strategies for Chemotherapy of Tuberculosis

Pattisapu R.J. Gangadharam

Until the middle of this century, tuberculosis had been a serious and often life-threatening disease. Starting from the late forties, following the introduction of specific drugs, the disease has been brought rapidly to a stage of a curable one. Even though some epidemiologists attribute this rapid success to the uplift of economic situation and improvement of living conditions in some affluent countries, by and large, availability of several powerful drugs and, more importantly, the evolution of optimal regimens and proper application of specific antituberculosis chemotherapy with these drugs should be given credit for this success. It should be stressed that all these drugs had been brought to clinical use by major inputs by the pharmaceutical industry by extensive investigations to develop the initial discovery from their own laboratories or by academic institutions. As discussed later, this sort of fortunate situation is not existing now!

Optimal regimens to make the best use of these drugs, to get rapid cure of the disease, have been arrived at following several controlled clinical trials conducted around the world. Credit should be given to many developing countries (e.g., India, Singapore, Hong Kong, Kenya, Rhodesia, Tanzania, and others) and various international organizations [e.g., the British Medical Research Council (BMRC), World Health Organization (WHO), the International Union Against Tuberculosis and Lung diseases (IUALTD), The Indian Council for Medical Research (ICMR), East African Tuberculosis Research Organization, tuberculosis associations in Hong Kong, Singapore, and other places] for this achievement. In the United States, the U.S. Public Health Service and the Veterans Administration–Armed Forces (VA-AF) Cooperative Studies played a major role. A continuous reduction of tuberculosis about 10% every year was noted in the United States until about the mid-eighties (1). Such a success was also noted in many other developed countries (e.g., the Netherlands, the Scandinavian countries, the United Kingdom, etc.). At about 1984, the annual reduction of the prevalence in the United States stopped (1) and even took an upward trend until 1994, after which period, during the past few years, there has been a 6% reduction (2). This increase during 1984 to 1994 has been attributed to the large influx of immigrants from countries with high incidence of tuberculosis, to "negligence" of the disease, and to the failure of the infrastructure (3). More unfortunately, most of the "excess" cases have been shown to have multidrug resistant (MDR) tubercle bacilli (i.e., bacilli which are resistant to at least two drugs, mostly isoniazid and rifampin).

The dramatic success of chemotherapy seen in the fifties and sixties has perhaps "overexcited" the health authorities in the United States and other affluent countries, which has resulted in a premature abandonment of their interest, let alone their concern over this disease. The pharmaceutical industry, which was playing a major role earlier, virtually eliminated their antituberculosis programs. The VA-AF who were conducting many cooperative trials in chemotherapy of tuberculosis and annual conferences have stopped these activities by 1972. More and more scientists in the academic institutions were diverting their interests to other areas. Even in the developing countries, most of which had not witnessed any dramatic improvement in the control of the disease similar to that seen in the advanced countries, interest in the disease declined rapidly. This attitude might have been prompted by the closing of sanatoria and shifting the treatment to outpatient settings.

World bodies got concerned continuously over this situation and the IUATLD devoted entire sessions in many of their global conferences to topics such as "How to inform the general public that tuberculosis is still not under control." Several pamphlets and brochures issued by WHO stress this aspect. More seriously, the global tuberculosis problem is aggravated by the human immunodeficiency virus (HIV) infection and acquired immune deficiency syndrome (AIDS). The emergent global impact of the two serious life-threatening infections has been a serious concern of many international authorities. Other chapters in this book deal extensively on the chemotherapy of tuberculosis in presence of HIV infections and the seriousness of these two major diseases in conjunction.

In such a grim and near "panic" situation, frantic attempts are sought to discover new drugs, as most of the available drugs are ineffective because of the high prevalence of MDR disease. The pharmaceutical industry, which in earlier times was ready to get involved in antituberculosis drugs, is now very reluctant to do so; in fact, most companies have abandoned their facilities for such expertise. It was stated that it costs approximately $200 million to develop an antituberculosis drug, whereas the global market (not considering the "discounts" negotiated by the international health agencies like the WHO and the IUATLD) is only around $150 million (3). Lack of financial incentives, thus, added to their reluctance to get involved in this area. On the other hand, attempts are made continuously to discover new drugs or to uncover new potentialities of existing or some old drugs to be able to offer some leads to tackle this situation. The attempts thus made are discussed under these various categories:

1. Conventional screening of several drugs from (a) plant and (b) synthetic origins. Compounds in the second category can be grouped as (i) fluoroquinolines, (ii) macrolides, (iii) nitroimidazoles, (iv) oxazolidones, and (v) poloxamer surfactants.
2. Rediscovery of some old drugs. These include (a) clofazimine, (b) paromomycin, and (c) allicin.
3. Structural analogs and derivatives of existing drugs—isoniazid, rifampin, and pyrazinamide.
4. Modification of permeability barriers in drug-resistant tubercle bacilli, so that the drugs would be made active on the bacilli which were resistant to the same drugs.
5. Metabolic inhibitors based on the available knowledge of mechanism of action and biochemistry of tubercle bacilli.
6. Approaches based on the available knowledge on molecular biology of tubercle bacilli.
7. Attempts to harness the maximum benefit from the existing drugs by circumventing the host influences on the drugs

In many of the studies, in vitro, macrophage and in vivo models have been used, whereas in some, only in vitro screening was done. At the other extreme, some studies have progressed to clinical investigations.

1. Conventional Screening

1.1. Plant Sources

Considering the seriousness of the disease, several attempts were made to discover antituberculosis drugs from plant sources. In many developing countries (e.g., India, China, Philippines, Mexico, etc.), the indigenous systems and traditional folk medicines are flourishing very well in close competition with Western medicine and have descriptions of several remedies in their pharmacopeias. Allicin, a compound isolated from *Allium sativum* (common garlic) which was shown a long time ago to possess powerful antituberculosis activity by Rao et al. (4) is being reinvestigated recently by in vitro studies (5). More recently, tryptanthrisis isolated from higher plants have been shown to have high in vitro activity against *M. tuberculosis* (M.tb) and the *M. avium* complex (MAC) (6).

1.2. Synthetic Drugs

FLUOROQUINOLINES

Even though quinolines were discovered as early as 1960, their clinical use commenced only in 1983, following the discovery of ciprofloxacin. Further growth rapidly ensued following the discovery of the DNA gyrase enzyme, which facilitated the introduction of several fluoroquinolines. Until recently, most of the work with these compounds dealt with gram-positive and gram-negative organisms and, only recently, they have been investigated for potential antimycobacterial activity. Among several fluoroquinolines screened, only four (Fig. 13.1)—ciprofloxacin, ofloxacin, levofloxacin, and sparfloxacin—showed considerable antimycobacterial activity (7,8). There are several other drugs, e.g., Bay 43118, DU 5859a, enoxacin, fleroxacin, and amifloxacin which have been shown to possess some in vitro activity even though they have not been explored to the same extent as the other four drugs.

The four active quinolines appear to be bactericidal against M.tb with minimal inhibitory concentrations (MICs) in the range 0.5–1.0 μg/ml and with minimal bactericidal concentrations (MBCs) approximately double their MICs. In vitro, sparfloxacin has been shown to be superior to the others. Both sparfloxacin and levofloxacin were active against intracellular tubercle bacilli (9), including MDR strains of tubercle bacilli. In the mouse model (10), ofloxacin at a dose of 40 mg/kg caused a significant reduction of the bacillary counts in the liver. Similarly, ciprofloxacin, at a 160-mg/kg dose, showed great reduction of the bacillary load in spleen and liver and it has a synergistic activity with isoniazid (11). Studies by the French investigators (12) have shown the high in vivo activity of sparfloxacin; at a 50-mg/kg dose, it displayed greater bactericidal activity than ofloxacin at a

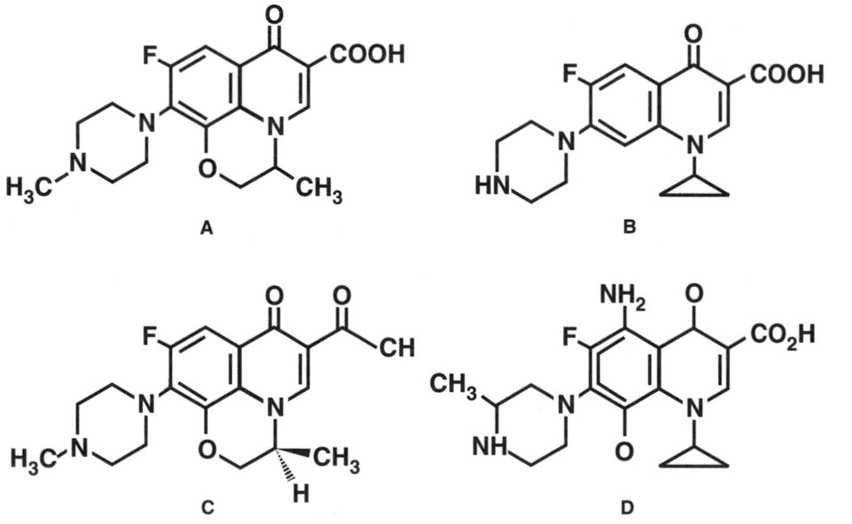

Figure 13.1. The chemical structures of quinolones: (A) ciproflaxacin, (B) ofloxacin, (C) levoflaxacin, (D) sparfloxacin.

300-mg/kg dose. Interestingly, these studies have shown that at a 100-mg/kg dose, sparfloxacin was more active than rifampin, and at a 500-mg/kg dose, it was significantly better than isoniazid and pyrazinamide. Levofloxacin at a 300-mg/ kg dose was bactericidal, and at a 200-mg/kg dose, it had more than twice the activity of ofloxacin (13,14).

Few clinical studies have dealt with ciprofloxacin and ofloxacin. Using the elegant technique of early bactericidal activity (EBA) originally introduced by Mitchison and colleagues (15) (see the Chapter 2 for a detailed discussion on this technique), Kennedy et al. (16) showed that ciprofloxacin at a dose of 750 mg/ kg, given for 7 days to smear-positive adult patients, was effective in that it caused a mean daily fall of 0.20 $\log_{10}$ CFU/ml of tubercle bacilli in the sputum. This activity compares favorably with that of isoniazid, which causes a 0.25 $\log_{10}$ CFU/ ml decrease.

Ofloxacin was investigated in two different studies (17,18) when given alone or in combination with rifampin and isoniazid. In the monotherapy study done in Japan by Tsukumara (17); ofloxacin (300 mg/kg) was given daily for 1 year to 16 patients who failed to respond to different regimens before this trial. Sputum bacterial counts decreased in all patients and culture negativity was seen in 5 (31%) of the 16 patients. Combination chemotherapy containing ofloxacin, isoniazid, and rifampin given to 17 previously untreated patients with pulmonary tuberculosis showed conversion to culture negativity in an average of 1.8 months, which was comparable to the other short-course regimens like isoniazid, rifampin, and ethambutol (18). However, it is doubtful whether any of the fluoroquinolines would find a place in the initial treatment of patients with drug-susceptible tubercle bacilli as in this study.

Interest in uncovering new agents from this area is still continuing, more so for drugs active against MAC disease. Recent drug discovery techniques, including quantitative structure activity relationships (QSAR) with some statistical procedures (e.g., multicase) are used in these approaches (19).

MACROLIDES (AZALIDES)

The search for drugs with antimycobacterial activity from this group of drugs was initiated only recently and, so far, this led to the introduction of at least three potential agents—clarithromycin, azithromycin, and roxythromycin. (Fig. 13.2). All three drugs are more active against MAC than against M.tb; in fact, clarithromycin has become a very important drug for the treatment of MAC disease. Although clarithromycin and roxithromycin are not active against M.tb, with an in vitro MIC of >64.0 μg/ml, they could enhance the activity of isoniazid (INH) and rifampin against MDR tubercle bacilli which are resistant to these drugs. Luna-Herrera et al. (20) have demonstrated synergistic activity of clarithromycin with INH and rifampin against MDR tubercle bacilli, more so in the macrophage

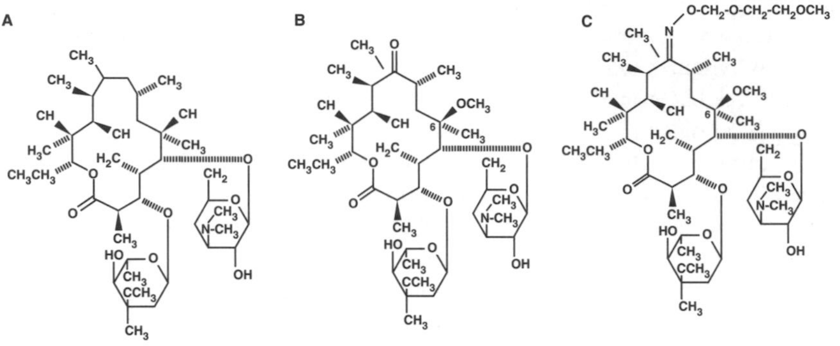

Figure 13.2. The chemical structures of the macrolides: (a) azithromycin, (b) clarithromycin, (c) roxithromycin.

and animal models than in the in vitro studies. The higher activity in these systems has been ascribed to the higher concentration of the drug in macrophages and tissues (e.g., lung).

NITROIMIDAZOLES

Metronidazole (Fig. 13.3), a nitroimidazole, was initially shown to be active against *Trichomonas vaginalis, Giardia lamblia,* and *Entamoeba histolytica,* and some obligate anaerobes like *Bacteriodes fragilis* (21). An interesting finding reported recently (22) was that it is highly active in vitro and more importantly on dormant and slow-growing tubercle bacilli in conditions simulating those existing in host lesions. It is readily absorbed after oral administration and yields a peak concentration of 6–11 μg/ml, after a dose of 250–500 mg/kg; its bactericidal activity against dormant bacilli is seen with 8 μg/ml. Since no antagonism was seen with other antituberculosis drugs, it would be of great value to see its action in combination with other drugs.

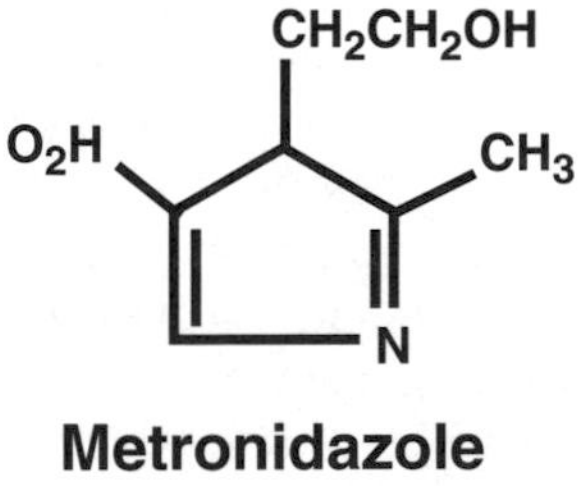

Metronidazole

Figure 13.3. The chemical structure of metronidazole.

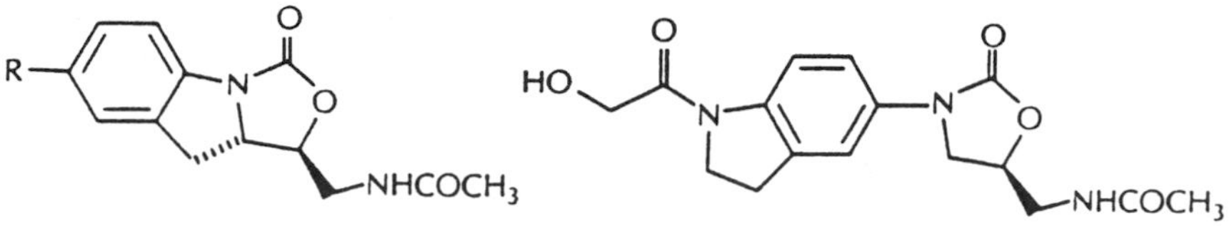

Figure 13.4. The chemical structure of oxozolidinone.

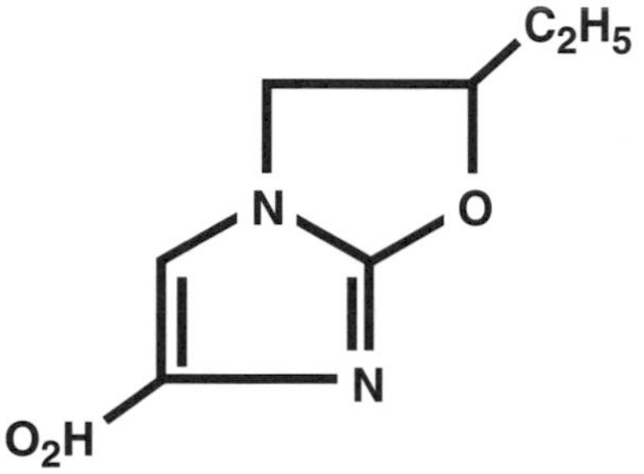

511-(CGI 17341)

Figure 13.5. The chemical structure of 511 (CGI17341).

OXAZOLIDINONES

Oxazolidinones (Fig. 13.4) are a relatively new class of drugs unrelated to other agents with activity against gram-positive and anaerobic bacteria and M.tb (23). They have similar MIC against MDR tubercle bacilli as with susceptible strains. One compound DUP721 (CGP17341) (Fig. 13.5) has an MIC of 0.1–0.3 μg/ml, with no cross-resistance with the antituberculosis drugs and, more importantly, has been shown to be highly protective against experimental tuberculosis infections in mice (24).

POLAXAMER SURFACTANTS

These compounds were earlier referred to as "defense stimulators." Several years ago, Conforth et al. (25) attempted to modify the susceptibility of mice to experimental tuberculosis using some surface-active agents (now called polaxamers) by condensing p-tetra-octylphenol with formaldehyde to give polynuclear compounds which have either a linear or macrocyclic structure.

Subsequently, it was shown that Tyloxpol (popularly known as Triton A-20), an important member in this series, was one-sixth as active in vitro as dihydro-streptomycin and that it acted synergistically with that drug (26). Another surface-active agent, α-methyl-α-n-dodecyl succinic acid (called B-53) was also found to be active in guinea pig tuberculosis. However, it was soon found that its activity was antagonized by serum and other body fluids and was toxic on long-term usage.

Due to the resurgence of tuberculosis, Chinnaswamy and associates (27) took interest in this group of compounds and have evaluated the in vitro, intracellular and in vivo activity of CRL8131, called poloxamer by them. It was shown to have a dose-related in vitro activity, with an MIC of 6.25 μg/ml and on intracellular bacilli, the MIC was 1.25 μg/ml, suggesting that the drug achieves greater intra-cellular concentration. In experimental chemotherapy studies, the CRL8131 group showed a 33% mortality by day 30 and 100% mortality by day 40, whereas thiacetazone produced 66% mortality by 30 and 40 days. Combination chemo-therapy of both drugs showed 0% mortality by 40 days and had a 1 $\log_{10}$ reduction in the CFU in lungs and spleens.

2. Rediscovery of Some Old Drugs

Tentatively, four agents—allicin, surface-active agents (poloxamers), clofazim-ine, and paromomycin—are to be considered in this section. The first two have been discussed earlier and the other two will be considered here.

2.1. Clofazimine and Its Analogs

The history of chemotherapy of tuberculosis abounds in several examples of drugs which were initially abandoned, based on few observations, and could enter the field later, quite often after a lapse of several decades. Pyrazinamide, thioac-etazone, and clofazimine are such examples. Although the former two could come back directly to the realm of antituberculosis chemotherapy, clofazimine had to take a rather circutious route, via leprosy, MAC disease, and then to tuberculosis. (See several interesting reviews on this drug in Refs. 28–31.)

The antituberculosis activity of clofazimine is now actively investigated by us using in vitro, macrophage, and animal models (31–33). In our attempts, besides clofazimine, several other analogs have been investigated with the objective of finding an agent which is (a) more active than clofazimine, (b) which gives higher serum levels than clofazimine, which does not show significant levels in the se-rum, but only high levels in tissues and macrophages, and (c) which causes no or reduced pigmentation. (Clofazimine administration results in reddish-brown pigmentation in the body, a feature which persists in the body for a long time because the drug has a long (>65 days) half-life.)

Of the several analogs tested, only five (B746, B4100, B4101, B4154, and B4157) showed *in vitro* activity greater than clofazimine (Fig. 13.6). B4100 was found to be toxic to macrophages and B4101 was similar in intracellular activity to clofazimine. In the experimental chemotherapy studies, B746 and B4157 showed similar or slightly better activity than clofazimine. More interestingly, both of these analogs caused less pigmentation in the visceral organs than clofazimine. Hyperpigmentation is dose related; when low doses are given, for example, via a depot preparation (as discussed later in this chapter), no pigmentation is seen. When used singly, all the three drugs—clofazimine, B746, and B4157— had similar *in vivo* activity as that of isoniazid and rifampin. Overall, drugs in this series have great chemotherapeutic potential and should be pursued with greater intensity.

2.2. Paromomycin

Paromomycin (Fig. 13.7), an aminoglycoside antibiotic related structurally to streptomycin, neomycin, and kanamycin, was discovered nearly 30 years ago (34). Besides its high activity against many parasitic infections in humans and cattle, it was also found to be active in vitro and in vivo against M.tb (34). However, it was not pursued further as an antituberculosis drug, because of the then-prevalent notion that tuberculosis was getting under control and that there was no necessity for introducing new drugs.

In their attempts to discover the antituberculosis activity of some old drugs, Kanyok et al. (35) have demonstrated its in vitro activity against a number of drug-susceptible and MDR strains of tubercle bacilli as well as several MAC strains. The MICs for M.tb ranged between ≤ 0.09 and $1.5 \, \mu g/ml$ and for MAC, between ≤ 1.56 and $12.5 \, \mu g/ml$. In the in vivo studies conducted by the same group (36), paromomycin was effective in preventing mortality and reducing the CFU counts relative to the mean log CFU in the lungs, spleens, and liver of M.tb-infected C57BL/6. However, at a 100- or 200-mg/kg dose, it was less active than isoniazid (25 mg/kg) in reducing the bacterial load in lungs, spleens, and liver, although it was more active against MDR tubercle bacilli. In studies against MAC-infected beige mice, paromomycin (200 mg/kg) was found to be as active as amikacin (50 mg/kg) in reducing the CFU counts in lungs, liver, and spleens. It was suggested that this drug should be evaluated in combination with other drugs against these infections.

3. Structural Analogs of Some Existing Drugs

In the frantic efforts to strengthen our armamentarium to treat tuberculosis, especially the MDR disease, attempts are being made by several groups of work-

RIMINOPHENAZINES

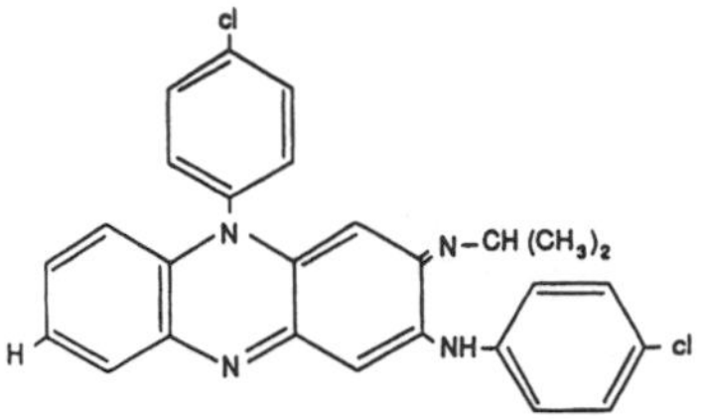

CLOFAZAMINE

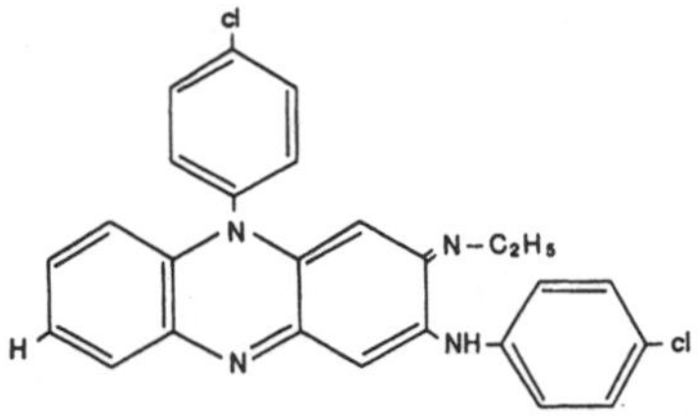

B 746

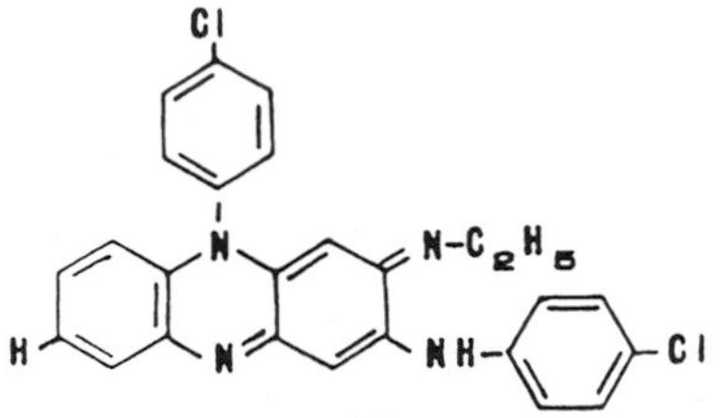

B4100

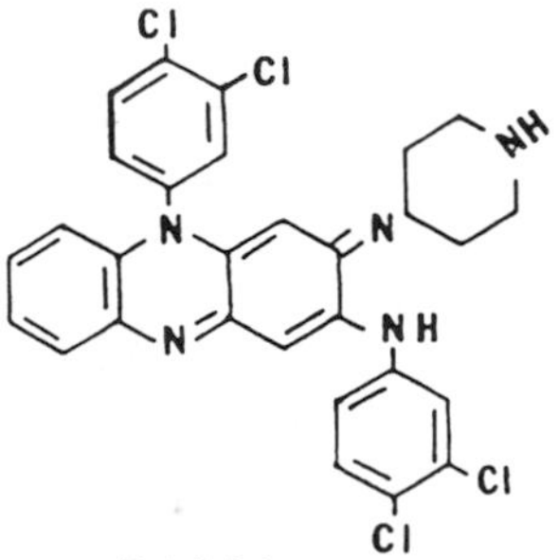

B4101

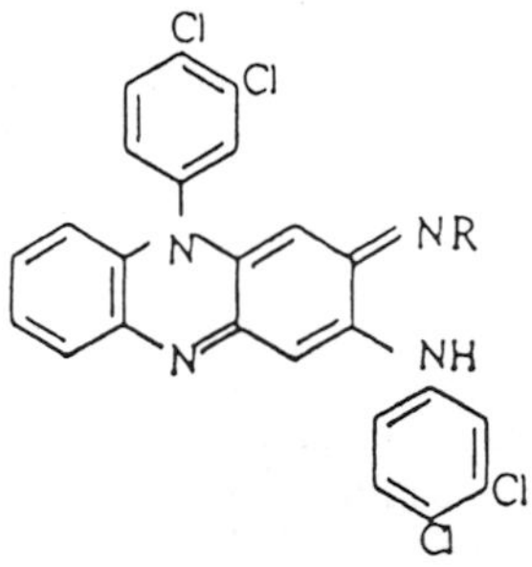

B4154   R= CH₂CH₂CH₂N(C₂H₅)₂

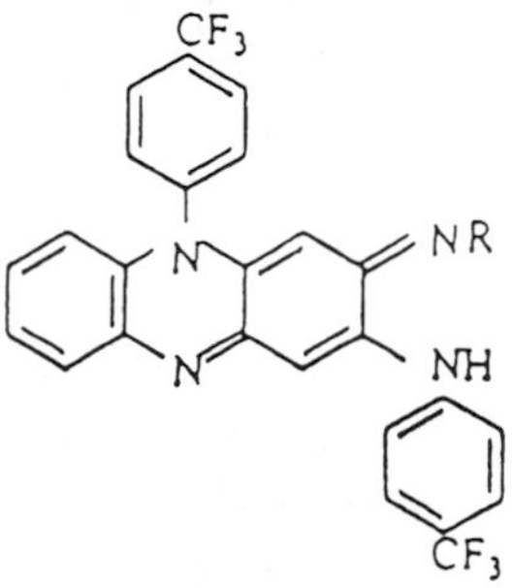

B4157   R= C₂H₅

Figure 13.6. The chemical structures of riminophenazines (clofazimine and its analogs).

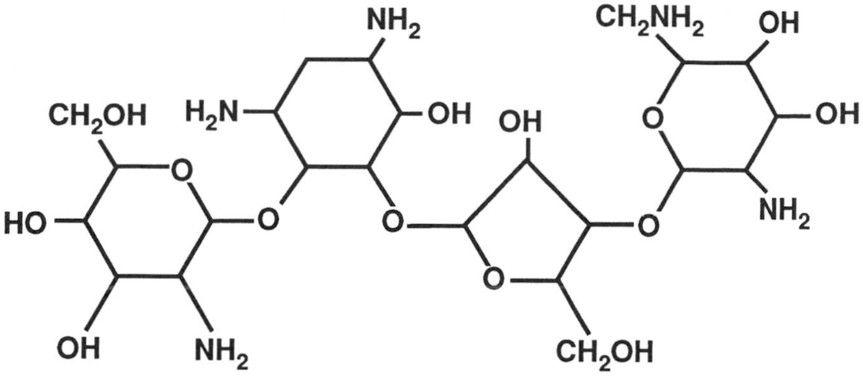

Figure 13.7. The chemical structure of paromomycin.

ers to search for analogs of the existing powerful drugs which are used in the treatment of the disease. Of the five first-line drugs, streptomycin is not pursued further, as its own stay is questioned and serious attempts to discontinue its production are considered. Of some 60 available analogs of ethambutol, none revealed any useful leads (37). However, an interesting property of ethambutol in enabling greater penetration of other drugs, which is based on the knowledge on its mechanism of action, is explored in developing newer strategies for the treatment of MDR tuberculosis. This aspect is discussed in a later section in this chapter. By far, most recent endeavors to uncover active analogs dealt with the other three first-line drugs—isoniazid, rifampin, and pyrazinamide—of which most of the effort was concentrated on rifampin.

3.1. Isoniazid Analogs

Only a few studies were reported concerning the screening for effective analogs of isoniazid. Rastogi and associates (38) tested several modifications of isoniazid that had a greater permeability into the MAC, which have the innate resistance to isoniazid, and have identified one compound, isonicotinyl-2-palmityl hydrazine. In subsequent studies (39), they mentioned that fluorophenyl alanine enhances the activity of this compound against MAC in macrophages.

Another report dealt with polymeric derivation of isoniazid with polysuccinamide (40). These were prepared by condensation of isoniazid with polysuccinamide to obtain water-soluble and water-insoluble compounds in order to get a "less toxic" drug, by using some artificial stomach wall lipid membrane models. However, most of these studies dealt with pharmacological aspects on the release of the drug in simulated gastric juice or plasma rather than identifying a new analog of isoniazid. The readers are referred to the earlier work by Barry and associates (41), who obtained similar results by condensing isoniazid with thiosemicabazone and periodate oxidized starch. This preparation, called "Hincon starch," was shown to give a slow release of isoniazid. (Extensive discussion on the sustained release of isoniazid using depot preparations after impregnation in biodegradable polylactic-co-polyglycolic acid (PLGA) polymer is given in a later section of this chapter.)

3.2 Rifampin Analogs

Exploring the same potential source of the Streptomyces group of fungi, many pharmaceutical industries have developed analogs of rifamycin with greater activity. Some (e.g., rifabutin and rifapentine) have recently been approved for clinical trials in the United States and some (e.g., KRM1648) are in Phase I and Phase II clinical trials. At the other extreme, there are some (e.g., the CGP series) which have been rather prematurely discontinued, consequent to the lack of interest in antituberculous drugs.

RIFAPENTINE

This drug (Fig. 13.8), previously called cyclopentyl rifampin or DL473, was investigated initially by Tsukumara (42) in several comparative in vitro studies

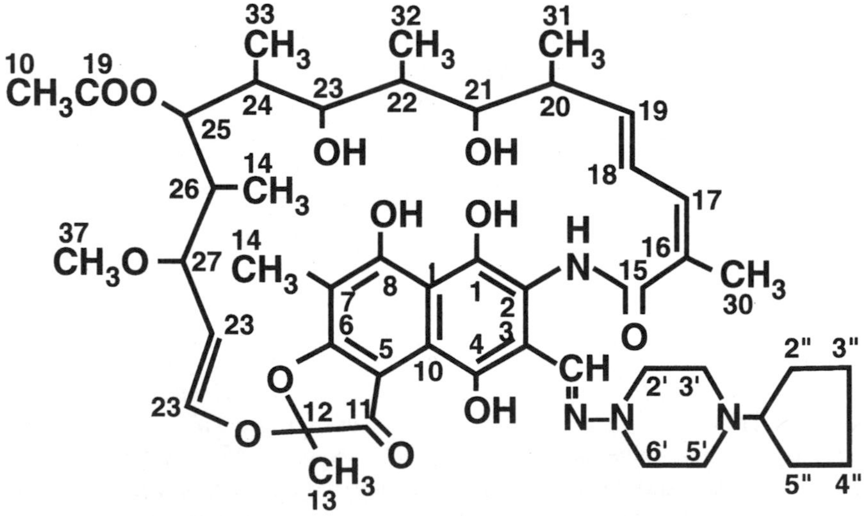

Figure 13.8. The chemical structure of rifapentine.

with rifampin. The MIC_{90} was shown to be to 0.06 μg/ml. The MBC for both rifapentine and rifampin was the same (0.5 μg/ml). Further detailed studies showed the following: (a) the antimycobacterial spectra of rifapentine and rifampin were almost identical; however, the growth-inhibitory activity of rifapentine against M.tb, *Mycobacterium kansasii* and *M. szulgai* was twice that of rifampin. (b) All the rifampin-susceptible strains were susceptible to rifapentine, and all rifampin-resistant strains were resistant to rifapentine. (c) About 75% of MAC strains isolated from patients who had not received rifampin but were resistant to the drug (naturally resistant strains) were resistant to rifapentine, and the remaining 25% were susceptible to rifapentine. (d) Serotypes 1 and 2 of MAC strains were often resistant to rifapentine, but those of serotype 16 were often susceptible to this drug.

Pharmacologically, rifapentine is claimed to be superior to rifampin in that it has a prolonged half-life in mice (43) and in humans (44). Animal studies have shown that intermittent administration of rifapentine (10 mg/kg) twice a week was as effective as the administration of rifampin (10 mg/kg) 6 days a week. Administration of the drug once a week still showed activity against experimental tuberculosis in mice. Based on these preclinical data, this drug is now being investigated in a multicenter study for the initial treatment of tuberculosis.

Rifapentine has also been shown to have another advantage in that it is taken up by macrophages to a greater extent than rifampin (45). It was shown that its intracellular concentration is 60 times that in the extracellular fluid; rifampin; on the other hand, accumulates intracellularly only fivefold. More detailed studies were done on the comparative uptake and intracellular distribution of radiolabeled drugs into the subcellular organelles of mouse peritoneal macrophages (46). Using analytic fractionation on sucrose density gradients, it was found that the uptake of radiolabeled rifapentine in serum-free medium was thrice that of rifampin; in presence of serum, the uptake of both the drugs was low and equal. Almost all of the cell-associated drug was found in the cytosol and no difference in distribution was detected between the two rifampins. It was concluded that the prolonged action of rifapentine compared to rifampin in the chemotherapy of tuberculosis cannot be attributed to its better penetration into the macrophages (46).

RIFABUTIN

Rifabutin (Fig. 13.9), also called LM427 or Ansamycin, is chemically a spiropiperidyl derivative of rifamycin and was developed by Sanfillippo and associates (47) of Farmatalia Carlo Erba, Milan, Italy. It has high in vitro activity against MAC and M.tb strains. In fact, its activity against MAC stimulated considerable interest and resulted in numerous contributions, which finally paved the way for clinical use. Its in vitro activity against M.tb was two to four times greater than that of rifampin, with a mean MIC of 0.022 μg/ml (48). Riframpin-resistant

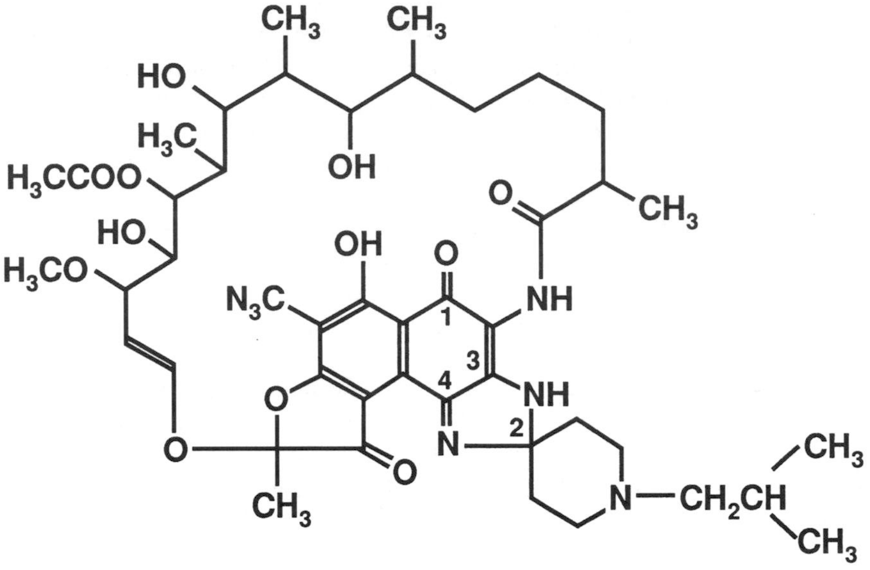

Figure 13.9. The chemical structure of rifabutin.

mutants isolated from the H37Rv strain were 160 times more resistant to rifampin and 128 times more resistant to rifabutin than the parent strain. The responses to rifabutin of rifampin-resistant strains isolated from patients under rifampin therapy differed from strain to strain; four of the eight strains tested were resistant to 40 μg/ml of rifabutin, but the other four strains were less resistant, and one of them was susceptible to 1.25 μg/ml of rifabutin. The response of naturally occurring rifampin-resistant MAC strains was different from those of M.tb strains. Of the 40 strains tested, 32 showed natural resistance to 40 μg/ml of rifampin and 19 were susceptible to 1.25 μg/ml of rifabutin. Thus, the natural resistance to rifampin did not always accompany resistance to rifabutin. Although a clear-cut relationship between the serotype and resistance to rifabutin had not been established, MAC strains with serotypes 4, 8, and 16 are always susceptible to rifabutin. This is encouraging, as most MAC strains from AIDS patients belong to serotypes 4 and 8. Rifabutin is more active than rifampin against M.tb, *M. bovis, M. kansasii, M. Marinum, M. xenopi, M. haemophilum, M. scrofulaceum, M. nonchromogenicum,* and *M. terrae.* Among these, the greatest differences have been noted with *M. marinum, M. xenopi,* and *M. haemophilum.*

Extensive in vitro studies, post-antibiotic effect, dynamic studies, including some done under conditions simulating those existing in vivo, macrophage studies, and studies in experimental chemotherapy in the beige mice model have established the chemotherapeutic role of rifabutin by itself and in combination with other drugs against MAC disease (49–51).

With experimental tuberculosis, Ji and associates (52) showed that rifabutin (10 mg/kg) given twice a week for 6 weeks resulted in a reduction of the CFU in lungs and spleens, similar to that observed with rifampin (10 mg/kg) given 6 days a week for the entire experimental period. Luna-Herrera et al. (53) have confirmed its chemotherapeutic role in the treatment of tuberculosis and against MDR tubercle bacilli. They found that rifabutin exhibited similar or greater in vitro activity than rifampin, as judged by its MIC, MBC, and MIC/MBC ratios, as well as on continuous exposure and its post-antibiotic effects. More importantly, it was shown to be active against some rifampin-resistant strains. In macrophage studies, rifabutin was found to be more effective than rifampin in killing intracellular bacilli after continuous exposure or after only 24 h contact. In animal studies (54) where rifabutin was compared with KRM1648 (discussed later) and rifampin, rifabutin was shown to be effective in preventing mortality and reduction of CFU counts, similar to KRM1648; both drugs were superior to rifampin and caused complete sterilization of lungs. However, resistant organisms started appearing 2 weeks after stopping treatment, whereas they did so after 16 weeks after stopping treatment with KRM1648.

Concerning clinical investigations, besides the few earlier studies reported by the manufacturer, two other reports dealt with MAC and M.tb disease, although the later one is only of a preliminary nature. Earlier, O'Brien et al. (55) reported the finding of a large-scale study on the role of rifabutin in the treatment of MAC patients with or without AIDS and MDR patients. Preliminary clinical studies (56) in AIDS and non-AIDS patients with pulmonary tuberculosis treated with rifabutin in combination with isoniazid and ethambutol given once a day showed that rifabutin may be an effective substitute for rifampin. About 50–60% of the M.tb strains that developed resistance to rifampin also developed resistance to rifabutin.

All the findings suggest that rifabutin has a place in the treatment of tuberculosis, especially in certain types of drug-resistant cases. However, the possibility of cross-resistance with rifampin should be considered. Its main usefulness is in the treatment of MAC disease, for which it has already been approved for clinical use.

FCE 22250

This compound, (Fig. 13.10), which is clinically 3-azo methyl rifamycin, has been shown to be a long-active derivative of rifabutin (57). It has been shown to have a good oral absorption and has a long half-life. It has a broad antibacterial spectrum, including high activity against mycobacteria. In experimental tuberculosis, using M.tb H37Rv challenge, FCE22250 showed an efficacy 14 times

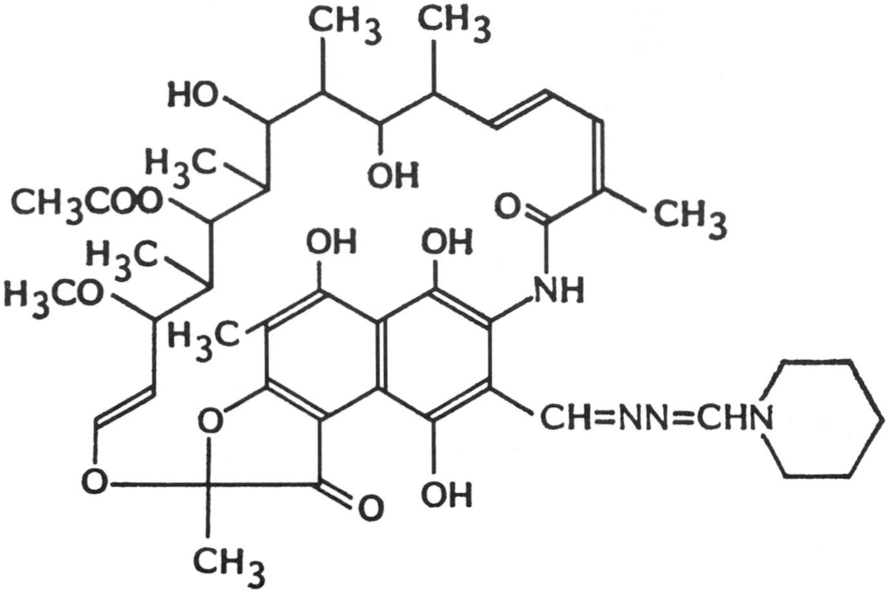

Figure 13.10. The chemical structure of FCE22250.

greater than that of rifampin and is therapeutically active when administered once every 3 weeks.

THE CGP SERIES OF RIFAMPIN ANALOGS

Three rifampin derivatives CGP7040, CGP27557, and CGP29861 (Fig. 13.11) were developed by the Ciba-Geigy Pharmaceutical Company. The biological half-lives of these three drugs were considerably longer than that of riampin (58). The half-life in hours for CGP7040 was 47 in mice, over 120 in rabbits, and 80 in dogs; the corresponding figures were 20, 70, and 15 for CGP27557, 45, 100, and 15 for CGP29861, and only 6 in mice and 4 in the rabbits for rifampin. Drug concentrations in these animal species increased approximately proportionally to the dose administered. The serum binding capacity of these compounds was similar to rifampin and the drugs accumulated in the liver to a lesser extent than rifampin.

Similar data were obtained with human volunteers; the half-life of CGP7040, CGP27557, and CGP29861 were 30, 8, and 40 h, respectively, as compared to 4 h for rifampin. Unlike rifampin, food had no influence on the absorption of CGP7040 and CGP297557, whereas the absorption of CGP29861 was twice as

Chemical structures of rifamycin derivatives

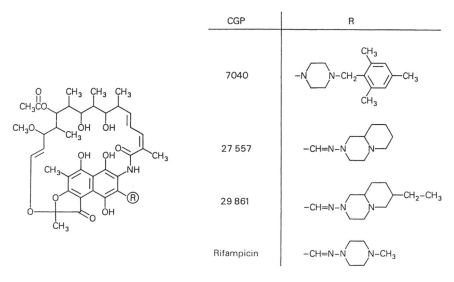

Figure 13.11. The chemical structures of the CGP series of rifampin derivatives.

high when taken after breakfast. All the three drugs showed slower absorption and lower peak serum concentrations than pure rifampin.

In vitro and in vivo activities of these compounds also showed superiorities over rifampin (58). *In vitro,* all three compounds had similar activities against gram-positive and gram-negative bacteria as well as tubercle bacilli; all were similar to rifampin, but CGP7040 was superior to rifampin against nontuberculous mycobacteria with an MIC of 0.003 μg/ml against MAC. In experimental tuberculosis in the mouse model, with short duration of treatment, the ED50 of CGP7040 was comparable to that of rifampin, whereas CGP27557 and CGP29861 were five times more active than rifampin. With longer periods of treatment and different dosage intervals, CGP27557 and CGP29861 proved to be several times more active than rifampin, as evidenced by enumeration of tubercle bacilli in the lungs. CGP29861 was also investigated in experimental short-course chemotherapy of tuberculosis, wherein the drug was given either in the initial or in the continuation phase of treatment. In spite of these indications of potential usefulness, these compounds were not explored further, perhaps due to the general lack of financial incentives.

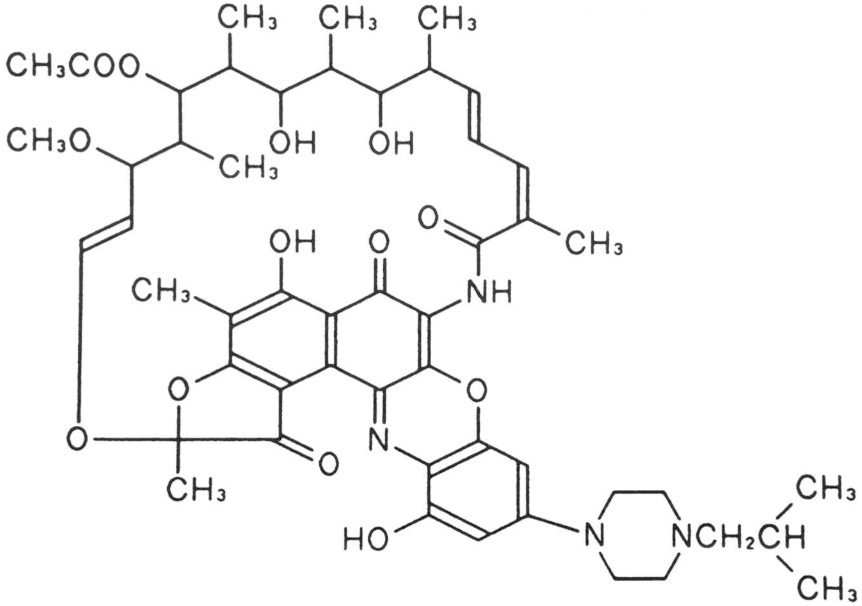

Figure 13.12. The chemical structure of KRM1648.

BENZOXAZINORIFAMYCIN (KRM1648)

Kaneka Corporation of Japan has been investigating several modifications of the isobutyl, propyl, and secondary butyl groups of benzoxazinorifamycin (e.g., KRM1648, KRM1657, KRM1668, and KRM1687) for antimycobacterial activity (59). Among them, only KRM1648 emerged as the best, based on several in vitro and other screening tests (60) (Fig. 13.12). Besides the extensive studies done in Japan (61,62), this compound has been studied in the United States (54,63) and is now undergoing Phase I and Phase II clinical testing.

Activity of KRM1648 Against MAC.   The MIC_{90} of MAC strains from AIDS patients was 0.25 μg/ml and it was bactericidal for MAC isolates with a reduction of viability of one to four orders of magnitude in 72 h. At 1.0 μg/ml, it was bactericidal to MAC in human macrophage systems. In experimental chemotherapy studies in beige mice (64) and in rabbits (65), KRM1648 (40 mg/kg) was effective in reducing CFU in blood, liver, and spleen. In combination with clarithromycin or ethambutol, KRM1648 exhibited an in vivo synergistic effect against experimental MAC disease in beige mice.

Antituberculosis Activity of KRM1648.   Soon after its introduction, several Japanese investigators have established its in vitro and in vivo activity against tuberculosis. In the United States, detailed investigations done by Luna-Herrera et al. (63) and Reddy et al. (54) have dealt with in vitro macrophages and animal model studies. In the in vitro studies, its activity was assessed against several drug-susceptible and MDR strains of tubercle bacilli, many of which are resistant to rifampin. For rifampin-susceptible strains, the MIC_{90} for KRM1648 was ≤0.015 μg/ml, whereas for rifampin, it was 0.25 μg/ml; in fact, there were some strains which were inhibited by KRM1648 [ranging between 0.007 and 0.03 μg/ml, which was much lower than that of rifampin (range 0.5–1.0 μg/ml)]. Post-antibiotic-effect studies with KRM1648 showed a rapid reduction in the CFU counts with an exposure time of 24 h or more, and its sterilizing effect was maintained even up to 21 days thereafter. Under parallel conditions, rifampin showed a less significant effect, with a faster recovery of growth, and failed to sterilize the organisms even after a 72-h exposure.

In the macrophage studies using J774 A.1 cells, KRM1648 at 0.125 and 0.25 μg/ml caused complete inhibition of intracellular growth of M.tb after a 48-h exposure (63). After a similar exposure time, rifampin at 0.25 μg/ml concentration, caused complete inhibition of growth, but at 0.12 μg/ml, it caused only a 50% reduction of growth, compared with that of the controls at 7 days. With a 24-h pulsed exposure of the intracellular tubercle bacilli, KRM1648 at 0.25 μg/ml caused complete inhibition of growth, whereas rifampin caused only moderate inhibition of intracellular growth.

In vivo studies against experimental tuberculosis of C57BL/6 mice showed that KRM1648, similar to rifabutin and rifampin, prevented mortality up to 28 days, by which time all untreated mice died (54). Analysis of CFU showed superior therapeutic effect of KRM1648 and rifabutin as compared to rifampin against infection with drug-susceptible tubercle bacilli under early treatment protocol (i.e., where treatment is commenced 1 day after infection). Twelve weeks' treatment with KRM1648 or rifabutin caused complete sterilization of the lungs. However, residual organisms started appearing in the spleens 6 weeks after cessation of treatment with rifabutin and 16 weeks after KRM1648 treatment. In mice infected with a MDR strain of M.tb which was susceptible to KRM1648 in vitro, the drug did not appear to have any activity. Because the MDR organisms did not multiply in vivo and did not cause any mortality even in untreated mice up to 28 weeks, a state of semidormancy of the organisms which might prevail in vivo could be responsible for the refractoriness to the treatment with KRM1648. Overall, the available data with KRM1648 offer strong hopes of its potential value as a powerful antimycobacterial drug.

ANTIMYCOBACTERIAL ACTIVITY OF A NEW DERIVATIVE, 3-(4-CINNAMYL PIPERAZINYL IMINOMETHYL RIFAMYCIN SV (T9)

The Chemical Pharmaceutical Research Institute, Sofia, Bulgaria has investigated several analogs of rifampin and have identified 3-(4-cinnamyl piperazinyl

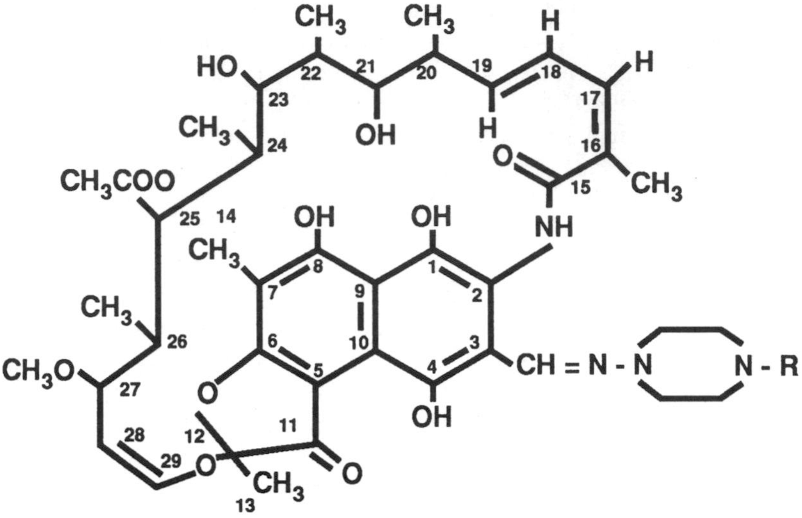

$R = ($ methyl $) - CH_3$ Rifampin

$R = ($ cinamyl $) \; C_6H_5CH=CHCH_2-$ T9

Rifampin   = 3-(4-Methyl-1-piperazinyl imino methyl) rifamycin SV

T9   = 3-(4-cinamyl-1-piperazinyl imino methyl) rifamycin SV

Figure 13.13. The chemical structure of T9, a new rifamycin derivative.

iminomethyl) rifamycin SV, called T9, to be most active (66) (Fig. 13.13). Several investigations conducted by the developers have indicated good therapeutic activity, lack of toxicity, and favorable pharmacokinetics and bioavailability in experimental animals (67). Reddy et al. (68) have confirmed and extended the earlier studies using detailed in vitro macrophage and animal models. The MIC_{90} of T9 for drug-susceptible M.tb strains was found to be ≤ 0.25 $\mu g/ml$, whereas for rifampin tested under the same conditions, it was ≤ 0.5 $\mu g/ml$. Interestingly, T9 had lower MICs against some rifampin-resistant M.tb strains. It had better activity against MAC strains with an MIC_{90} of ≤ 0.125 $\mu g/ml$, whereas for rifampin, it was ≤ 2.0 $\mu g/ml$. It also had superior MBC activity compared to rifampin.

With regard to its intracellular activity, T9 at both 0.125- and 0.5-$\mu g/ml$ concentrations caused significant reduction of intracellular growth of M.tb strain H37Rv and MAC strain 101, not only after continuous exposure for 4 days but also after only 24 h pulsed exposure. Rifampin under similar conditions showed much less activity.

More interesting leads emerged in these studies with experimental tuberculosis in C57BL/6 mice (68). With a 40- and a 20-mg/kg dose of T9, no organisms could be detected at 4 weeks either in spleen or in lungs. With a 10-mg/kg dose, the organisms were detected in only three of five mice at the fourth week. Even with a 5-mg/kg dose, T9 caused significant reduction of CFU counts in both lungs and spleens, compared with the 20-mg/kg dose of rifampin. All these studies have thus shown a high potential for antimycobacterial activity of T9 against M.tb and MAC infections.

3.3. Pyrazinamide Analogs

Attempts were made by two groups of workers to search for analogs of pyra-zinamide, the third most important drug used in the short-course chemotherapy of tuberculosis (69,70). Unlike the studies with the other two drugs, these attempts were based on the available knowledge on the mode of action of the drug. It is postulated that pyrazinamide exerts its antimycobacterial activity after conversion to pyrazinoic acid within the bacterial cell by a bacterial enzyme, pyrazinamide deamidase (71). Several analogs which act on the same or similar pathway were tested. Thus, Cynamon and associates (69) screened several compounds for their anrimycobacterial activity and used quantitative structure activity relationships (QSAR) to identify compounds with potential usefulness against tubercle bacilli and MDR strains. So far, these studies dealt with in vitro testing only.

The other series of studies reported by Yamamoto and associates (70) also dealt with in vitro testing. Among 39 analogs of pyrazinamide and/or pyrazinoic acid tested, only four exhibited strong in vitro activity. These are (i) pyrazine thiocar-boxamide, (ii) *N*-hydroxymethyl pyrazine thiocarboxamide, (iii) pyrazinoic acid *n*-octylester, and (iv) pyrazinoic acid pivaloyloxymethyl ester. All four drugs showed high bacteriostatic and bactericidal activities against M.tb and MAC in conditions under which pyrazinamide showed little or no activity. As pyrazin-amide is more active at pH 5.5 than at pH 6.8, these four compounds were tested at both pHs. It was found that the relative antimycobacterial activities of these drugs compared with pyrazinamide were essentially the same in medium with a pH of 6.8 as in medium with a pH of 5.5. Further studies using macrophage and animal models are warranted before the chemotherapeutic potential of these compounds is established.

4. Investigations on Potential Agents Which Bypass the Permeability Barriers in MDR Bacilli

These can be broadly divided into two categories: (a) penicillinase inhibitors and (b) other agents which influence the permeability of the bacterial cell.

4.1. Penicillinase Inhibitors

Several years ago, Kasik and associates (72) investigated the possible chemo-therapeutic activity of penicillin against tuberculosis. Although some reports indicated its in vitro activity, at least at high concentrations, many could not confirm it (73). It was soon realized that mycobacteria possess a high penicillinase activity, which make them less susceptible to this drug. Subsequently, as the several hundreds of β-lactam compounds were discovered, it was also shown that mycobacteria were resistant to these drugs as well; it was soon recognized that this was due to the high β-lactamase content in these organisms. Even though not enough knowledge has accumulated concerning β-lactamases of mycobacteria compared to the enormous literature with other organisms, some (74) have attempted to categorize this enzyme in mycobacteria as intracellular, constitutive, or noninductive types. However, it was also questioned if β-lactamase activity cannot by itself be considered responsible for the resistance of these organisms to β-lactam antibiotics. For instance, MAC, which are highly resistant to these drugs and several other antimycobacterial drugs, do not have the β-lactamase enzyme.

Interest in exploring antimycobacterial drugs from this group did not exist until recently; only after the resurgence of MDR tuberculosis did several investigators explore this area. Knowledge which is accumulating on the biochemistry of these enzymes in other microbial species and the discovery of specific inhibitors of the β-lactamase enzyme (e.g., clavulinic acid) have also been rapidly applied for screening drugs for their antimycobacterial activity. Intensive work on the nature of such enzymes in mycobacteria is also progressing, and several specific penicillin-binding proteins have been identified. Likewise, the inducibility of β-lactamase enzyme in M.tb has also been speculated.

Several investigators (75,76) have assessed the in vitro antituberculosis activity of the combination of clavulinic acid and amoxicillin (the best known combination is Augmentin). Other combinations [e.g., clavulinic acid and ticarcillin, ampicillin and a sublactam (unasyn)] showed much less in vitro activity than Augmentin™ (77). In contrast to the in vitro data, macrophage studies using J774 cells and in vivo studies in C57BL/6 mice did not show any activity of Augmentin™ (77). This is perhaps due to the inability of the drugs, particularly clavulinic acid, to penetrate intracellularly. On the other hand, a clinical report (78) claimed that two patients with MDR tuberculosis had been successfully treated with this combination.

4.2. Other Agents Influencing Permeability Barriers

Two agents, ethambutol and dimethyl sulfoxide, were investigated in this series using in vitro and macrophage models, and animal models as well for ethambutol.

ETHAMBUTOL

Ethambutol (EMB), is by itself, a powerful antituberculosis drug; it has also the property of enhancing the permeability of other drugs, especially in MDR bacilli (79). This property is due to its inhibition of the cell-wall synthesis of mycobacteria by acting on the mycolic acid and arabinogalactan biosynthesis (80,81). Treatment with ethambutol increases the permeability of many drugs which otherwise cannot penetrate easily. Initial studies by Rastogi and associates (82), which were based on this phenomenon, had shown that in the J774 cell macrophage system, isoniazid was more active on bacilli resistant to it when it was given in conjunction with ethambutol. Encouraged by these background data, Chinnaswamy et al. (83) have extended these investigations using in vitro and macrophage models. In vitro studies indicated that exposure of isoniazid, rifampin, or streptomycin individually along with sub-MIC concentrations of ethambutol to four MDR strains resistant to these drugs enhanced susceptibility to these drugs. In contrast, exposure to the sub-MIC level of ethambutol to drug-susceptible M.tb H37Rv or to an EMB-resistant strain did not alter the susceptibility pattern. Macrophage studies in J774 cells indicated similar outcome, although the effect was more evident with isoniazid and rifampin than with streptomycin.

Interesting results were obtained in experimental chemotherapy studies using an MDR strain of tubercle bacilli (84). These studies were done using isoniazid (25 mg/kg), rifampin (25 mg/kg), or ethambutol at normal (120 mg/kg), half the normal (60 mg/kg), and one-fourth the normal (30 mg/kg) dose. The animals were followed for mortality and CFU counts in spleens and lungs at 4, 8 and 12 weeks. With spleens, a combination of ethambutol at one-fourth the dose (30 mg/kg) plus isoniazid (25 mg/kg) showed considerably greater activity than either isoniazid alone or the regular dose of ethambutol. The reduction in mean CFU counts was seen all through, showing the greatest difference at 12 weeks. With lungs, slightly different results were seen. Synergism was seen only with the normal (highest dose—120 mg/kg) of ethambutol, and the synergistic activity was gradually reduced as the dose of ethambutol was reduced to 60 or 30 mg/kg. Moreover, the activity was most evident at 8 weeks, but at 12 weeks, the activity was less evident.

DIMETHYL SULFOXIDE

Dimethyl sulfoxide (DMSO), which was proved to be fairly nontoxic to rabbits and humans (85), was shown by many (86,87) to be capable of enhancing the permeability of tubercle bacilli and "sensibilize" MDR tubercle bacilli both in vitro and, more importantly, in experimental tuberculosis in the guinea pig model. Most of these reports, however, were published several years ago in languages other than English. Muller and Urbanczik (88) studied the role of DMSO in altering the susceptibility of MDR strains to isoniazid, ethambutol, rifampin, streptomycin, para-aminosalicylic acid (PAS), and thiacetazone and could find reversal in some but not in all strains. They suggested that the continuous presence of DMSO is essential for the expected reversal action. Some recent unpublished reports indicated that DMSO could enhance penetration of many drugs in humans (Urbanczik, personal communication, 1995).

Chinnaswamy et al. (83) investigated, in in vitro and macrophage models, whether DMSO could alter the susceptibility of MDR strains to isoniazid, rifampin, and streptomycin to which the bacilli were initially resistant. Preliminary studies titrated to arrive at the MIC of DMSO which was found to be 5%; at 2.5%, only partial inhibition of the strains could be seen. Using the 2.5% concentration as the sub-MIC, an eightfold increase in the susceptibility of MDR stains to isoniazid was seen; the increase was less evident with rifampin and streptomycin. Similar to the in vitro studies, enhancement of the susceptibility to isoniazid of intracellular MDR strains was seen in the macrophage model (83).

5. Metabolic or Competitive Growth Inhibitors

Based on the available knowledge to date, only three areas are considered in this section: (a) mycolic acid synthesis inhibitors, (b) vitamin K and ubiquinone analogs, and (d) calmodulin antagonists.

5.1. Inhibitors of Mycolic Acid Synthesis

Mycolic acid has been demonstrated to be the most important constituent in the cell-wall architecture of mycobacteria. Recently, a mycolic acid synthetase pathway has been worked out. One compound, methyl 4-(2-octodecyl cyclo pro-pen-lyl)butanoate, which is a structural analog of an important precursor in the mycolic acid biosynthesis pathway in *M. smegmatis,* has been shown to be an inhibitor of that enzyme (89). No further studies on the activity of this compound or its analogs on M.tb have been reported.

5.2. Gangamicin

Dual analogs of vitamin K types of compound, which are present in the cell wall of mycobacteria, and ubiquinone (coenzyme Q_{10}), which is known to stimulate mycobacterial growth, were tested for their antimycobacterial activity (90,91). In this process, the powerful activity of 6-cyclo octylamino-5,8-quinoline quinone (CQQ) was discovered. This was patented by the National Jewish Center for Immunology and Respiratory Medicine, Denver, Colorado as Gangamicin (GM) (U.S. patent No. 4,963,563). In vitro, its MIC is between 0.1 and 2.0 μg/

ml for M.tb, including those resistant to isoniazid and rifampin and between 2.0 and 8.0 μg/ml against MAC. At concentrations of 1.0 and 2.0 μg/ml, it inhibited the intracellular M.tb, including drug-resistant strains and MAC in mouse peritoneal and human monocyte-derived macrophages and the J774 A macrophage cell line. In experimental chemotherapy studies, using doses of 20 and 40 mg/kg given orally or subcutaneously in C57BL/6 and beige mice models for M.tb and MAC, respectively, it has demonstrated high activity both in preventing mortality and in reduction of CFU counts from spleen and lungs. In general, its activity is more pronounced in spleen than in lungs, by subcutaneous rather than by oral treatment, and more against M.tb than MAC. Of more significance is its similar activity against infection with drug-resistant M.tb where it exhibited equal activity as with susceptible strains. Besides M.tb and MAC, it is highly active against *M. kansasii* and *M. leprae*. The drug has been found to be nontoxic to several animal species. All these experimental findings warrant strong hopes for the chemotherapeutic potential of this drug against M.tb, drug-resistant M.tb, and MAC.

5.3. Calmodulin Antagonists

Demonstration of calmodulin like proteins in mycobacteria (92) and inhibition of in vitro growth of M.tb by calmodulin antagonists led to investigation on the antimycobacterial activity of trifuoperazine (TFP) (93,94). The MIC of TFP ranged between 12.5 and 25.0 μg/ml for MAC, between $\leq$3.0 and 25.0 μg/ml for drug-susceptible strains of M.tb, and between 6.25 and 12.5 μg/ml for MDR strains. The MBC against M.tb strains ranged from 12.5 to 50.0 μg/ml. Trifluoperazine showed no in vitro synergistic activity with INH or rifampin. The drug was active against intracellular MTB (H37Rv) in U937 cells at 4.0 μg/ml. At subinhibitory concentrations, TFP enhanced intracellular activity of INH and rifampin against H37Rv.

However, it should be stressed that the concentrations at which TFP shows antimycobacterial activity in vitro and in macrophages are too high to be achieved in normal circumstances. For instance, with a 5.0-mg/kg dose, the maximum concentration reached in plasma is about 1.4 μg/ml, and with a 20-mg/kg dose, the level can go between 0.9 and 4.0 μg/ml (95). The effect of TFP at subinhibitory concentrations in enhancing the in vitro and intracellular activity of INH and rifampin may be due to increased accumulation and retention of antituberculosis drugs, as has been earlier demonstrated with anticancer drugs (96,97). Although it is difficult to achieve an in vivo antimycobactericidal concentration of TFP, at least it has the potential to enhance the efficacy of other drugs. The development of less toxic analogs of TFP or lipososmal preparations could be a profitable alternate approach.

6. Molecular Biological Approaches

Recent advances in molecular biology of mycobacteria have identified some genes which are instrumental in the action of some drugs, particularly INH. The first discovery by Zhang et al. (98) on the *katG* gene and, subsequently, the *inh-a* gene by Bannerjee et al. (99) have given some strong leads of understanding the mechanisms of development of drug resistance. It has been hoped that these landmark discoveries will pave the way for the discovery of newer drugs, although, so far, no definitive advances have been published. Following the development of these genes for INH, genes for resistance for other drugs particularly rifampin, streptomycin, and fluoroquinolines (100–102) have been developed. (These aspects have been discussed extensively in the Chapters on drug resistance and on molecular biology of mycobacteria in Volume 1.).

7. Approaches to Harness the Maximum Benefit of Available Drugs

This aspect is attaining greatest importance in recent times, not only in tuberculosis but in almost all other diseases as well. This is due to the increasing realization of several aspects in medicine, which are influenced to a great extent of some important realities.

1. As discussed earlier, it is becoming increasingly difficult for the pharmaceutical industry to get involved in introducing new drugs, mainly due to financial considerations. This inertia is evident in spite of strong evidence of highly encouraging preclinical data, as has been reviewed in the earlier sections of this chapter.
2. Recent advances in clinical pharmacology and detoxification mechanisms of drugs stress that most drugs are "short-lived" in the body and are destroyed and excreted rapidly before their full therapeutic values are utilized.
3. The last and perhaps most serious is the problem of "noncompliance" of the patient in taking the full course of treatment. In diseases like tuberculosis, such premature dicountinuation of treatment would cause failure, quite often resulting in drug resistance of the organisms. Such an unfortunate situation would not only harm the patient but, more seriously, would act as source of resistant bacilli, to his contacts wherein also these drugs would be inactive. (These aspects have been discussed in the chapters on drug resistance in Volume 1 and on chemotherapy of drug-resistant tuberculosis in this volume.)

Recent advances in knowledge of pharmaceutics and chemical engineering have facilitated introduction of several new leads to solve some of these problems. Thus, several new and innovative methodologies have given scope of new strategies of drug delivery, whereby the drug can bypass to a great extent the metabolic insults of the body. More importantly, these drug-administration tactics target most of the free drug to the required site of action in vivo. Such targeted drug delivery systems are being attempted for the chemotherapy of mycobacterial diseases as well.

Other strategies deal with controlled and sustained drug delivery. There, developments are possible, thanks to the enormous progress made in recent years in biopharmaceutical and chemical engineering knowledge of the polymers as drug carriers. These attempts are giving immense opportunities to circumvent the man-made problem of noncompliance. Before discussing these various achievements with respect to the treatment of tuberculosis, some comments on the drug–parasite–host interactions will be useful. Such considerations will also correlate other important aspects of pathogenesis, immunology, and chemotherapy of tuberculosis, which are discussed extensively in other chapters in these two volumes.

7.1. Drug–Parasite–Host Interactions

Paul Ehrlich, the "father of chemotherapy," had defined chemotherapy as a positive balance in drug–parasite–host interactions, in favor of the host. He stressed the concept of "selective toxicity," which was further elucidated by subsequent authors. For convenience of discussion of this important topic with special relevance to the newer strategies of chemotherapy and with reference to several other topics of chemotherapy, which have been elucidated in other chapters in the two volumes in this series, it will be advantageous to conceptualize the drug–parasite–host relationships in a three-way triangle (Fig. 13.14).

Each of the two-way relationships is designated as an arrow. Arrow 1 is the interaction of the parasite (tubercle bacilli) with host (man). This is the process of infection, followed by disease, in the rare events where the host's resistance cannot control it. This aspect is elaborated in chapter 5 on Pathogenesis by Dr. Grange and in Chapter 6 on Transmission by Dr. Falkinham in Volume I of this series. Normal healthy hosts could easily control the spread of infection in vivo and progress to the state of overt disease. This aspect (arrow 2), which can be called "immunity," is reviewed by the chapter on immunology of tuberculosis by Drs. Reddy and Andersen and on the immunological cause, prevention and treatment of tuberculosis by Dr. Stanford, both in the first volume. A compound is called a drug if it is specifically active, and this aspect, which is depicted by arrow 3, is covered by Chapters 2 and 3, and 5 and 6 on chemotherapy of tuberculosis in developed and developing countries respectively. Likewise, the mechanisms by which the parasite evades the action of drug, the so-called "drug resistance," is reflected in arrow 4 and is covered by the chapter by Dr. Gangadharam with the same title in volume 1. The toxicity of the various drugs (arrow 5) is covered by Chapters 5 and 6.

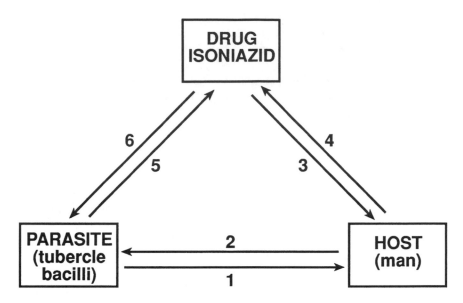

Figure 13.14. A schematic representation of drug–parasite–host interactions.

The subject indicated by arrow 6, which is the influence of host on the drug, is not included anywhere. In fact, this aspect had escaped the concern of Ehrlich when he defined the basic concepts of chemotherapy. He might have, perhaps, assumed that the host will only be the direct beneficiary of the action of the drug on the parasite and will only be an innocent participant in this game. He might not have considered that the host would confuse and complicate the direct beneficial action of the drug on the parasite. Now, thanks to the advances in our knowledge in clinical pharmacology, drug metabolism, in vivo fate of drugs administered by various routes, and the various detoxication mechanisms, the influence of the host on the drug has become a subject of great importance. In simplistic terms and for convenience of discussion, these aspects can be broadly divided as (a) involuntary and (b) voluntary influences.

7.2. Involuntary Influences

These are dictated by the nature of the host and, in many cases, by nutrition and other factors like the general health, metabolic status, and the species variation, and in some cases, by the race, genetic makeup, and other concommitant infections (e.g., HIV). All these factors, as the name indicates, are not in the control of the host. Only intervention by other means can attempt to rectify or at least lessen some of these influences. Some like those influenced by race, and genetic makeup, cannot thus be altered. Reviewing the available literature on the involuntory influences of the host on the drug, three areas can be considered as most important: absorption, metabolism, and excretion. Of these, very little information is available on the excretion of the drugs, except the knowledge on routes by which the drugs are removed, but such data have not been shown to have any influence on the drugs' action on the parasite in vivo.

ABSORPTION OF THE DRUGS

Considering the absorption of the drugs, with relevance to their in vivo chemotherapeutic action, which is reflected by the pharmacokinetics, peak and sustained serum levels, and maintenance of sufficient drug levels, knowledge is available on only a few drugs: isoniazid, rifampicin, and cycloserine. However, recent regulatory requirements for all new drugs insist on the availability of such data. With isoniazid, it had been reported that fatty foods could influence the attainable serum levels. By far, a great amount of knowledge had been obtained with rifampin, where it had been shown that the levels are significantly higher when the drug is given on an empty stomach than when given after food. In contrast, drugs like cycloserine are prescribed after food, lest the higher levels which are obtained when it is given on an empty stomach should reach toxic limits.

Recent studies from the National Jewish Center, Denver, have made the interesting observation that AIDS patients in general have a reduced capacity to absorb certain orally administered drugs (103). Their studies have also suggested that there may be a correlation between the severity of immunosuppression and the degree of malabsorption. Earlier, such an experience was noted only in patients with specific malabsorption syndromes. Species variation in drug absorption is best explained by clofazimine, which is poorly absorbed by some animal species (e.g., guinea pigs and monkeys), in contrast to mice and hamsters and perhaps man.

METABOLISM

For all practical purposes, any drug, however powerful it might be in curing any ailment, would be considered as a "foreign compound" by the host. With this approach, the body tries to "get rid" of the drug as it is or, in most cases, after alteration by metabolism to derivatives which could be handled more easily by the host's excretory organs (e.g., kidneys, bowel, etc.). This aspect of metabolism is more aptly called "detoxification" and knowledge in this area has progressed considerably in recent times so that accurate predictions of the pathways of metabolism of the drugs with known chemical structures can be made. The metabolites of most of the drugs are inactive against the parasite, and therefore the metabolism is commonly referred to as "inactivation." This aspect has been extensively worked with isoniazid, and more recently with other drugs. Although all the metabolites of isoniazid are "inactive," justifying the term "inactivation" for the metabolism, the principle metabolite of rifampin, desoxyrifamycin, is still active against tubercle bacilli. Likewise, the chief metabolite of clarithromycin, 14-hydroxy clarithromycin, is still active against *M. avium complex.* However, most importance has been given to the study of isoniazid metabolism due to several reasons.

First, isoniazid is easily considered the best antituberculosis drug, with least toxicity. It is inexpensive and easily available. Second, over the years, an enormous amount of knowledge has accumulated on its metabolism, which is racially influenced and genetically controlled. Humans can be divided into two distinct groups, the slow and rapid inactivators, with different clinical and toxicological outcomes. Third, the action of the drug on the parasite and the altered properties of isoniazid-resistant tubercle bacilli have shown many interesting biological properties, like loss of several key enzymes, reduction of pathogenicity, and so on. Parallel to these developments, attempts have also been made to control or modify the metabolism of this important drug with a view to achieve greater therapeutic outcome.

The various attempts to alter the metabolism of isoniazid can be grouped as follows: (i) concomitant administration of other agents, which also act as specific substrates to the principal metabolic enzyme, *n*-acetyl transferase, so that more of the free isoniazid is available for therapeutic action; (ii) to slow down the release of the drug in the rapid inactivators by matrix preparations so that higher levels are achieved; Several methods are being made to evolve a sustained release (depot) type of formulations, which facilitate prolonged coverage with therapeutically active levels of the drug (discussed later in the chapter); and (iii) to encapsulate the drug in liposome preparations so that the drug can escape the metabolic influence of the liver and can go directly to macrophages.

Concomitant Administration of Other Agents.   The first extensive study in this regard was with PAS. For several years, an isoniazid–PAS combination was prescribed as a standard double-drug regimen for the treatment of tuberculosis. In both slow and rapid acetylators, it was shown that PAS elevated the levels of isoniazid. It was claimed that PAS, when given with isoniazid, has two functions, one to prevent emergence of resistance to isoniazid and the other to protect it from being fully metabolized. Subsequent studies have shown that this competitive metabolism is due to the fact that PAS can also function as a substrate for the *n*-acetyltransferase enzyme, which metabolizes isoniazid. Subsequent studies on this enzyme have shown that a sulfonamide derivative, sulfamethazine (also called sulfadimidine), is a more potent substrate for this enzyme. In fact, based on this observation, the inactivation of isoniazid is assessed by sulfamethazine acetylation (104). Concomitant administration of sulfamethazine along with iso-niazid either simultaneously or 15 min prior to isoniazid has been shown to prolong the metabolism of isoniazid in both rapid and slow inactivators (105). Even though at one time it was hoped that this strategy could be used in routine clinical practice, it was not pursued further due to the fact that sulfamethazine has some of its own immunomodulatory properties and sulfa drugs are, in general, contraindicated in AIDS patients.

Matrix Preparations. With a view to slowing the metabolism of isoniazid, especially in rapid inactivators, attempts were made to entrap isoniazid in some matrix preparations (106). By doing so, it was shown that higher doses of isoniazid could be administered (17–20 mg/kg instead of 5–8 mg/kg) and that the levels of free isoniazid could be seen for slightly prolonged periods than with conventional administrations. Although these studies have shown a pharmacological advantage, clinical studies have shown more toxic manifestations, mainly in the shape of giddiness, which prompted discontinuation of this approach.

Liposome Encapsulation. By far the most popular attempt in recent times to circumvent the involuntary influences of the host is with liposomes. These have been extensively used in several systems of medicine and in the cosmetic industry. With mycobacterial diseases, only recently some investigations have begun; most of them, however, dealt with MAC disease, and only few with M.tb.

In the pharmaceutical industry, liposomes are one of the most actively used drug delivery systems and are popularly known as "vesicles." They can be prepared from phospholipids with or without cholesterol, although inclusion of cholesterol increases the in vivo stability and retard the drug release. Liposomes can be prepared in a wide range of sizes, from 20 nm to 10 μm, and, structurally, have one (unilamellar) or more (multilamellar) lipid layers, separated by aqueous compartments, surrounding an aqueous core. Liposomes are capable of entrapping a wide range of materials, hydrophilic drugs are dissolved in the aqueous regions, and hydrophobic drugs are associated with the lipid bilayers. Liposomes are avidly phagocytosed by macrophages and are, therefore, naturally targeted to the sites of mycobacterial existence in vivo. Because they are directly targeted to the cells, by and large, they escape the metabolic pathways of the liver and other tissues in the body. Using the encapsulation in unilamellar and multilamellar liposomes several drugs (e.g., amikacin, streptomycin, gentamicin, and clofazimine) have been investigated in the macrophages and beige mouse models for the efficacy against MAC disease (107–111). All these studies have shown that (a) increased activity of the liposomal encapsulated drugs in macrophages compared to the free drugs and (b) efficacy, with a fraction (1/10–1/5) of a dose given in 3 or 4 doses, comparable to the outcome when the full dose was given daily for the entire experimental period.

Most of the in vivo studies demonstrated poorer activity in the lungs than in the other tissues. This has been explained by the fact that unilamellar or multi-lamellar liposomes deliver most (>95%) of the drug to the macrophages directly and that only a small amount of the drug can be transported elsewhere, as these liposomes are not stable in serum. This defect is now being corrected by the introduction of the stearically stabilized liposomes, the so-called "stealth" liposomes, which (a) deliver only 40–50% of the drug to the macrophages, the remaining being transported elsewhere, as these liposomes are stable in serum and (b) more importantly they have strong affinity to the lung tissue, partly because their lipids are similar to those existing in the lung surfactant (112,113). An operational advantage is that the "stealth" liposomes can be administered subcutaneously, giving equal response, whereas the unilamellar and multilamellar liposomes should be given only intravenously. The possibility thus exists to develop inermittent treatment with a fraction of the daily doses, to achieve equal success. Another important advantage of the stealth liposomes is the feasibility of delivering the drugs aerogenically. Such studies are still in infancy and, if successful, will offer a new strategy of treating tuberculosis.

7.3. *Voluntary Influences of the Host*

All the voluntary influences of the host can be summarized in one word, "noncompliance." Noncompliance is a well-recognized limitation for the success of any disease, and its influence was recognized long ago (114). Although the seriousness is not easily felt in other diseases because the duration of treatment, in general, is short, during which period most patients will be compliant, in diseases like tuberculosis the consequences of noncompliance are serious. Because symptoms act as an incentive for taking the drugs, most people discontinue their treatment soon after their symptoms wane. Disappearance of symptoms is quite rapid, thanks to the powerful drugs we now have. It is therefore not uncommon for patients with tuberculosis to discontinue their treatment quite early after initiation. This sort of behavior is not influenced by age, sex, race, economic status, educational background, or any other known human qualifications (115). On the other hand, there are marked differences in the background reasons for noncompliance between patients in developed and developing countries. Whereas most of the above-mentioned criteria do apply in developed countries, other factors, like convenience of operation of the clinics, distance for travel, lack of conveyance, and so forth, also play a major role for noncompliance in developing countries. Excellent reviews of these factors appear Refs. 116 and 117, 121.

Several direct and indirect methods for identifying the noncompliant patient have been discussed. These include pill counts, pill calenders, medication monitors, and so on, which are the well-adapted inferential methods, and the urine or serum tests for the drugs or their metabolites, as the definitive methods (115). All these attempts will only identify the defaulters with the hope that the defaulting patients will be warned or advised by the physician or health care worker to mend his ways. In spite of this, defaulting still occurs.

Therapeutic attempts have also been made to ensure compliance. These include shortening the chemotherapy or giving combination tablets. All these attempts are also not guaranteeing complete success. The only recourse taken was to give the drug under direct supervision. This concept, which was brought out in the early studies as "supervised chemotherapy" from Madras, is now reintroduced with great emphasis as directly observed treatment (DOT) or directly observed treatment short course (DOTS), which is discussed later.

In all early studies on supervised chemotherapy, it was soon realized that daily administration of chemotherapy under direct supervision is difficult in many places and, therefore, intermittent administration was sought (118). A controlled study done in Madras showed that a twice-weekly administration of supervised intermittant chemotherapy gives equal response as daily unsupervised treatment (118). Subsequently, several studies confirmed this conclusion (119,120). Elaborate discussions on these regimens are given in Chapters 2 and 4–6.

Twice-weekly intermittent chemotherapy could be successfully administered in an urban setting, as has been excellently demonstrated in a study in Bohemia (120). However, such an application is not feasible in rural areas as is common in many developing countries (121). For successful application of supervised chemotherapy, an intermittency of once a month or so is advantageous. On the other hand, stretching the intermittency with conventional drug administration, even to once a week, will not yield satisfactory results, as had been shown by another study done in Madras (122). In this context, depot preparations are developed, as discussed later.

DOT OR DOTS

The recent upsurge in tuberculosis in the United States has reawakened supervised chemotherapy as directly observed treatment (DOT) or, as is called by the World Health Organization (WHO), as directly observed treatment short course (DOTS). This strategy has been applied in all vigor to solve the severe problem of tuberculosis in New York (123), and the success achieved there has been taken as a role model for all other places (124). A drastic reduction in the incidence of tuberculosis and, more importantly, MDR disease in New York is considered by many as a miraculous testimony of the success of the DOT strategy. As is mentioned by Iseman et al. (125), DOT strategy is not to be ignored. Besides many affluent countries, DOTS has given fruitful results in many developing countries like Tanzania, China, and Korea. The WHO and World Bank are putting their maximum efforts in this approach, which is bound to achieve the desired success.

There may be a few difficulties in the widespread application of DOTS, however. These will deal not with patients compliance but with the proper function ("compliance") of the providers. There are administrative problems which have to be solved for a successful application of DOTS universally (126,127). Furthermore, lack of communication, conveyance, transport facilities etc. will also limit the direct DOT. For this reason also, an intermittent DOT, for example with an interval of once a month or so, will be optimal. If one can deliver the whole treatment of tuberculosis in one or few doses of drugs given as DOT, it will be ideal. Such thoughts can be approached with some success, thanks to the recent developments in biotechnology. This is the background for the new strategy of depot drug delivery for the treatment of tuberculosis.

DEPOT DRUG DELIVERY

Encouraged by the development in biotechnology and the introduction of bio-degradable and nondegradable polymers which are in wide use in some fields of medicine (e.g., family planning), attempts have been made to explore the use of biodegradable polymers as vehicles to contain and carry the antituberculosis drugs (128). Of all the available biodegradable polymers, the polylactic polyglycolic acid (PLGA) polymer was chosen; it has several advantages of maneuverability, least toxicity, and the only one approved by the Food and Drug Administration (FDA). By delivering the drug incorporated in the optimal concentrations of the PLGA polymer, significant levels of the drug could be demonstrated in serum, urine, and tissues for prolonged periods of time, with no abnormally high levels released immediately. Such findings were shown with isoniazid, rifampin, pyrazinamide, and streptomycin in mice and rabbits (128–131). More importantly, the single administrations of these depot preparations of the drugs have shown chemotherapeutic efficacy by indirect and direct methods. In the indirect methods [also called in vitro–in vivo methods by Conalty (132)], the serum, filter-sterilized urine or tissue honogenates obtained at 4–6 weeks after the implant, could demonstrate high antimycobacterial activity in vitro and in macrophages, as assessed by the conventional or radiometric (BACTEC) methods (128,133). More significantly, direct chemotherapeutic activity could be demonstrated against experimental tuberculosis in the mouse model. Such efficacy was shown with isoniazid, rifampin, streptomycin, and combinations of those drugs, as well as isoniazid and pyriazinamide (130,134,135). Whereas substantial evidence is available in the animal models in the use of this depot drug strategy, clinical trials which are being conducted will give the final proof. Although this strategy is being experimented with the existing drugs, it should be vital for any new drug to be discovered in the future, as noncompliance will be an anticipated universal complication of any drug, especially on prolonged use as in needed in the treatment of tuberculosis.

8. Conclusions

In this chapter, the various approaches to discovering new drugs for the treatment of tuberculosis are discussed. Besides the many new leads, some approaches have centered around the modifications of the existing drugs. However, it is realized that development of the drugs by the pharmaceutical industry is as not forthcoming as it used to be. As such, new strategies to harness the maximum benefit from the existing drugs are sought. In this endeavor, the involuntary and voluntary host factors in the control of chemotherapy are considered. In the involuntary factors, liposome incorporation has been considered very important. In the involuntary factors, one has to accept inevitable consequences and resort to DOT. The other promising approach is to give the drugs in the depot formulation, ensuring sustained drug delivery by one or few administrations.

References

1. Bloch AB, Rieder HL, Kelly GD, Canthan GM, Hayden CW, Snider DE (1989) The epidemiology of tuberculosis in the United States. Semin Respir Inform Infec. 4:159.
2. Infection Expanded Tuberculosis Surveillance and Tuberculosis Morbidity—United States. (1993) Morbid Mortal Wkly Rep 43:361.
3. Bates J (1995) Tuberculosis chemotherapy. Am J Respir Crit Care Med 151:942.
4. Raghunath Rao U, Srinivas Rao S, Natarajan S, Venkataraman PR (1946) Inhibition of *Mycobacterium tuberculosis* by garlic extract. Nature 157:441.
5. Deshpande RG, Khan MB, Bhat DA, Navalkar RG (1993) Inhibition of *Mycobacterium avium* complex isolates from AIDS patients by garlic (*allium sativum*). J. Antimicrob Chemother 32:623.
6. Baker WR, Mitscher LA, Feng B, Cai S, Clark M, Leung T, Towell JA, Derwish I, Stover K, Kreiswirth B, Moghazeh S, Henriquez T, Resconi A, Arain T (1995) Part II. Antitubercular agents from plants: antimicrobial activity of azaindoquinazolinidiones. Novel alkaloids active against sensitive and multidrug resistant tuberculosis. Interscience Conference on Antimicrobial Agents and Chemotherapy, p. 116.
7. Berlin OGW, Young LS, Brucner CA (1987) In vitro activity of six fluorinated quinolones against *Mycobacterium tuberculosis*. J Antimicrob Chemother 19:611.
8. Ji B, Truffot-Pernot C, Grosset J (1991) In vitro and in vivo activities of sparfloxacin (AT04140) against *Mycobacterium tuberculosis*. Tubercle 72:181.
9. Skinner PS, Furney SK, Kleinert DA, Orme IM (1995) Comparison of activities of fluoroquinolones in murine macrophages infected with *Mycobacterium tuberculosis*. Antimicrob Agents Chemother 39:750.
10. Tsukumura M (1985) Antituberculosis activity of ofloxacin (DL8280) on experimental tuberculosis in mice. Am Rev Respir Dis 915:144.

11. Sanders CC, Sanders WE, Goering RV (1987) Overview of preclinical studies with ciprofloxacin. Am J Med 82(Suppl 4A):2.

12. Lalande V, Truffot-Pernot C, Paccaly-Moulin A, Grosset J, Ji B (1993) Powerful bactericidal activity of sparfloxacin (AT4140) against *Mycobacterium tuberculosis* in mice. Antimicrob Agents Chemother 37:407.

13. Ji B, Lounis N, Truffot-Pernot C, Grosset J (1995) In vitro and in vivo activities of levofloxacin against *Mycobacterium tuberculosis*. Antimicrob Agents Chemother 39:1341.

14. Klemens SP, Sharpe CA, Rogge MC, Cynamon MH (1994) Activity of levofloxacin in a murine model of tuberculosis. Antimicrob Agents Chemother 38:1476.

15. Jindani A, Aber VR, Edwards EA, Mitchison DA (1980) The early bactericidal activity of drugs in patients with pulmonary tuberculosis. Am Rev Respir Dis 121:939.

16. Kennedy N, Fox R, Kinyombe M, Aloyce O, Saruni S, Uiso LO, Ramsay AR, Ngowi FI, Gillespie SH (1993) Early bactericidal and sterlizing activities of ciprofloxacinin pulmonary tuberculosis. Am Rev Respir Dis 148:1547.

17. Tsukumura M, Nakamura E, Yoshii S, Amano H (1985) Therapeutic effect of a new antibacterial substance ofloxacin (DL280) on pulmonary tuberculosis. Am Rev Respir Dis 131:352.

18. Leysen DC, Haemers A, Pattyn SR (1989) Mycobacteria and the new quinolones, Antimicrob Agents Chemother 33:1.

19. Klopman G, Li JY, Wang S, Pearson AJ, Chang K, Jacobs MR, Bajaksouzian S, Ellner JJ (1994) In vitro antimycobacterium avium activities of quinolones: predicted activities structures and mechanistic considerations. Antimicrob Agents Chemother 38:1794.

20. Luna-Herrera J, Reddy MV, Danneluzzi D, Gangadharam PRJ (1995) Antituberculosis activity of clarithromycin Antimicrob Agents Chemother 39:2692.

21. Burchard GD, Mirelman D (1988) *Entamoeba hystolica:* virulance potential and susceptibility to metronidazole. Exp Parasitol 66:231.

22. Wayne LG, Sramek HA (1994) Metronidazole is bactericidal to dormant cells of *Mycobacterium tuberculosis*. Antimicrob Ag Chemother, 38:2054.

23. Kilburn J, Glickman S, Brickner SM, Manninen P, Ulanowicz D, Lovatz K, Zurenko C (1994) In vitro antimycobacterial activity of novel multicyclic, fused ring oxazolidinones In: Programs & Abstracts of the 33rd Interscience Conference on Antimicrobial Agents and Chemotherapy. Washington, DC: American Society of Microbiology.

24. Ashtekar DR, Costa-Periera R, Srinivasan T, Ivyer R, Vishwanathan N, Rittel W (1991) Oxazolidinones, a new class of synthetic antituberculosis. Agent Diagn Microbiol Infect Dis 14:465.

25. Cornforth JW, Hart P, Rees R, Stock J (1951) Antituberculosis effect of certain surface active polyoxyethlene ethers in mice. Nature 168:150.

26. Cornforth JW, Hart P, D'Arcy Hart P, Nicholls GA, Rees RJW, Stock JA (1955)

Antituberculous effects of certain surface active polyoxyethylene ethers. Br J Pharmacol 10:73.

27. Chinnaswamy J, Allaudeen HS, Hunter RL (1995) Activities of polaxmer CRL8131 against *Mycobacterium tuberculosis in vitro* and *in vivo*. Antimicrob Agents Chemother 39:1349.

28. Barry VC, Belton JG, Conalty ML, Denneny JM, Edward DW, O'Sullivan JF, Twomey D, Winder F (1957) A new series of phenazines (rimino-compounds) with high antituberculosis activity. Nature 179:1013.

29. Barry VC, Conalty ML (1965) B663 in the treatment of leprosy: Lepr Rev 36:3–7.

30. Gangadharam PRJ, Candler ER (1977) Activity of some antileprosy compounds against *Mycobacterium intracellulare in vitro*. Am Rev Respir Dis 115:705.

31. Reddy MV, O'Sullivan JF, Gangadharam PRJ (1996) Riminophenazines (minireview). Antimicrob Agents Chemother 1997 (in press).

32. Reddy MV, Geeta N, Danelluzi D, O'Sullivan JF, Gangadharam PRJ (1996) Antituberculosis activities of clofazimine and its new analogs B4154 and B4157. Antimicrob Agents Chemother 40:633.

33. Chinnaswamy J, Reddy MV, Kailasam S, O'Sullivan JF, Gangadharam PRJ (1995) Chemotherapeutic activity of clofazimine and its analogues against *Mycobacterium tuberculosis: in vitro,* intracellular and *in vivo* studies. Am Rev Respir Crit Care Med 151:1083.

34. Kucers A, Bennett NM (1987) Neomycin, framycetin and paromomycin. In: The Use of Antibiotics, 4th Ed., with of R.J. Kemp published Lippincott C, London: Heineman.

35. Kanyok TP, Reddy MV, Chinnaswamy J, Danziger LH, Gangadharam PRJ (1994) Activity of aminosidine (paramomycin) for *Mycobacterium tuberculosis* and *Mycobacterium avium*. J Antimicrob Chemother 33:323.

36. Kanyok TP, Reddy MV, Chinnaswamy J, Danziger LH, Gangadharam PRJ (1994) In vivo activity of paramomycin against susceptible and multidrug resistant *Mycobacterium tuberculosis* and *M. avium*. Antimicrob Agents Chemother 38:170.

37. Gangadharam PRJ, Reddy MV (1993) Unpublished observations.

38. Rastogi N, Moreau B, Capmau ML, Goh KS, David HL (1988) Antibacterial action of amphipathic derivatives of isoniazid against the *Mycobacterium avium* complex. Zbl Bakt Hyg A268:456.

39. Rastogi N, Goh KS (1990) Antibacterial action of 1-isonicotinyl-2-palmitoyl hydrazine against the *Mycobacterium avium* complex and the enhancement of its activity by *m*-fluoro-phenylalanine. Antimicrob Agents Chemother 34:2061.

40. Gaetano G, Giannola LI, Carlish B (1989) Synthesis of polymeric derivatives of isoniazid: characterisation and in vitro release from a water-soluble adduct with polysuccinamide. Chem Pharm Bull 37:1106.

41. Barry VC (ed) (1964) In: Chemotherapy of Tuberculosis: The Development of the Chemotherapeutic Agent for Tuberculosis, p. 46. London: Buttersworth.

42. Tsukumura M, Mizuno S, Toyoma H (1986) In vitro antimycobacterial activity of rifapentinine (compared with rifampicin). Kakkaku 63:144.

43. Asondi A, Batti B, Christina T (1984) Pharmacokinetics of rifapentine, a new long lasting rifamycin in rat, the mouse and the rabbit. J Antibiot 37:1066.

44. Lee HS, Shin HS, Han SS, Rol JK (1992) High performance liquid chriomatographic determination of rifapentine in serum using column switching. J Chromatogr 574:175.

45. Wylie GL, Scoging A, Lowrie DB (1986) Uptake and intracellular distribution of rifamycin DL 473 and rifampicin in mouse macrophages. Bull Int Union Against Tuberc 61:11.

46. Gangadharam PRJ (1988) In: Peterson PK, Verhoef J, eds. Antimycobacterial Drugs. New York: Elsevier Science Publ.

47. Sanfilippo A, Della Bruna C, Marsili L, Morvillo E, Pasqualucci CR, Schioppacassi G, and Ungheri D (1980) Biological activity of a new class of rifamycins, spiropiperdyl-rifamycins. J Antibiot 33:1193.

48. Tsukumura M, Mizuno S, Toyoma H, Ichiyama S (1986) Comparison of *in vitro* antimycobacterial activities of ansamycin and rifampicin. Kekkakku 61:497.

49. Perumal VK, Gangadharam PRJ, Heifets LB, Iseman MD (1985) Dynamic aspects of the in vitro chemotherapeutic activity of ansamycin on *Mycobacterium intracellulare*. Am Rev Respir Dis 132:1278.

50. Perumal VK, Gangadharam PRJ, Iseman MD (1987) Effect of rifabutin on the phagocytosis and intracellular growth of *Mycobacterium intracellulare* in mouse resident and activated peritoneal and alveolar macrophages. Am Rev Respir Dis 136:334.

51. Gangadharam PRJ, Perumal VK, Jairam BT, Rao PN, Nguyen AK, Farhi DC, Iseman MD et al (1987) Activity of rifabutin alone or in combination with clofazimine or ethambutol or both against acute and chronic experimental *Mycobacterium intracellulare* infections. Am Rev Respir Dis 136:329.

52. Ji S, Truffot-Perot C, La Croix C, Raviglione MC, O'Brien R, Ofliaro P, Roscigno G, Grosser J (1993) Effectiveness of rifampin, rifabutin, and rifapentine for preventive therapy of tuberculosis in mice. Am Rev Respir Dis 148:1541.

53. Luna-Herrera J, Reddy MV, Gangadharam PRJ (1995) In-vitro and intracellular activity of rifabutin on drug-susceptible and multidrug resistant (MDR) tubercle bacilli. J Antimicrob Chemother 36:355.

54. Reddy MV, Luna-Herrera J, Daneluzzi D, Gangadharam PRJ (1996) Chemotherapeutic activity of benzoxirifamycin KRM 1648, against *Mycobacterium tuberculosis*. Tubercle Lung Dis 77:154.

55. O'Brien RJ, Lyle MA, Snider DE (1987) Rifabutin (ansamycin LM427): a new rifamycin-S derivative for the treatment of mycobacterial diseases. Rev Infect Dis 9:519–530.

56. Dautzenberg B, Truffot C, Mignon A, Rozenbaum W, Katlama C, Pronne C, Parroth R, Grosset J (1991) Rifabutin in combination with clofazimine, isoniazid and

ethambutol in the treatment of AIDS patients with infections due to opportunistic mycobacteria. Tubercle 72:168.

57. Della Bruna C, Ungeri D, Sebben G, Sanfillipo A (1985) Laboratory evaluation of a new long acting 3-azinomethylrifamycin FCE22250. J Antibiot 38:779.

58. Vischer WA, Gowrishankar R, Ashteker DR, Costapereira R, Subramanyan D, Kump W, Traxler P (1986) Antitubercular activity in vitro and in vivo of new long acting rifamycin derivatives. Bull Int Union Against Tuberc 61:8–10.

59. Yamane T, Hashizume T, Yamashita K, Konishi E, Hosoe K, Hidaka T, Watanabe K, Kawaharada H, Yamamoto T, Kuze F (1993) Synthesis and biological activity of 3'-hydroxy-5'aminobenzoxazinorifamycin derivatives. Chem Pharm Bull 41(1):148.

60. Tomoika H, Saito H, Sato K, Yamane T, Yamashita K, Hosoe K, Fujii K, Hidaka T (1992) Chemotherapeutic efficacy of newly synthesized benzoxirifamycin KRM-1648, against *Mycobacterium avium* complex infection induced in mice. Antimicrob Agents Chemother 36:387.

61. Saito H, Tomioka H, Sato K, et al. (1991) In vitro antimicrobial activities of newly synthesised benzoxirifamycins. Antimicrob Agents Chemother 35:542.

62. Saito H, Tomioka H, Sato K, Kawahara S, Hidaka T, Dekio S (1995) Therapeutic effect of KRM 1648 with various antimicrobials against *Mycobacterium avium* complex infection in mice. Tuberc Lung Dis 76:51.

63. Luna-Herrera J, Reddy MV, Gangadharam PRJ (1995) In vitro activity of benzoxirifamycin KRM 1648 against drug susceptible and multidrug resistant tubercle bacilli. Antimicrob Agents Chemother 39:440.

64. Bermudez LE, Kolonoski P, Young LS, Inderlied CB (1994) Activity of KRM 1648 alone or in combination with ethambutol or clarithromycin against *Mycobacterium avium* complex. Antimicrob Agents Chemother 38:1844.

65. Emori M, Saito H, Sato K, Tomoika H, Setogawa T, Hidaka T (1993) Therapeutic efficacy of the benzoxazinorifamycin KRM-1648 against experimental *Mycobacterium avium* infection induced in rabbits. Antimicrob Agents Chemother 37:722.

66. Dimova V, Dobrev P, Kalfin E, Vlasov V (1994) Therapeutic effect of 3/4-cinnamyl-1-piperazinyl/iminomethyl rifamycin SV on generalized tuberculosis in guinea pigs. In: Recent Advances in Chemotherapy, Proceedings of the 18th International Congress of Chemotherapy. Washington, DC: American Society for Microbiology.

67. Dimova V, Stefanova P, Valova N (1994) Pharmacokinetic studies in experimental animals on cinnamyl rifamycin derivative (T9). In: Recent Advances in Chemotherapy, Proceedings of the 18th International Congress of Chemotherapy. Washington, DC: American Society for Microbiology.

68. Reddy MV, Geeta N, Daneluzzi D, Dimova V, Gangadharam PRJ (1995) Antimycobacterial activity of a new rifamycin derivative, 3-(-4-cinnamylpiperazinyl iminomethyl) rifamycin SV (T9). Antimicrob Agents Chemother 39:2320.

69. Cynamon MH, Gimi R, Gyenes F, Sharpe CA, Bergmann KE, Han HJ, Gregor LB, Rapolu R, Luciano G, Welch JT (1995) Pyrazinoic acid esters with broad spectrum in vitro antimycobacterial activity. J Med Chem 38:3902.

70. Yamamoto S, Toida I, Watanabe N, Ura T (1995) In vitro antimycobacterial activities of pyrazinamide analogs. Antimicrob Agents Chemother 39:2088.

71. Konno L, Feldmann FM, McDermott W (1967) Pyrazinamide susceptibility and amidase activity of tubercle bacilli. Am Rev Respir Dis 95:461.

72. Kasik JE (1979) Mycobacterial beta-lactamases. In: Hamilton JMT, Smith JT, eds. Beta-lactamases, pp. 339–350. London: Academic Press.

73. Kasik JE (1965) The nature of mycobacterial penicillinase. Am Rev Respir Dis 91:117–119.

74. Casal M (ed) (1986) New *in vitro* antimicrobial possibilities in the treatment of tuberculosis. In: Mycobacteria of Clinical Interest, p. 155. Amsterdam: Elsevier Science Publishers, B.V.

75. Cynamon MH, Palmer GS (1983) In vitro activity of amoxicillin in combination with clavulinic acid against *Mycobacterium tuberculosis*. Antimicrob Agents Chemother 24:429.

76. Casal MJ, Rodriguez FC, Luna MD, Benavente MC (1987) In vitro susceptibility of *Mycobacterium tuberculosis, Mycobacterium africanum, Mycobacterium bovis, Mycobacterium avium, Mycobacterium fortuitum* and *Mycobacterium chelonae* to Ticarcillin in combination with clavulinic acid. Antimicrob Agents Chemother 31:132–133.

77. Reddy, MV, Luna-Herrera J, Gangadharam PRJ (1995) Unpublished observations.

78. Nadler JP, Berger J, Nord JA, Cofsky R, Saxena M (1991) Amoxicillin–clavulinc acid for treating drug-resistant *Mycobacterium tuberculosis*. Chest 99:1025.

79. Hoffner SE, Svenson SB, Kallenius G (1987) Synergistic effects of antimycobacterial drug combinations on *Mycobacterium avium* complex determined radiometrically in liquid medium. Eur J Clin Microbiol 6:530.S.

80. Takayama K, Armstrong EL, Kunugi KA, Kilburn JO (1979) Inhibition by ethambutol of mycolic acid transfer into the cell wall of *Mycobacterium smegmatis*. Antimicrob Agents Chemother 16:240.

81. Takayama K, Kilburn JO (1989) Inhibition of synthesis of arabinogalactan by ethambutol in *Mycobacterium smegmatis*. Antimicrob Agents Chemother 33:1493.

82. Rastogi N, Potar MC, David HL (1987) Intracellular growth of pathogenic mycobacteria in the continuous macrophage cell line J-774: ultrastructure and drug susceptibility studies. Curr Microbiol 16:79.

83. Chinnaswamy J, Reddy MV, Gangadharam PRJ (1995) Enhancement of drug susceptibility of multidrug resistant strains of *Mycobacterium tuberculosis* by ethambutol and dimethylsulfoxide. J Antimicrob Chemother 35:381.

84. Reddy MV, Gangadharam PRJ. Unpublished.

85. Jacobs S, Bishel M, Herschler RJ (1964) Dimethylsulfoxide (DMSO): a new concept in pharmacotherapy. Curr Therap Res 66:134.

86. Szydlowska T, Pawlowska I (1974) In vivo studies on reversion to sensitivity of INH-resistant tubercle bacilli under the influence of dimethylsulfoxide (DMSO). Arch Immunol Therap Exp (Warszawa) 22:559.

87. Szydlowska T (1972) Studies on the role of dimethylsulfoxide in resensibilisation of antibiotic resistant bacterial strains. Arch Immunol Therap. Exp (Warezawa) 20:193.

88. Muller U, Urbanczik R (1979) Influence of dimethylsulfoxide (DMSO) on restoring sensitivity of mycobacterial strains resistant to chemotherapeutic compounds. J Antimicrob Chemother 5:326.

89. Wheeler PR, Besra GS, Minnikin DE, Ratledge C (1993) Inhibition of mycolic acid biosynthesis in a cell wall preparation from *Mycobacterium smegmatis* by methyl 4-(2-octadecylcyclopropen-1-yl) butanoate, a structural analogue of a key precursor. Lett Appl Microbiol 17:33.

90. Gangadharam PRJ, Pratt PF, Damle PB, Porter TR, Folkers K (1978) Inhibition of *Mycobacterium intracellure* by some vitamin K and ubiquinone analogues. Am Rev Respir Dis 118:467.

91. Gangadharam PRJ (1994) The antimycobacterial activity of gangamicin, a potential drug. In: Global Congress on Lung Health 28th World Conference of IUATLD.

92. Falah AMS, Bhatnagar R, Bhatnager N, Singh Y, Sidhu GS, Murthy PS, et al. (1988) On the presence of calmodulin like protein in mycobacteria. FEMS Microbiol Lett 56:89.

93. Ratnaker P, Murthy PS (1992) Antitubercular activity of trifluoperazine, a calmodulin antagonist. FEMS Microbiol Lett 76:73.

94. Reddy MV, Geeta N, Gangadharam PRJ (1996) In vitro and intracellulare antimycobacterial activity of trifluoperazine. J Antimicrob Chemother 37:196.

95. Baselt RC, Cravey RH (1989) Disposition of Toxic Drugs and Chemicals in man. Chicago, IL: Year Book Publishers Inc.

96. Ganapathi R, Kuo T, Teeter L, Grabowski D, Ford J (1991) Relationship between expression of *p*-glycoprotein and efficacy of trifluoperazine in multidrug resistant cells. Molec Pharmacol 39:1.

97. Gollapudi S, Reddy MV, Gangadharam PRJ, Tsuruo T, Gupta S (1994) *Mycobacterium tuberculosis* induces expression of *p*-glycoprotein in pro-monocytic U1 cells chronically infected with HIV type, 1. Biochem Biophys Res Commun 199:1181.

98. Zhang Y, Heym B, Allen B, Young D, Cole S (1992) The catalase–peroxidase gene and isoniazid resistance of *Mycobacterium tuberculosis*. Nature 358:591.

99. Banerjee A, Dubnau E, Qumard A, Balasubramanian V, Um KS, Wilson T, de Lisle G, Jacobs WR (1994) *inha,* a gene encoding a target for isoniazid and ethionamide resistance in *Mycobacterium tuberculosis* Science 263:227.

100. Williams DL, Waguespack C, Eisenach K, Crawford JT, Portaels F, Salfinger M, Nolan CM, Abe C, Sticht-Groh V, Gillis TP (1994) Characterisation of rifampin resistance in pathogenic mycobacteria. Antimicrob Agents Chemother 38:2380.

101. Honroe' N, Cole ST (1994) Streptomycin resistance in mycobacteria. Antimicrob Agents Chemother 38:238.

102. Takiff H, Salazar C, Phillip W, Huang WM, Kreiswirth B, Cole ST, Jacobs WR Jr, Telenti A (1994) Cloning and nucleotide sequence of *Mycobacterium tuberculosis*

gyrA and *gyrB* genes and detection of quinolone resistance mutations. Antimicrob Agents Chemother 38:773.

103. Peloquin CA, Macphee AA, Berning SE (1993) Malabsorption of mycobacterial medications. N Engl J Med 329:1122.

104. Rao KVN, Mitchison DA, Nair NGK, Prema K, Tripathy SP (1970) Sulphadimidine acetylation test for classification of patients as slow or rapid inactivators of isoniazid. Br Med J 3:495.

105. Gangadharam PRJ, Bautista EM (1972) Competitive acetylation between sulfamethazine and isoniazid in slow and rapid inactivators of isoniazid. In: Proceedings of Fifth International Congress on Pharmacology.

106. Ellard GA, Gammon PT, Lakshminarayan S, Fox W, Aber VR, Mitchison DA, Citron KM, Tall R (1972) Pharmacology of some slow release preparations of isoniazid of potential use in intermittent treatment of tuberculosis. Lancet 340.

107. Düzgunes N, Perumal VK, Debs RJ, Gangadharam PRJ (1988) Enhanced effect of liposome-encapsulated amikacin on *Mycobacterium avium-intracellulare* complex infection in biege mice. Antimicrob Agents Chemother 32:1404.

108. Kesavalu L, Goldstein JA, Debs RJ, Düzgunes N, Gangadharam PRJ (1990) Differential effects of free and liposome encapsulated amikacin on the survival of *Mycobacterium avium* complex in mouse peritoneal macrophages. Tuberc Lung Dis 71:215–218.

109. Düzgunes N, Ashtekar DR, Flasher DL, Ghori N, Debs RJ, Friend DS, Gangadharam PRJ (1991) Treatment of *Mycobacterium avium-intracellulare* complex infection in beige mice with free and liposome encapsulated streptomycin. Role of liposome type and duration of treatment. J Infect Dis 164:143.

110. Gangadharam PRJ, Ashtekar DR, Ghori N, Goldstein JA, Debs RJ, Düzgunes N (1991) Chemotherapeutic potential of free and liposome encapsulated streptomycin against experimental *Mycobacterium avium* complex infections in beige mice. J Antimicrob Chemother 28:425.

111. Ostro MJ (1992) Drug delivery via liposomes. Drug Therap vol. 22(4):61.

112. Allen TM (1989) Stealth TM liposomes: avoiding reticuloendothelial uptake. In: Liposomes in the Therapy of Infectious Diseases and Cancer, p. 405, New York: Alan R. Liss, Inc.

113. Irma AJM, Woudenberg B, Lokerse AF, Marian T, ten Kate, Storm G (1992) Enhanced localisation of liposomes with prolonged blood circulation time in infected lung tissue. Biochim Biophys Acta 1138:318.

114. Evans L, Spelman M (1983) The problem of non-compliance with drug therapy. Drugs 25:63.

115. Fox W (1962) Self administration of medicaments Bull Int Union Against Tuberc 32:307–331.

116. Fox W (1983) Compliance of patients and physicians; experience and lessons from tuberculosis. Br Med J 287:33.

117. Sumartojo E (1993) When tuberculosis treatment fails. A social behaviour account of patient's adherence. Am Rev Respir Dis 147:1311–1320.

118. Tuberculosis Chemotherapy Center, Madras (1964) A controlled comparison of intermittent (twice-weekly) isoniazid plus streptomycin and daily isoniazid plus PAS in the domiciliary treatment of pulmonary tuberculosis. Bull WHO 31:247.

119. Sbarbaro JA, Johnson S (1968) Tuberculosis chemotherapy of recalcitrant outpatients administered directly twice-weekly. Am Rev Respir Dis 97:895.

120. World Health Organization Collaborating Center for Tuberculosis Chemotherapy, Prague (1971) A comparative study of daily and twice-weekly combination regimens of tuberculosis chemotherapy, including a comparison of two durations of sanatorium treatment. 1. First report: the results of 12 weeks. Bull WHO 45:573.

121. Pamra SP (1980) Problems of tuberculosis in developing countries. In: Stead WW, Dutt AK, eds., p. 265. Clinics in Chest Medicine. W.B. Saunders Co.

122. Tuberculosis Chemotherapy Center, Madras (1970) A controlled comparison of a twice-weekly and three once-weekly regimens in the initial treatment of pulmonary tuberculosis. Bull WHO 43:143.

123. Frieden TR, Fujiwara PL, Washko RM (1995) Tuberculosis in New York City— turning the tide. N Engl J Med 333:229.

124. Weis SE, Slocum PC, Blais FX, King B, Num M, Malney GB, Gomez E, Foresman BK (1994) The effect of directly observed therapy on the rates of drug resistance and relapse in tuberculosis. N Engl J Med 330:1179.

125. Iseman MD, Cohn DL, Sbarbaro JA (1993) Directly observed treatment of tuberculosis: we can't afford not to try it (editorial). N Engl J Med 328(8):576.

126. Gangadharam PRJ (1994) Chemotherapy of tuberculosis under program conditions, with special relevance to India (editorial). Tuberc Lung Dis 75:241.

127. Olle-Goig JE (1995) Noncompliance with tuberculosis treatment: patients and physicians. Tubercle 76:277.

128. Gangadharam PRJ, Ashtekar DR, Farhi DC, Wise DL (1991) Sustained release of isoniazid *in vivo* from a single implant of a biodegradable polymer. Tubercle 72:115.

129. Kailasam S, Daneluzzi D, Gangadharam PRJ (1994) Bio-availability of rifampin after a single implant of a biodegradable polymer. Poster presented at the 93rd ASM General Meeting.

130. Kailasam S, Daneluzzi D, Reddy MV, Gangadharam PRJ (1994) Bioavailability and chemotherapeutic activity against experimental tuberculosis of streptomycin in mice after a single subcutaneous biodegradable polymer implant. Poster presented at the 34th Interscience Conference on Antimicrobial Agents and Chemotherapy (ICAAC).

131. Gangadharam PRJ, Kailasam S (1993) Sustained release of pyrazinamide in mice from a single implant of a biodegradable polymer. Poster presented at the 8th International Congress of Chemotherapy.

132. Conalty ML (1964) Methods of preclinical evaluation of antituberculosis drugs. In: Barry VC, ed. Chemotherapy of Tuberculosis, p. 150. London: Butterworths.

133. Kailasam S, Daneluzzi D, Gangadharam PRJ (1994) Maintenance of therapeutically

active levels of isoniazid for prolonged periods in rabbits after a single implant of biodegradable polymer. Tuberc Lung Dis 75:361.

134. Gangadharam PRJ, Kailasam S, Srinivasan S, Wise DL (1994) Experimental chemotherapy of tuberculosis using single dose treatment with isoniazid in a biodegradable polymer. J Antimicrob Chemother 33:265.

135. Gangadharam PRJ, Geeta N, Danelluzzi D (1996) Unpublished.

14

Epidemiology of Leprosy

S.K. Noordeen

1. Introduction

Leprosy, from an epidemiological point of view, is a disease with several interesting features. Among these, one should mention its high infectivity and low pathogenicity, its prolonged incubation period, and its high tendency for spontaneous healing, in addition to its wide range of manifestations.

The study of the epidemiological situation in leprosy is beset with several problems. To begin with, there is the compulsion to base the evaluation entirely on clinical disease without any measurement possible of subclinical infection. Further, clinical leprosy itself occurs in a wide spectrum of forms, from the early minimal, self-healing, single cutaneous patch with limited evidence of disease, to disseminated progressive multibacillary disease with billions of organisms all over the skin and elsewhere. The significance of the wide spectral occurrence of leprosy in relation to the immune response of individuals to *M. leprae* infection and the potential of the various forms in the spectrum for transmission are not fully understood. Until we have a more precise understanding of the exact mode or modes of transmission of *M. leprae,* some of these questions will remain unanswered. Finally, evaluation of leprosy control necessitates follow-up of large populations over long periods in view of the low incidence and chronicity of the disease.

Thus far, the studies on the epidemiology of leprosy are based on the occurrence of clinical leprosy and on the assumption that the occurrence of the clinical leprosy is directly correlated with the occurrence of earlier infection. This assumption is not fully justified in the absence of studies on infection status of populations together with the absence of any evidence that only a specific and constant proportion of infected individuals get clinical disease.

Another important constraint while dealing with clinical leprosy for epidemiological evaluation is the lack of uniform criteria for defining the disease, its various types and stages. This has made geographical comparisons and comparisons over a period of time difficult. Further, the existing definitions for the end point of disease are rather vague and varied, particularly for epidemiological and operational purposes.

2. Global Leprosy Situation

Over the past 40 years or more, the leprosy situation in different countries as well as globally has been described through statistics on the number of existing registered leprosy patients at any particular point or period of time and, sometimes, by the number of new cases detected during a particular period. In addition, limited statistical information has been available on the distribution of the types of leprosy and the occurrence of deformities. All this information has been based on leprosy registers maintained by the different programs which did not always reflect the complete picture. The definition of what constituted "leprosy" was often variable. Notwithstanding these problems, information on registered cases has been used systematically to evaluate the situation, and over the years, this information has apparently become more and more dependable. The definition of "a case of leprosy" by the Who Expert Committee in its Sixth Report (1987) (1) as "a person requiring chemotherapy" has improved the situation still further. All the same, it is clear that registered cases reflect only a proportion of the true number and that such proportions vary with individual countries. There are inherent difficulties in making estimates, through sample surveys, of the number of cases in any country because of the relatively low frequency of the disease which requires extremely large samples, the very uneven geographic distribution characteristic of leprosy, and the difficulties of discriminating between early self-healing disease and established disease at one end of the evolution of the disease, and between active and inactive disease at the other end. In spite of these difficulties, estimates of the number of leprosy patients in the world, made in the past by WHO (2) and others, had remained constant at around 10–12 million for several years. Although this assumption of a static situation was, to some extent, justifiable during the period prior to the introduction of multidrug therapy (MDT), it can no longer be justified in the post-MDT period, as large numbers of patients are being cured and thus eliminated from the prevalence pool. As such, based on a detailed review of available information, WHO estimates (3) the number of leprosy patients in 1996 in the world at 1.3 million. This is also clear from the steep decrease in the number of registered cases since 1985 in spite of improved registration. Although the number of registered cases had increased from 2.8 million in 1966 to 5.4 million in 1985, it has since dropped to 0.9 million in 1996. This reduction of 82% since 1985 is, however, not uniformly distributed in

Table 14.1. Registered cases over the years by WHO region

WHO region	Number of registered cases (000s)				Change between 1985 and 1996
	1966	1976	1985	1996	
Africa	1,686	1,398	988	93	−90.6%
America	178	241	306	163	−46.7%
Eastern Mediterranean	41	63	75	18	−76.0%
Europe	20	20	17	1	−94.1%
Southeast Asia	791	1,748	3,737	635	−83.0%
Western Pacific	117	128	246	31	−87.4%
Total	2,832	3,600	5,368	941	−82.5%

Leprosy in the world – 1996

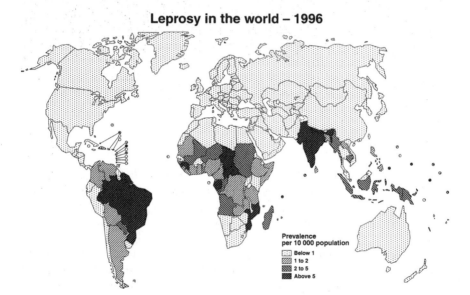

Figure 14.1. Current distributing of registered cases of leprosy.

Table 14.2. Estimated cases (in thousands) by continent 1963–1994

	1963	1996	Change between 1963 and 1996
Africa	3,868	200	−94.8%
America	358	170	−52.5%
Asia and Oceania	6,508	890	−86.3%
Europe	52	1	−98.1%
World Total	10,786	1261	−88.3%

the different WHO regions as shown in Table 14.1. The current distribution of registered cases by countries is shown in Fig. 14.1. The downward trend of leprosy is also seen in the number of estimated cases in the world as published by WHO. Table 14.2 gives this information by continent.

It is clear that although there has been a substantial decrease in registered cases in the WHO regions of Southeast Asia and the Western Pacific following the introduction of MDT, in the African region the decrease had started in the mid-seventies even before the introduction of MDT. There is no clear explanation for this steady decline in Africa, although, since 1985, a proportion of the decrease is attributable to MDT. Among the other contributing factors to this decline in Africa, mention should be made of (a) the late effects of strong dapsone-based leprosy campaigns of the fifties and sixties in several countries of Africa, (b) the effect of widely applied Bacille–Calmette–Guérin (BCG) vaccination acting as a prophylactic against leprosy, and (c) secular trends not contributed by any specific intervention. The leprosy situation on the American continent has shown a deterioration due largely to ineffective control programs in some countries. The steady increase in registered cases in the eastern Mediterranean region is probably due more to improved case detection than to a real increase in the disease. As far as Europe is concerned, leprosy, which at present is confined to a few countries and that too in extremely small numbers, is well on its way to extinction.

2.1. Age Distribution

Leprosy is known to occur at all ages ranging from early infancy to very old age. The youngest case seen by the author was in an infant of 2½ months, where the diagnosis of tuberculoid leprosy was confirmed by histopathology. The occurrence of leprosy, presumably for the first time, is not uncommon, even after the age of 70.

Figure 14.2 shows the age-specific incidence rates in a part of South India, where leprosy is highly endemic (4). The pattern is very similar to that seen in many high endemic areas, where there is a clear peak at age 10–14, followed by a depression which, in turn, is followed by a rise and then a plateau covering ages 30–60. The bimodal curve in high endemic areas suggests the possibility of two distinct experiences, one among children and the other among adults. In the absence of specific immunological tools to measure subclinical infection, one can only speculate on the assumption that the disease occurrence parallels the acquiring of infection. Even so, it is difficult to accept that a large number of persons in high endemic areas acquire infection and disease for the first time at a late age. There are two possible explanations for this. One is that the incubation or latent period is very long in a proportion of infected individuals, resulting in manifestation of disease late in their lives, possibly somewhat similar to endogenous reactivation in tuberculosis. The other explanation is that leprosy in adult life in

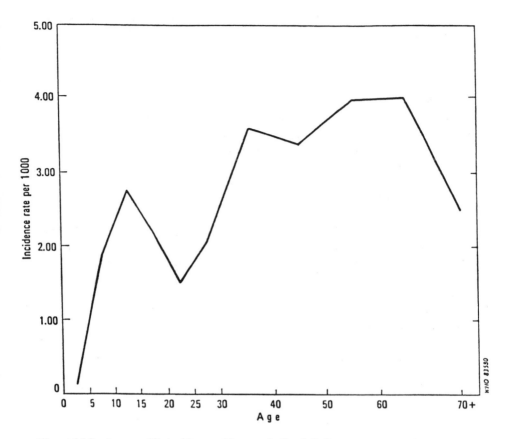

Figure 14.2. Age-specific incidence of leprosy in South India.

endemic areas is often the result of reinfection or superinfection among individuals who had previously been infected and whose immune response to leprosy had become inadequate as they grew older. In either case, in the absence of a specific method for identifying subclinical infection and strain variations of *M. leprae,* the hypotheses will remain untested.

2.2. Sex Distribution

Although leprosy affects both sexes, males are affected more frequently than females in most parts of the world, often in the ratio of 2 : 1. This preponderance of males is observed in as diverse geographic situations as India, the Philippines, Hawaii, Venezuela, and Cameroon. Doull et al. (5), from their studies in the Philippines, have also pointed out that the difference was a true difference due to higher incidence among males, not due to the differing duration of the disease for the two sexes.

2.3. Clustering of Leprosy

The more frequent occurrence of leprosy in certain clusters, particularly family clusters, is well recognized. However, the most debated point is whether this is due to the clusters sharing the same environment or the same genetic predisposition, or a combination of both. The occurrence of leprosy in clusters has been particularly observed in low endemic areas.

2.4. Time Trends in Leprosy

Because leprosy as a disease has a chronic course, it is often assumed that the epidemiological situation in any area remains static. In fact, the epidemiological situation is capable of a considerable amount of dynamic changes, and the factors that influence these changes are many. Both long-term and short-term trends have been studied with regard to the occurrence of leprosy.

In northern Europe, continental United States, Venezuela, Japan, and Hawaii, there have been well-documented studies on the decline in incidence of leprosy, leading gradually to the virtual disappearance of the disease in the native-born population. In northern Europe, the peak was reached in medieval times, with the decline occurring last in Norway during the nineteenth century (6). Careful analyses of declining incidence rates in Norway, Hawaii, and Japan reveal several features which are similar to those of tuberculosis under similar circumstances, including (a) a gradual increase in the mean age at onset of disease over time, (b) a decrease in age-specific incidence rates within successive cohorts associated with a fall in the mean age at onset, and (c) a gradually increasing proportion of the lepromatous type over a period of time among incidence cases.

Among short-term trends, the well-documented leprosy epidemic at Nauru Island in the Pacific is unique in many ways (7). It showed that, although leprosy is generally an endemic disease, occasionally it is capable of reaching epidemic proportions when conditions are favorable. The disease was probably introduced into Nauru for the first time in 1912 by a patient from the nearby Gilbert Island. By 1920, there were four known cases, and by 1924, at least 24% of the population of 1200 were known to have been affected. The sudden increase followed an epidemic of influenza. The disease started declining after 1927, and by 1952, only 4% of the population were affected and this had declined to less than 1% by 1981. Less dramatic outbreaks have been reported from eastern Nigeria, New Guinea, and the Pacific Islands of Phonpe and Truk.

The short-term outbreaks reported so far have certain common features. They include occurrence of disease in an unselected manner throughout the community, irrespective of age, sex, and household contact status, and the type of leprosy, which was mostly tuberculoid with a high tendency for spontaneous healing.

2.5. Occurrence of Deformities

The occurrence of deformities in leprosy is one of the important concerns about the disease. Up to one-third of leprosy patients develop deformities of varying degrees. Deformity in leprosy is not only permanent, but in many instances, it is also progressive, even after the disease has become inactive. This is largely due to the component of sensory loss that occurs with the disease. The proportion of deformity is higher in lepromatous leprosy than in nonlepromatous leprosy resulting from the progressive nature of the former type. In addition to physical deformities, and mainly as a result of them, leprosy patients in many societies suffer from an additional burden of social disability due to the stigma attached to the disease.

3. The Prevalence Pool

The prevalence pool of leprosy in a population in general is in a constant flux resulting from inflow and outflow of patients. The inflow is due to the occurrence of new cases, relapse of cured cases, and immigration of cases. The outflow is mainly through cure, death, and emigration of cases. Of the various factors that influence the prevalence pool, the importance of spontaneous inactivation of disease and mortality are less well recognized.

3.1. Inactivation of the Disease

Where treatment facilities exist, inactivation or cure due to specific treatment is an important mode of elimination of cases from the prevalence pool. Even in the absence of specific treatment, a majority of patients, particularly of the tuberculoid and indeterminate types, tend to get cured spontaneously. A study in Culion Island in the Philippines showed that, among children, self-healing occurred in 77.7% of cases (8). A later study in South India involving long-term follow-up of a high endemic population (9) showed that, among newly detected tuberculoid cases of all ages and both sexes, the rates of inactivation was 10.9% per year, the bulk of inactivation in the study being spontaneous (Table 14.3).

3.2. Mortality in Leprosy

Mortality in leprosy is often considered unimportant because the disease is rarely an immediate cause of death. However, leprosy patients are exposed to

Table 14.3. Inactivation of tuberculoid leprosy by age

Age group in years	Total cases	No. of cases inactive	Inactivation rate (% per year)
0–9	47	24	14.6
10–19	72	24	9.5
20–29	41	18	12.5
30–39	45	14	8.9
40–49	30	11	10.5
50 and over	35	12	9.8
TOTAL	270	103	10.9

increased mortality risks due to the disease's indirect effects. In a study in Cebu, the Philippines (10), it had been found that the mortality rate for lepromatous patients was four times more than that of the general population and that the situation for nonlepromatous patients was very similar to that of the general population. A comparative study of lepromatous patients, nonlepromatous patients, and general population from the same rural area in South India (11) showed that the standardized death rate for lepromatous patients was 3½ times that of the general population, the nonlepromatous themselves having a mortality risk which was twice that of the general population. In that population, leprosy was found to contribute to about 1% of all deaths.

4. Transmission Factors

4.1. General Considerations

There are several constraints in studying the transmission of leprosy. Unlike many other communicable diseases, in leprosy there is considerable difficulty in identifying the three reference points that are involved in the transmission of the disease, these being the onset points of exposure, infection, and disease. The problem with the onset point of exposure relates mainly to the clear identification of the source of infection, which is not always easy. The problem with the onset point of the disease is related mainly to the insidious nature of the onset of the disease in most instances. The identification of the point of onset of infection is the most important and most difficult problem in the study of transmission. Although the future in this area appears to be very promising with the availability of specific and sensitive tests, at present there is no test dependable enough to measure subclinical infection with sufficient sensitivity and specificity for use in epidemiological studies. Until such a test becomes available, the epidemiological picture of leprosy will remain incomplete.

4.2. Reservoir of Infection

The only known reservoir of infection in leprosy is the human being. However, a naturally occurring disease with organisms indistinguishable from *M. leprae* has also been detected among wild armadillos in parts of the southern United States, although the epidemiological significance of the animal is generally considered to be negligible. Among human beings, it is the lepromatous cases that carry the largest load of organisms, with the maximum load reaching over 7 billion organisms per gram of tissue. Patients of nonlepromatous leprosy carry a very much smaller bacillary load, probably not exceeding 1 million organisms in total. In addition to clinically identified cases, the occurrence of acid-fast bacilli (AFB) in the skin (12,13) and nasal mucosa of healthy subjects (14) have also been reported. The evidence that the AFB found on such "carriers" is *M. leprae* is not conclusive, although there is some evidence that persons who carry such AFB have a higher chance of developing the disease, as was found during their follow-up (15).

4.3. Portal of Exit of M. leprae

The two portals of exit of *M. leprae* often described are the skin and the nasal mucosa. However, the relative importance of these two portals is not clear. It is true that the lepromatous cases show large numbers of organisms deep down the dermis. However, whether they reach the skin surface in sufficient numbers is doubtful. There is no doubt that when lepromatous patients have ulcers from the breaking down of nodules or when they have other breaks in their skin, large numbers of organisms could be discharged. It is also possible that, apart from breaks in the skin, small numbers of organisms escape to the surface of the skin along with sweat and sebaceous secretions. Regarding the nasal mucosa, its importance had been recognized as early as 1898 by Schaffar (16), particularly that of the ulcerated mucosa. The quantity of bacilli from nasal mucosal lesions in lepromatous leprosy has been demonstrated by Shepard (17) to be as large, with counts ranging from 10^4 to 10^7.

4.4. Portal of Entry

The portal of entry of *M. leprae* into the human body is not definitely known. However, the two portals of entry seriously considered are the skin and the upper respiratory tract.

With regard to the respiratory route of entry of *M. leprae,* the evidence in its favor is in the increase, in spite of the long held belief that the skin is the exclusive portal of entry. Rees and McDougall (18) have succeeded in experimental transmission of leprosy through aerosols containing *M. leprae* among immune-suppressed mice, suggesting a possible similarity among humans.

4.5. Subclinical Infection in Leprosy

Although reliable tools for a routine study of subclinical infection in leprosy have yet to be made available, limited studies based on measuring immune response in healthy subjects have indicated that a much larger proportion of persons exposed to leprosy than those seen with the clinical expression of the disease acquire infection.

Godal and Negassi in 1973 (19) were the first to measure cell-mediated immunity (CMI) response through the lymphocyte transformation test (LTT) among different categories of persons exposed to leprosy. They found that the test was showing a gradation of response among Europeans visiting Ethiopia according to the period of their stay and their proximity to leprosy patients. Contacts of leprosy patients also showed a high rate of response to LTT.

With regard to the humoral antibody response, Abe et al. (20) had applied his indirect florescence (FLA-ABS) test among different categories of the population of Okinawa. The test was found not only to be positive in 100% of polar lepromatous and borderline lepromatous patients, 88% of borderline tuberculoid patients, and 77% of polar tuberculoid patients, but also positive to the extent of 92% among household contacts. None of the healthy noncontacts or patients with pulmonary tuberculosis were positive to the test. However, there are problems with repeatability of the test. More recently, several tests which measure *M. leprae*-specific antibodies and antigens have been developed (21). Several tests based on *M. leprae*-specific antigens or epitopes are under development, including those defined by the recently produced *M. leprae* monoclonal antibodies and synthetic peptides derived from DNA sequences of the corresponding genes.

Phenolic glycolipid-1 (PGL-1) was the first antigen specific to *M. leprae* to be identified and its antigenic moiety chemically synthesized (22). In limited studies, high antibody titres have also been reported in some household contacts of multibacillary (MB) patients as well as in other inhabitants of endemic areas, confirming that infection is more frequent than overt disease. The predictive value of PGL-1 antibody tests for the detection of individuals at high risk of developing the disease is not yet clear.

With the advent of *M. leprae*-specific monoclonal antibodies and improved T-cell cloning technology, the mapping of protein epitopes of cloned *M. leprae* gene products has generated a battery of tests based on chemically defined antigens. Further definition of the sensitivity and specificity of these tests is required.

Although we have more than enough evidence that the number of people who are subclinically infected is far greater than the number who contract the disease, our understanding of the infection of *M. leprae* in endemic populations in space and time has not improved very much, largely because (a) we still do not have an ideal immunological test which meets all requirements of sensitivity, specificity, and applicability and (b) the number of well-studied populations for leprosy or "population laboratories" for leprosy is very limited. When such an ideal test or tests become available, widespread application in field studies is bound to greatly expand the understanding of leprosy. With regard to the requirements for this ideal test, both sensitivity and specificity are, no doubt, important. However, because of the low frequency of the disease in most populations, for this purpose, specificity will be relatively more critical than sensitivity. For instance, even if the specificity of 98% of the test is what in many other situations would be considered as satisfactory, it may not provide meaningful information if the positivity rate in an endemic population is just around 4%, because, in this instance, about half of the positive results would represent false positives. This problem is compounded further when the test has, in addition, a low sensitivity. Thus, it is clear that large-scale field applications of the tests currently developed will not be possible until the tests have been evaluated and found quite satisfactory from the points of view of both specificity and sensitivity in addition to applicability.

In addition to the above, skin tests with various preparations of lepromin, and more recently with soluble antigens from *M. leprae,* have also provided useful information on the occurrence of subclinical infection, although the specificity of these tests, particularly of the integral lepromin, has been rather questionable. Zuniga et al. (23), using the soluble skin test antigen prepared by a method developed by Convit, have found that the skin test positivity in a part of Venezuela was 19% among nonhousehold contacts and 48% among household contacts.

4.6. Method of Transmission of Leprosy

The exact mechanism of transmission of leprosy is not known. At least, until recently, the most widely held belief was that the disease was transmitted by contact between cases of leprosy and healthy persons. More recently, the possibility of transmission by the respiratory route is gaining ground.

The term "contact" in leprosy is generally not clearly defined. All that we know at present is that individuals who are in close association or proximity with leprosy patients have a greater chance of acquiring the disease. However, it is the definition of contact by early workers with qualifications such as "skin to skin," "intimate," "repeated," and so forth that has made it appear as if the disease could be acquired only under such conditions and that the transmission involved some kind of "inunction" or "rubbing in" of the organisms from the skin of affected persons into the skin of healthy subjects.

There is considerable evidence that household contacts of leprosy are at high risk of infection and of disease. A large population-based study in the Philippines was the first to provide age-standardized attack rates for clinical leprosy per 1000 persons-years of observation according to type of primary case. In noncontacts and contacts of "neural" (nonlepromatous) and "cutaneous" (lepromatous) cases the attack rates were 0.83, 1.6, and 6.23 per 1000 person-years of observation,

Table 14.4. Incidence by contact state

Contact state	Number	New cases in 5 years	Incidence per 1000 per year	Relative risk
Noncontacts	186,047	1,723	1.85	1
Contacts of nonlepromatous cases	11,173	379	6.78	3.7
Contacts of lepromatous cases	1,025	90	17.56	9.5
Contacts of both types	12,198	469	7.69	4.2

respectively (5). Later studies have confirmed this trend as was seen in South India (Table 14.4) (24).

An interesting observation with regard to risk for contacts is the exceptional situation in Europe where immigrant cases and Europeans, who had returned home after acquiring leprosy in endemic countries, have failed to produce secondary cases among their contacts. There is, as yet, no plausible explanation for this. The other interesting observation in many studies is the observance of a relatively low rate of conjugal transmission.

In endemic areas, the observance of high risk for contacts should not lead to underestimation of the importance of the noncontact population in terms of their contribution to the total yield of new cases. Even with a relatively low risk, the noncontact population contributes to a larger share of new cases solely because of its large size in comparison with the contact population. Even in highly endemic areas, the contact population contributes to less than 15% of the total population, and even with the increased risk, its contribution to the total new cases is less than 25%, the rest of the 75% or so of new cases coming from the noncontact population which has a relatively low risk.

With regard to contacts of nonlepromatous cases, although they have a low risk relative to contacts of lepromatous cases, their risk is still higher than that for noncontacts. Even with a relatively low infectivity, nonlepromatous cases contribute to as many or more new cases as lepromatous cases. This is because of the much larger proportion of nonlepromatous cases which, therefore, contribute to a much larger share of the total contact population. Thus, the collective potential of nonlepromatous cases as sources of infection should not be underestimated.

5. Factors Determining Clinical Expression After Infection

There is sufficient evidence in leprosy to show that all people who get infected do not develop the disease. The factors that determine clinical expression after infection appear to be as important as the factors that determine infection after exposure. Of the many possible factors that determine clinical expression of disease, the more important are genetic predisposition, route of infection, and, possibly, reinfection.

5.1. Genetic Predisposition

Although the relative contribution of genetic host factors versus environmental factors are still far from clear, both twin and family studies indicate an important contribution of host genetics to the type of disease developing after infection. Whether genetic factors also contribute to differential susceptibility to infection with *M. leprae* or to the development of clinical leprosy irrespective of the type is less clear. There is now ample evidence that HLA-linked genes influence the development of tuberculoid leprosy (25) and evidence has also been presented for HLA-linked control of lepromatous leprosy (26). These HLA-linked genes do not seem to control susceptibility to clinical leprosy per se, but rather to determine the type of disease to develop.

5.2. Route of Infection

Studies by Shepard et al. (27) in the mouse footpad model suggest that the route of entry of the organism may, to some extent, determine the occurrence of leprosy. This is based on the observation that whereas intradermal administration of killed *M. leprae* sensitizes the animal, intravenous administration of killed *M. leprae* tends to tolerize the animal, as studied through skin test reactivity. This also raises the possibility of tuberculoid and lepromatous leprosy being the result of different routes of entry of the organisms.

5.3. Reinfection

The occurrence of leprosy, presumably for the first time, in older individuals in endemic areas has raised the possibility of reinfection in these individuals, as it is difficult to believe that they remained uninfected for such a long period in an endemic area. However, this occurrence in the older age can also be explained by the possibility that the disease in these persons represents reactivation of old, undetected primary disease following waning of previous acquired immunity. As there is no evidence of a distinct primary disease occurring in leprosy as in tuberculosis, the hypothesis of reinfection gains some importance. Further, the occurrence of relapse in lepromatous leprosy also suggests, at least in a proportion of relapsed individuals, the possibility of reinfection. There is nothing against these immune-deficient inactive patients living in endemic areas from succumbing to fresh infection. In the absence of a method for identification of strain variations of *M. leprae,* the hypothesis on reinfection will remain untested.

Table 14.5. Major field trials with BCG in leprosy

Country and no. of study projects	Control			BCG			Protection (%)
	Person-years of follow-up	New cases	Incidence per 1000 per year	Person-years of follow-up	New cases	Incidence per 1000 per year	
Burma (28,220)	151,060	831	5.5	151,415	663	4.4	20.4
New Guinea (5,544)	27,100	172	6.3	29,300	100	3.4	46.0
Uganda (10,990)	42,800	192	4.5	43,300	37	0.9	80.9
India (181,400)	240,000	2,301	9.6	488,00	3,602	7.4	23.0

5.4. Prior Infection with Other Mycobacteria

There is some evidence that prior infection with the atypical environmental mycobacteria and, possibly, M. tuberculosis influence the occurrence of leprosy. BCG vaccination itself is known to provide a degree of protection against leprosy as shown in Table 14.5 (28–31). This is possibly due to the antigenic overlap between M. leprae and other mycobacteria. The varying degrees of protection given by BCG against leprosy in different geographic areas and the limited protection seen among natural tuberculin-positive reactors in the BCG study in Uganda support this possibility. Rook et al. (32) have gone further and have suggested that the protective efficacy of BCG in different areas may be enhanced or diminished depending on the local environmental mycobacteria, some acting synergistically with BCG and some antagonistically.

5.5. HIV Infection and Leprosy

It is well known that HIV infection has created a serious situation with regard to the occurrence of tuberculosis, particularly in African countries. Several studies have shown that an increase in pulmonary tuberculosis in Africa is attributable to HIV infection. This is also true for atypical mycobacteriosis. Therefore, there was considerable apprehension that HIV infection in a similar way might have deleterious effects on the leprosy situation in endemic countries. Early reports based on a not so well-controlled study (33) suggested that the prevalence of HIV infection was significantly higher among leprosy patients compared with blood donors or surgical patients. However, later studies as reviewed by Lucas (34) did not confirm this. In well-organized case control studies involving at least 100 people per group, no significant differences in HIV prevalence were found between leprosy patients and control nonleprosy subjects. The conclusion from studies carried out so far is that there is no convincing evidence that there is any causal association between HIV and leprosy.

6. Conclusions

A critical review of the epidemiology of the disease reveals several features unique to leprosy. These include study of the disease by traditional leprologists often isolated from the mainstream of developments elsewhere, the relative scarcity of hard information in published literature, the extensive use of ill-defined and nonstandardized tools such as the lepromin test for epidemiological reasoning, and the widespread confusion in the application of terminology relating to disease states. Nonetheless, the fact remains that leprosy is one of the most challenging of diseases both from the points of view of its understanding and control.

The progress in basic research in leprosy in recent years has opened or promised to open a wide vista of opportunities for the epidemiologist. With the imminent availability of dependable and easily applicable immunological tools to measure humoral as well as CMI response, the time has come now to formulate appropriate hypotheses to be tested under a variety of conditions utilizing standardized methodology. The choice of hypotheses to be tested, at least to begin with, should focus on those directly relevant to disease control.

Although measurement of infection state through immunological tools, as against measurement of disease state, could be very valuable in identifying risk factors for infection, it should not be forgotten that in leprosy disease determinants after infection are probably equally or even more important than determinants for infection. Again, there could be a degree of interaction between these two sets of determinants.

The relationship between infection and disease in leprosy does not appear to be a constant one, as seen from the finding that age-specific incidence of infection does not appear to parallel age-specific incidence of disease. For instance, the occurrence of new disease among significant numbers of older adults in high endemic areas cannot be explained simply as resulting from recent infection, as is the case among younger children in the same area. It is difficult to accept that the older individuals who develop the disease had remained uninfected for long periods in such high endemic areas where *M. leprae* infection is so ubiquitous and opportunities for exposure so frequent. Issues such as these need to be studied with appropriate tools to explain better the natural history of the disease and to explore possibilities such as reinfection.

In the past, the study of disease determinants appears to have focused more on genetic predisposition than on other factors. In this context, the experience of the host with regard to exposure to other mycobacteria prior to infection with *M. leprae* and the subsequent experience of the host with regard to repeated doses of *M. leprae* after primary infection are questions which need to be studied in some depth. Studies on these need to be carried out not only in high endemic areas, as has been the common practice hitherto, but also in areas where leprosy has a low endemicity and in areas, as in parts of Europe, where autochthonous leprosy fails to occur in spite of the presence of active sources of infection in the immigrant population.

A very interesting feature of leprosy is the variety and gradation of response of the host to *M. leprae*. This response extends from subclinical infection as demonstrated by in vitro lymphocyte tests, skin test conversion, serum antibodies, and occurrence of AFB on healthy skin at the one end, to lepromatous leprosy at the other end. In between, one observes the early monomacular self-healing lesions as well as well-characterized disease states such as tuberculoid and borderline leprosy. However, the factors that contribute to this wide gradation of response are not clear, whether they are mainly genetic or environmental or a combination of both. Although genetic predisposition has been demonstrated both for the tuberculoid and lepromatous leprosy through HLA markers, its importance vis-à-vis the environmental influences is still to be determined. In this context, the occurrence of divergent types of leprosy among monozygotic twins, at least among some, is a case in point.

Regarding the transmission process itself, although direct man-to-man transmission is the well-accepted view in leprosy, whether through respiratory or skin route, the possibility of extra human reservoirs existing in close proximity to man cannot be excluded. In any case, there appears to have been very few attempts to search for these. The contribution of an extra human reservoir can possibly explain some of the unexplained features of leprosy such as the very uneven geographic distribution of the disease, the uneven risk of leprosy in different geographic situations even for household contacts, the rapid rise and fall of leprosy in certain situations, and the nonoccurrence of secondary cases among contacts of immigrant leprosy cases in parts of Europe.

A major problem facing leprosy research is how to evaluate the efficacy of tools for intervention such as vaccines in a reasonable period of time. The present approach of measurement of outcome through disease occurrence is not only an indirect measure of transmission of infection but also one that requires follow-up of populations for very long periods of time. Therefore, there is an urgent need to develop tools that could serve as dependable intermediate markers in the measurement of outcome in such intervention trials. In addition, there is a need for the development of appropriate epidemiometric models to predict and compare the different methods of intervention, as had been demonstrated by Lechat et al. (35) earlier.

It is clear that even as we are making headway in the control of leprosy, we do not have a full understanding of the epidemiology of leprosy for want of dependable tools and the need for very long-term follow-up of populations. However, one can reasonably conclude that *M. leprae* is a relatively less efficient pathogen, trying to survive in the human host through different adaptations and, from the evolutionary point of view, possibly on the way toward extinction. If it manages to thrive in certain human environments, it is largely because of opportunities provided for it through very poor socioeconomic conditions. In spite of this, it is possible to get rid of *M. leprae* from human sources through MDT and hope to achieve elimination of leprosy as a public health problem, as reflected in a landmark resolution of the World Health Assembly in May 1991 committing WHO to the goal of elimination of leprosy as a public health problem by the year 2000, defining elimination as attaining a level of prevalence below 1 case per 10,000 population.

References

1. World Health Organization (1988) Report of the Expert Committee on Leprosy—6th report, p. 14. Geneva: World Health Organization.
2. Bechelli LM, Martinez Dominguez V (1966) The leprosy problem in the world. Bull. WHO 34:811–826.
3. Anonymous (1994) Progress towards eliminating leprosy as a public health problem. WHO Wkly Epidemiol Rec 20:145–151; 21:153–157.
4. Noordeen SK (1994) Epidemiology of leprosy. In: Hastings RC, ed. Leprosy, pp. 29–45. London: Churchill Livingstone.
5. Doull JA, Guinto RS, Rodriguez JN, Bancroft H (1942) The incidence of leprosy in Cordova and Talisay, Cebu, P.I. Int J Leprosy 10:107–131.
6. Irgens LM (1980) Leprosy in Norway—an epidemiological study based on a national patient registry. Leprosy Rev 51(Suppl 1):1–130.
7. Wade HW, Ledowsky V (1952) The leprosy epidemic at Nauru: a review with data on the status since 1937. Int J Leprosy 20:1–29.
8. Lara CB, Nolasco JO (1956) Self-healing or abortive and residual forms of childhood leprosy and their probable significance. Int J Leprosy 24:245–263.
9. Noordeen SK (1975) Evolution of tuberculoid leprosy in a community. Leprosy India 47:85–93.
10. Guinto RS, Doull JA, de Guia L, Rodriguez JN (1954) Mortality of persons with leprosy prior to sulfone therapy, Cordova and Talisay, Cebu, Philippines. Int J Leprosy 22:273–284.
11. Noordeen SK (1972) Mortality in leprosy. Indian J Med Res 60:439–445.
12. Figueredo N, Desai SD (1949) Positive bacillary findings in the skin of contacts of leprosy patients. Indian J Med Sci 3:253–265.
13. Chatterjee BR (1976) Carrier state in leprosy. Leprosy India 48:643–644.
14. Chacko CJG, Mohan M, Jesudasan K, Job CK, Fritschi EP (1979) Primary leprosy involvement of nasal mucosa in apparently healthy household contacts of leprosy patients. Abstracts of the XI Biennial Conference of the Indian Association of Leprologists.
15. Chatterjee BR, Tayler CE, Thomas J, Naidu GN (1976) Acid fast bacillary positivity

in asymptomatic individuals in leprosy endemic villages around Jhalda in West Bengal. Leprosy India 48:119–131.

16. Schaffer (1898) Uber de Verbreitung der Lepra bacillen von den oberen Luftwegen aus (The spread of leprosy bacilli from the upper parts of the respiratory tract). Arch Dermatol Syphil (Berlin) 44:159–174.

17. Shepard CC, Walker LL, Van Landingham RM, Ye SZ (1982) Sensitization or tolerance to *Mycobacterium leprae* antigen by route of infection. Infect Immun 38:673–680.

18. Rees RJW, McDougall AC (1977) Airborne infection with *Mycobacterium leprae* in mice. J Med Microbiol 10:63–68.

19. Godal T, Negassi K (1973) Subclinical infection in leprosy. Br Med J 3:557–559.

20. Abe M, Minagawa F, Yoshino Y, Ozawa T, Saikawa A, Saito T (1980) Flourescent leprosy antibody absorption (FLA-ABS) test for detecting subclinical infection with *Mycobacterium leprae*. Int J Leprosy 48:109–119.

21. Anonymous (1986) Serological tests for leprosy. Lancet 1:533–535.

22. Hunter SW, Brennen PJ (1981) A novel phenolic glycolipid from *Mycobacterium leprae* possibly involved in immunogenicity and pathogenicity. J Bacteriol 147:728–735.

23. Zuniga M, et al. (1982) Immunoprophylaxis trial in Venezuela. Preliminary results. Paper presented at the Conference and Workshop on Pathogenesis and Immunotherapy of Leprosy, National Institute of Dermatology (Pan American Center for Research and Training in Leprosy and Tropical Diseases, CEPIALET).

24. Noordeen SK (1978) Infectivity of leprosy. In: Chatterjee BR, ed. A Window on Leprosy, pp. 59–63. New Delhi: Ghandi Memorial Leprosy Foundation.

25. de Vries RRP, Van Eden W, Van Rood JJ (1981) HLA-linked control of the course of *M. leprae* infections. Leprosy Rev 52:109–119.

26. Van Eden W (1983) HLA and leprosy—a model for the study of genetic control of immune response in man. Thesis Drukkerij JH, Pasmans BV, Gravenha GE, (Netherlands).

27. Shepard CC, Walker LL, Van Landingham RM, Ye SZ (1982) Sensitization or tolerance to *Mycobacterium leprae* antigen by route of infection. Infect Immun 38:673–680.

28. Sundaresan T (1982) BCG vaccination of children against leprosy. Twelve years' findings of the controled WHO trial in Burma. Paper presented at the Sixth IMMLEP Scientific Working Group Meeting.

29. Scott GC, Russell DA, Wigley SC (1982) The Karimuri trial of BCG as a leprosy prophylactic. Paper presented at the Sixth IMMLEP Scientific Working Group Meeting.

30. Stanley SJ, Howland C, Stone MM, Sutherland I (1981) BCG vaccination of children against leprosy in Uganda: final results. J Hyg (Camb.) 87:233–248.

31. Tripathy SP (1983) ICMR Leprosy Prevention Trial. Paper presented at the Joint Indian and IMMLEP Scientific Meeting on Immuno-epidemiology of Leprosy.

32. Rook GAW, Bahr GM, Stanford JL (1981) The effect of two distinct forms of cell-mediated response to mycobacteria on the protective efficacy of BCG. Tubercle 62:63–68.

33. Meeran K (1989) Prevalence of HIV infection among patients with leprosy and tuberculosis in rural Zambia. Br Med J 298:364–365.

34. Lucas S (1993) Human immunodeficiency virus and leprosy (editorial). Leprosy Rev 64:97–103.

35. Lechat MF, Mission CB, Bouckaert A, Vellut C (1977) An epidemiometric model of leprosy: a computer simulation of various control methods with increasing coverage. Int J Leprosy 45:1–8.

15

Treatment of Leprosy

Baohong Ji

1. Introduction

Leprosy existed for thousands of years in the preantibiotic era and was treated in many different ways. Among the various treatments, chaulmoogra oil was most widely used. The clinical response, especially among lepromatous cases, to the treatment of chaulmoogra oil was inconsistent and relapse was common, indicating that its therapeutic effect was modest, and this has been confirmed by the modest activity of chaulmoogric acid against *M. leprae* in the mouse footpad system (1).

Chemotherapy for leprosy began in the 1940s with the introduction of sulfone (2), although sulfonamides were tried earlier but the overall results were rather poor (3). Instead of the parent compound dapsone, as it was thought to be too toxic, Promin (glucosulfone sodium) was the first sulfone tried in leprosy and very promising results were obtained, but it had to be given intravenously. Further work led to the development of Diasone (sulfoxone sodium), Promacetin (acetosulfone), and a number of other sulfones which could safely be given orally. In the late 1940s, a reevaluation of dapsone (4,5) led to the gradual realization that lower, relatively nontoxic dosages of dapsone were equally effective and became almost the only sulfone for the treatment of leprosy. Because dapsone is inexpensive, effective, virtually without toxicity in the dosages used, and can be given by mouth, dapsone monotherapy was the standard treatment for leprosy worldwide until the early 1980s and had made a significant contribution to the leprosy control activities in certain areas.

Between 1960s and early 1970s, additional antimicrobial agents such as clofazimine (6), rifampicin (7,8) and thioamide (ethionamide/prothionamide) (9) were introduced for the treatment of leprosy. At the beginning of 1960s, the mouse footpad technique was developed by Shepard (10), and represented the first major breakthrough to reproduce the infection of *M. leprae* in animal model. The mouse footpad technique plays a key role in identifying the anti-*M. leprae* activities of various antimicrobial agents both in mouse and in human and provides the possibility for testing the drug susceptibility of *M. leprae* strains.

2. Dapsone-Resistant Leprosy

During the first 15–20 years of the sulfone era, dapsone-resistant *M. leprae* virtually did not exist, despite the millions of leprosy patients treated by long-term, sometimes lifelong, dapsone monotherapy. However, by the mid 1970s, it was clear that attempts to control leprosy by lifelong dapsone monotherapy was failing because of the rapid increase of dapsone resistance (11). Multiple surveys conducted in a number of leprosy-endemic areas clearly demonstrated that both secondary and primary dapsone-resistant leprosy have been found wherever they have been sought (12); the prevalence of secondary dapsone resistance, which was close to 40% in certain countries, increased with time if dapsone monotherapy continued, even if patients were treated regularly and treatments were well supervised (12); the prevalence of primary dapsone-resistant leprosy, which was virtually nonexistent before 1977 (13), had become alarmingly high, about one-third of newly diagnosed multibacillary patients were resistant (12). It must be assumed that primary dapsone resistance, unlike secondary resistance, occurs in at least as high a proportion of paucibacillary leprosy as multibacillary leprosy, although one cannot demonstrate directly the resistance in paucibacillary patients through inoculation of mice, as they have too few bacilli in their skin biopsy specimens.

From what was known about the treatment of tuberculosis, it gradually became understood that drug-resistant *M. leprae* occurs spontaneously, as the result of a mutational event in wild strains when the populations reach a certain level. It has been well documented that the mutants of *M. leprae* resistant to a particular antimicrobial remain susceptible to other antimicrobials. Because *M. leprae* cannot be cultivated, the frequencies of mutants resistant to various drugs have not been measured directly; but by analogy with the experience of *M. tuberculosis,* it is estimated that in a wild strain of *M. leprae,* the frequency of resistant mutants is 10^{-7} to rifampicin and 10^{-6} to dapsone, clofazimine, or thioamide (ethionamide or prothionamide). In an untreated lepromatous patients, the maximal number of *M. leprae* could be 10^{11} or even 10^{12}, and, in general, 1%–10% of the bacterial population, or 10^9–10^{11}, are viable. Therefore, among such cases, there are two major subpopulations of *M. leprae:* a drug-susceptible subpopulation, close to 10^9–10^{11}, and a drug-resistant subpopulation, which includes about 10^3–10^4 mutants resistant to rifampicin and 10^4–10^5 to either dapsone, clofazimine, or thioamide. During the course of dapsone monotherapy, the drug-susceptible subpopulation and mutants resistant to other antimicrobials were gradually eliminated, but the dapsone-resistant mutants survived and multiplied selectively, eventually dominating the bacterial population and relapsing; thus, the patient became a secondary dapsone-resistant case. If an individual was infected by dapsone-resistant organisms and developed the disease, he/she became a primary dapsone-resistant case. Likewise, the selection of rifampicin-resistant mutants during monotherapy of rifampicin, or adding rifampicin to dapsone in patients who have already developed resistance to dapsone, has been well documented (14).

Because of the serious situation of dapsone resistance, a WHO study group believed that urgent actions must be taken, and the only way to prevent the emergence of dapsone resistance and the spreading of dapsone-resistant leprosy was to use multidrug therapy (MDT) (15). The objectives of the MDT are (i) to ensure the elimination of the drug-resistant mutants, especially rifampicin-resistant mutants and (ii) to substantially reduce the number of drug-susceptible viable organisms to a level which will not cause an unacceptable relapse rate after stopping treatment. From the experience of dapsone monotherapy, it was recognized that lifelong treatment is impractical in the field, and MDT should be administered for only a limited period; therefore, only bactericidal drugs should be considered as candidates for the MDT regimens.

3. Established Chemotherapeutic Agents for Leprosy

By the time the WHO study group met in 1981, only four established antileprosy drugs with different bactericidal mechanisms [i.e., dapsone, clofazimine, rifampicin, and thioamide (ethionamide/prothionamide)] were available.

3.1. Dapsone

Dapsone (DDS) is available as 25-, 50-, and 100-mg tablets. The routine dosage is 100 mg daily for adults and 1–2 mg/kg body weight daily for children. In terms of the minimal inhibitory concentration (MIC), *M. leprae* is extremely susceptible to dapsone the MIC being only 3 ng/ml. Dapsone acts as a synthetase inhibitor in the folate synthesizing enzyme system of *M. leprae,* and 100 mg daily displays weak bactericidal activity against *M. leprae* in human leprosy (17). It is rapidly and nearly completely absorbed when taken orally (18) and is eliminated relatively slowly, with a half-life averaging a little over 24 h. After ingestion of a single dose of 100 mg of dapsone, the peak blood level is about 1–2 μg/ml, some 500-fold in excess of its MIC against *M. leprae,* and measurable amounts can be found in the blood even 10 days later. It is well distributed throughout the body and ultimately is excreted for the most part in the urine.

A variety of side effects have been attributed to dapsone, including hemolytic anemia and a reduced life span of red blood cells, skin rashes, gastrointestinal complaints, agranulocytosis, hepatitis, psychosis, and peripheral neuropathies. Most of these effects are very rare. The so-called DDS syndrome, which is also very rare, usually develops within the first 6 weeks after the start of therapy and consists of exfoliative dermatitis and/or other skin rashes, generalized lymphadenopathy, hepatosplenomegaly, fever, and hepatitis. When this syndrome does occur, dapsone should be discontinued immediately and corticosteroids should be given. The most common side effect with dapsone is anemia. However, this is usually very mild, unless the patient has a complete glucose-6-phosphate dehydrogenase (G6PD) deficiency. The safety of dapsone in pregnancy has been fairly well established, and no evidence of teratogenicity has been observed.

As mentioned earlier, dapsone resistance is a result of selective multiplication of resistant mutants during dapsone monotherapy, and it can be classified into three levels of resistance: low degree (resistance equivalent to a dosage up to 1 mg daily), intermediate degree (resistance equivalent to a dosage up to 10 mg daily), and high degree (resistance equivalent to a dosage up to 100 mg daily). Until now, the vast majority of secondary dapsone-resistant leprosy cases have high-degree resistance, whereas most of the primary dapsone-resistant leprosy cases have low-degree resistance (12). The reason for the difference is still unclear. It is expected that in cases with low- and even intermediate-degree resistance, the patient may respond to treatment with dapsone in full dosage. Therefore, even though currently dapsone resistance is wide spread in combination with other antileprosy drug(s), there is still justification for employing dapsone in the treatment of leprosy.

3.2. Rifampicin

Rifampicin (rifampin) is a semisynthetic broad-spectrum antibiotic and is available as 150- and 300-mg capsules. It acts by inhibiting the DNA-dependent RNA polymerase of the organisms, thereby interfering with bacterial RNA synthesis. Rifampicin is rapidly absorbed from the gastrointestinal tract and distributed throughout the body, and about two-thirds of the absorbed drugs is ultimately excreted via the gastrointestinal tract. It is much more potent and rapidly bactericidal than other antileprosy drugs. *M. leprae* recovered from skin biopsy specimens taken 3 or 4 days after a single 600–1500-mg dose of rifampicin failed to multiply in mice (7), indicating that at least 99% of the bacilli are killed by the dose. It is because of this that rifampicin is the most important component of MDT for both paucibacillary and multibacillary leprosy patients (15). Because no significant difference in bactericidal activity against *M. leprae,* as measured by serial mouse footpad inoculations, has been detected between the daily and monthly administration of rifampicin, the WHO study group recommended that rifampicin be administered once monthly under supervision at dosage of 600 mg for adults (15). For children, the dose can be calculated at 10 mg/kg body weight. The monthly rifampicin administration not only greatly reduces the costs but also reduces the frequency and severity of side effects as compared with daily administration of the drug.

The major side effect of rifampicin is hepatotoxicity, but this is extremely rare with monthly administration unless it is combined with other hepatotoxic drugs such as thioamides (22–25). Although the "flu" syndrome and other syndromes consisting of shock, hemolytic anemia, and renal failure had been reported in the treatment of tuberculosis with once-weekly or twice-weekly rifampicin administration (19), very few "flu" syndromes have been observed in leprosy patients with the monthly administration of 600-mg rifampicin. Other side effects include skin rashes and mild gastrointestinal symptoms. Rifampicin can be safely used during pregnancy.

Secondary rifampicin-resistant leprosy has been detected after rifampicin monotherapy in multibacillary patients or after administration of rifampicin with dapsone for treatment of relapses following initial dapsone monotherapy (14). A great majority of the latter cases end up with resistance to both dapsone and rifampicin if there is further relapse. Rifampicin should, therefore, always be combined with other antileprosy drugs capable of preventing the development of rifampicin resistance.

3.3. Clofazimine

Clofazimine (Lamprene; B663) is a rimino-phenazine dye. The drug is available as 50- and 100-mg capsules. It is weakly bactericidal against *M. leprae* (20) and exhibits anti-inflammatory activity; it is, therefore, also effective in controlling the erythema nodosum leprosum (ENL or Type 1 reaction). It is absorbed to the extent of 70% via the gastrointestinal tract and is deposited mostly in fatty tissues and cells of the reticuloendothelial system, including the skin. For treatment of multibacillary leprosy, the routine dosage recommended by the WHO study group is 50 mg daily plus 300 mg once monthly (15); however, because of its extraordinarily long half-life in man, approximately 70 days (21), monthly administration of clofazimine 1200 mg is as effective for the treatment of lepromatous leprosy as the standard dosage (20). The dosage for the treatment of ENL is much higher, beginning with 100 mg two or three times daily. The most common side effect is a reddish black coloration of the skin which normally develops within 4–8 weeks after the start of the therapy in patients with active skin lesions. The severity of the coloration is dose dependent and is intensified in the areas of leprosy lesions. Among patients who have been treated with clofazimine for 24 months, the coloration diminishes gradually in 6–24 months after discontinuation of the drug. Coloration may also be seen in the mucous membrane. Sebum, sweat, feces, and urine also show a reddish color. Because of the coloration, some light-skinned patients refuse to take the drug. Gastrointestinal symptoms, such as nausea, vomiting, crampy abdominal pain, and diarrhea, are common but mild in patients treated with 50 mg daily. However, in patients receiving higher doses, usually over 100 mg daily, these symptoms may appear more serious. Another common side effect is ichthyosis, resulting from the anticholinergic activity of the drug.

3.4. Ethionamide/Prothionamide

The two thioamides are virtually interchangeable and show cross-resistance with each other. Both drugs are available as 250-mg tablets. They are weakly bactericidal against *M. leprae* (9) to about the same extent as are dapsone and clofazimine. The routine dosage for adults is 250–375 mg daily, and 5–7.5 mg/kg body weight daily for children. The drugs are absorbed from the gastrointestinal tract and are excreted mainly in the urine. The peak blood concentration after a dose of 375 mg in adults is 60-fold its MIC against *M. leprae,* about 0.05 μg/ml (15).

Ethionamide and prothionamide were recommended by the WHO study group as alternatives to clofazimine in the MDT regimen for multibacillary cases unable to take clofazimine (15). However, as it has been found that hepatotoxicity was quite common when ethionamide or prothionamide was combined with rifampicin, even if it was administered monthly (22–25), the WHO Expert Committee on Leprosy withdrew the recommendation for using thioamide in the MDT regimen, unless it was absolutely necessary (16). When used under exceptional situations, the drugs must be given with great caution and under close monitoring of hepatotoxicity. Other gastrointestinal symptoms, such as nausea, vomiting, abdominal pain, and anorexia, are common; therefore, the patients' compliance to self-administration is poor.

4. Standard Regimens of Multidrug Therapy

Because thioamide (ethionamide/prothionamide) should not be included as a component of MDT regimen (16), consequently only three drugs (i.e., dapsone, rifampicin, and clofazimine) are available to be used as components of the MDT regimens.

Based on clinical, bacteriologic, histologic, and immunologic findings, leprosy can be classified into five groups, TT, BT, BB, BL, and LL, as proposed by Ridley and Jopling (26). This classification has been widely accepted for research purposes but is too complicated to be applied under field conditions because of limited facilities. Taking into account the bacterial loads, in terms of the degree of skin smear positivity of the patients, the WHO study group proposed to classify leprosy into two different categories: paucibacillary (PB) and multibacillary (MB) leprosy (15); such classification was endorsed by the WHO Expert Committee on Leprosy with minor modifications (16). Paucibacillary or multibacillary leprosy refers to the initial skin smear-negative or initial skin smear-positive leprosy cases, respectively (16). The classification is essentially an operational categorization to facilitate the delivery of MDT (15). The design of the MDT regimens for paucibacillary and multibacillary leprosy were based on the considerations regarding the sizes and compositions of the bacterial populations, the immunological status of the patients, and the efficacies of available antileprosy drugs. It was estimated that as the bacterial population of viable organisms in paucibacillary leprosy is no more than 10^6, the possibility of these being drug-resistant mutants is low, and, therefore, theoretically, monotherapy, especially with rifampicin, should be enough. However, in order to avoid the possible risk of selecting rifampicin-resistant mutants in multibacillary patients who are wrongly diagnosed as paucibacillary, the study group recommended combined therapy with rifampicin plus dapsone for all paucibacillary patients. Although the bacterial load in a multibacillary patient is significantly greater than that in a paucibacillary patient, the chance to include a mutant resistant to two drugs is very remote because the frequency of such a mutant in a bacterial population should be 10^{-12} or 10^{-13}, and in theory, two drugs should be sufficient to eliminate all drug-resistant mutants. Nevertheless, taking into account the possibility of dapsone resistance, it was recommended that at least two additional drugs should be combined with dapsone for the treatment of multibacillary patients, and one of the two additional drugs should always be rifampicin because of its great potency (15):

Paucibacillary leprosy: Rifampicin 600 mg once monthly, supervised, for 6 months, plus dapsone 100 mg daily, self-administered, for 6 months.

Multibacillary leprosy: Rifampicin 600 mg once monthly, supervised, plus dapsone 100 mg daily, self-administered, plus clofazimine 300 mg once monthly, supervised, and 50 mg daily, self-administered. The combined therapy should be given for at least 2 years and be continued, whenever possible, up to skin smear negativity.

Because the great majority of national leprosy control programs discontinue the treatment of multibacillary leprosy after 2 years of MDT with very promising results, recently the duration of treatment among multibacillary leprosy has been fixed to 2 years at the second WHO study group meeting (27).

5. Achievements in Implementing Multidrug Therapy

The MDT regimens are accepted and are being implemented intensively in many endemic countries. Up to May 1994, 55% of the total registered leprosy cases in the world were being treated with MDT, and more than 5.7 million patients have already completed their treatment (28).

Both MDT regimens were well tolerated by the patients, except that the coloration of skin caused by clofazimine has been a problem in certain light-skinned patients. The side effects were extremely rare and mild. Both regimens are very effective, as the clinical response during treatment is satisfactory, and after stopping treatment, the cumulative relapse rates among paucibacillary or multibacillary were well below 1% (30). High motivation of patients as well as health workers has also been reported; in terms of attendance rate of the monthly supervised drug administration, the regularity of treatment is excellent (16).

In 1988, for the first time in the history of leprosy control, the global number of registered leprosy cases began to decline (16); it was 3.7 million in 1990 but reduced to 1.7 million in 1994 (28). Although such a decline could be partly due to the discharging of inactive patients during screening of all cases at the preparatory phase of implementing MDT (29), there is no doubt that the effectiveness of MDT and shortening the duration of treatment also played an important role. The very encouraging results led to the adoption in May 1991 of the World Health Assembly resolution WHA44.9: to eliminate leprosy as a public health problem (defined as the prevalence rate below 1 case per 10,000 population) in the world by the year 2000.

6. Needs for Improved Multidrug Regimens and New Bactericidal Drugs

Although the introduction and implementation of MDT is the most important development in the history of leprosy control, it must be emphasized that the current coverage rate of MDT at the global level, about 55%, is still far below the desirable level. Furthermore, many of the patients currently not covered by MDT are living in difficult access areas or areas with poor coverage of health infrastructures; therefore, further increasing the MDT coverage rate becomes more and more demanding.

From the operational point of view, the duration of MDT is still too long, especially 2 years for multibacillary leprosy, because the monthly supervised treatment for 2 years cannot always be applied in difficult access areas or areas with poor coverage of health infrastructures. If the duration of MDT for multibacillary leprosy can be significantly shortened, preferably to 1–6 months, it will greatly facilitate the implementation of MDT and increase the coverage rate. Nevertheless, to minimize the relapse rate after stopping treatment, especially relapse with drug-resistant organisms, treatment should be continued until drug-susceptible organisms have been substantially reduced and drug-resistant mutants have been eliminated. Among the three components of the current MDT regimen for multibacillary leprosy, only rifampicin displays very powerful and rapid bactericidal activity against *M. leprae* (7,31–33). The dapsone- or clofazimine-resistant mutants will be eliminated rapidly by treatment with rifampicin because, by definition, their response to rifampicin should be the same as drug-susceptible organisms. The major problem is the necessary duration of treatment to eliminate the rifampicin-resistant mutants. In the wild strains, however, the number of such mutants was estimated to be smaller than the mutants resistant to other drugs, but they can only be killed by dapsone and clofazimine in the MDT regimen. Both drugs are weakly bacteriocidal against *M. leprae,* and the duration of treatment with dapsone plus clofazimine to kill 10^4 viables organisms, the maximum number of the rifampicin-resistant mutants in an untreated multibacillary patient, remains unknown, but it seems unlikely that the duration may be substantially shorter than 2 years, because either of them took 3–6 months of monotherapy to kill 90% of viable organisms (17). Without the introduction of new, more powerful bactericidal drug(s) to the MDT regimens, it is unlikely that the duration of treatment can be substantially shortened.

The other weakness of the MDT regimen for multibacillary leprosy is the uncertainty of patients' compliance to the self-administered daily dapsone and clofazimine. The major objective to combined rifampicin with dapsone plus clofazimine is to prevent the selection of rifampicin-resistant mutants. However, in the field condition, it is very difficult to assess the compliance to or the regularity of the daily self-administration of dapsone and clofazimine. Unfortunately, like in other chronic diseases, noncompliance with the treatment is a very common phenomenon among leprosy patients. It has been well documented that close to 30% of leprosy patients in Karigiri, India, one of the best leprosy control programs in the world and where clofazimine is well accepted by the patients because of the dark skin, did not take their prescribed dapsone and clofazimine properly (34), suggesting that simply because of noncompliance, it is still possible to develop rifampicin-resistant leprosy in a program where MDT is implemented. The risk may be significantly reduced if a fully supervised MDT regimen is developed such that all components are administered once monthly under supervision. The basic requirement for a component of monthly administered regimen should be (i) its single dose displays a certain degree of bacteriocidal activity against *M. leprae* and (ii) the dosage is well tolerated by the patients (35,36). Dapsone is not suitable for once-monthly administration; and although clofazimine 1200 mg once monthly displayed a therapeutic effect similar to its standard daily dosage (20) gastrointestinal side effects are common following the high dosage. For these reasons the current MDT regimen cannot be administered monthly.

In brief, to significantly shorten the duration of treatment for leprosy and to develop a fully supervised monthly administered multidrug regimen, new antileprosy drugs are urgently needed. Ideally, a new drug should possess the following characteristics: (1) it displays powerful bacteriocidal activity against *M. leprae* but its antimicrobial mechanism is entirely different from those of existing drugs;

(2) its pharmacokinetic properties allow the treatment to be given no more frequently than once daily; (3) it can be administered orally; and (4) it is safe and well tolerated by the patients. The prospect as an antileprosy drug is bleak for a compound if it is required to be given by injection, because it is difficult to use under field conditions and the recent epidemiological trend of HIV infection further hampers the application of multiple-injection treatment in rural areas.

7. Strategies of Developing New Antileprosy Drugs

In theory, there are two different approaches: (1) synthesis of new compounds and (2) the screening of existing compounds.

With the development of experimental infection in the armadillo (37) and the method of purification of *M. leprae* from infected armadillo tissue (38), a limited amount of purified organisms have been obtained for studies of metabolism and physiology of *M. leprae* (39). The scientific progress in the understanding of the metabolism and physiology of *M. leprae* (39–42) and in the techniques for producing and studying the structures of the potential target enzymes have made it possible to exploit such knowledge and capabilities for synthesizing new compounds in a systematic way—the process becoming known as rational drug development. Unfortunately, rational drug design for leprosy is certainly not a high priority of the pharmaceutical industry because of the lack of commercial viability.

The laboratory screening of antileprosy drugs was started only after the mouse footpad model of *M. leprae* infection (10) had been established. Large numbers of different classes of antimicrobials are being developed. The recent discovery of strong bactericidal activity of pefloxacin (43,44,46), ofloxacin (45,46), sparfloxacin (47–49), minocycline (50–53), and clarithromycin (52–54) clearly demonstrated that the screening of existing compounds is still the most practical and productive approach. However, unless an in vitro screening method is developed, large-scale screening with the mouse footpad system is not possible due to limited facilities. The candidates should focus on compounds which show strong activities, in terms of MICs, against gram-positive organisms, including cultivable mycobacteria. Favorable pharmacokinetic properties, especially better absorption rate and longer half-life, are also critical in selecting the analogs of active compounds (e.g., newer fluoroquinolones and macrolides) for screening.

8. Methods in Screening Antileprosy Drugs

The mouse footpad technique is the only universally accepted experimental system for testing of drug activity against *M. leprae*. Three methods are employed for drug screening by this technique: (1) the continuous method (55); (2) the kinetic method (56,57); and (3) the proportional bacteriocidal method (58).

With the continuous method, the drug is administered from the day mice are inoculated until mice are sacrificed. Comparing the number of *M. leprae* harvested between the treated and untreated control mice, one may decide whether or not the drug displays anti-*M. leprae* activity, and therefore is the most sensitive screening technique. Having demonstrated that the drug is active, one may further determine the minimal effective dosage (MED) by testing the activity of the drug in a range of dosages. However, the continuous method cannot distinguish between merely bacteriostatic and bacteriocidal activity. Many drugs are bacteriostatic against *M. leprae,* whereas a few of them are bacteriocidal; only bacteriocidal drugs may be considered as potential components for the MDT regimens. Another disadvantage of the method is that it requires a large quantity, usually more than 10 g, of the drug for testing. Unless the compounds are being produced commercially or are undergoing active development. The pharmaceutical firms are unable to provide such a quantity.

In the kinetic method, the drug is administered for a limited period depending on the expected activity of the drug but usually far around 60 days, beginning about 60 or 70 days after inoculation when *M. leprae* is in early or mid-logarithmic growth (10). The activity of the treatment is assessed by the growth delay of the treated group, which is determined by comparing the growth curves between the treated and untreated control groups. Theoretically, a pure bacteriostatic drug will inhibit multiplication of the organisms as long as the drug is administered. Thus, the growth curve of *M. leprae* in mice treated with a purely bacteriostatic drug should be parallel to that in the control mice and should lag behind the control growth curve by a length of time no greater than that during which the drug is administered. The absence of bacteriocidal activity is reliably demonstrated by this approach. The failure of bacterial multiplication to resume immediately following cessation of drug administration may represent evidence that *M. leprae* are killed during treatment or prolonged bacteriostasis (also termed "bacteriopause"), which may indicate persistence of the drug in the tissues or within the organisms or it may reflect the recovery of the organisms that have been damaged reversibly. Thus, the kinetic method can distinguish between purely bacteriostatic and so-called "bacteriocidal-type" activity (59) but cannot distinguish between purely bacteriopausal and bacteriocidal activity, unless there is total failure of resumption of bacterial multiplication after drug administration has been stopped.

The proportional bacteriocidal method may clearly define whether the activity of a compound is bacteriostatic or bacteriocidal. Groups of mice are inoculated with 10-fold diluted *M. leprae* suspensions, ranged from 0.5 or 1 organism to 5,000 or 10,000 organisms per footpad. Except for control mice, they are treated for a period of time that varies, depending on the drug, from 1 to 60 days. The mice are then held for at least 1 year without treatment, a period of time theoretically sufficient for one surviving organism to multiply to 10^5. At the end of the year, harvests of *M. leprae* are performed from the individual footpads. If the harvest yields at least 10^5 organisms, the *M. leprae* are considered to have multiplied, indicating that the inoculum provided at least one viable organism. The number of *M. leprae* surviving the treatment are estimated by the most probable number (MPN) of viable organisms (60,61) or the median infectious dose (ID_{50}) (i.e., the number of organisms required to infect 50% of the animals) (62). Although the proportional bacteriocidal method requires more mice and more time than do the other methods, it is the most reliable method for detecting bacteriocidal activity. And permits the determination of the degree of activity. However, the method is incapable of detecting bacteriostatic or bacteriopausal activity and, therefore, is not routinely employed for initial screening.

All the antileprosy drugs have been demonstrated to exert at least bacteriostatic activity in the mouse footpad system, whereas no compound shown to be inactive in mice has been demonstrated to display definite therapeutic effect in leprosy patients. Nevertheless, the mouse footpad technique possesses several disadvantages. It is time-consuming and requires many mice and gram amounts of the compounds to be tested, and can, therefore, be employed to test only limited numbers of compounds that represent a very few selected classes. The search for active compounds from a wide variety of classes requires a rapid primary screening method that will yield results within days or, at most, a few weeks and that requires only milligram rather than gram amounts of the tested compounds. It is clear that in vitro primary screening methods are needed.

Despite the fact that *M. leprae* cannot be cultivated in vitro, the viable organisms still retain many of their metabolic functions for a limited period of time outside the host (63), and it is therefore possible to develop various techniques which may be able to demonstrate rapidly the in vitro activity of killing or impairing the key metabolic process. Within the last decade, many in vitro techniques have been developed and have been reviewed elsewhere (64). It is very encouraging to note that in a recent double-blind evaluation, the radiorespirometric assays, using either the BACTEC 460 or Buddemeyer $^{14}CO_2$ detection system, are capable of differentiating between antileprosy drugs (dapsones, rifampicin, ethionamide, and pefloxacin) and inactive substances following 2 weeks of incubation of freshly harvested viable leprosy bacillus under appropriate incubation conditions (65) and, therefore, are suitable to function as a primary drug screening system. Of course, the compounds found to be active in vitro should not be tested immediately in man. Besides the study of pharmacokinetics and toxicities, at least for the time being, their activities against *M. leprae* must be firmly established in the mouse footpad system prior to the initiation of a clinical trial.

9. Basic Methods in Monitoring the Therapeutic Effect of New Antileprosy Drugs and New Multidrug Regimens in Clinical Trials

The therapeutic effect of any promising new antileprosy drug should be evaluated in a clinical trial. Clinical assessment and evolution of the bacterial index (BI) of the skin smears during chemotherapy were the most important parameters in the earlier trials. Although a definite clinical improvement was observed among newly diagnosed lepromatous patients during treatment with any of the established or investigational antileprosy drugs, the assessment of clinical improvement is very much subjective and difficult to quantify for comparison. The BI represents the total bacterial load including both dead and viable organisms. Because the great majority of bacilli were dead even before treatment (44,46,53,70) and because the dead organisms persisted in the tissues and were eliminated by a process unrelated to the antimicrobial activity of the treatment, the reduction of BI was very slow and did not differ significantly between patients treated with dapsone monotherapy and MDT or other rifampicin-containing combined regimens, although it is well known that rifampicin is far more bactericidal than dapsone. The introduction of the morphological index (MI) was a major development in measuring the proportion of solid-staining, morphologically intact and, therefore, presumably viable *M. leprae* in the hosts (94). Nevertheless, the technique is difficult to standardize and to perform with accuracy; also because it is difficult to examine more than 50–100 organisms per site of skin smear, it may monitor a decrease in the proportion of viable organisms by no more than 90%, or one order of magnitude (17,86). Therefore, the clinical assessment, BI, and MI are still useful parameters in clinical trial, but are not sensitive enough to evaluate more accurately and precisely the bacteriocidal activities of treatment (86).

Serial mouse footpad inoculations have been applied as the most efficient technique for assessing the therapeutic effects of treatment. It needs only a small number of previously untreated lepromatous patients, as few as four to six patients per group, for demonstrating the anti-*M. leprae* activity, evaluating the nature of the activity, and monitoring the rate of the "initial killing" of *M. leprae* during treatment with an individual drug or combinations of drugs in short-term trials (17,20,33,44,46,53,70,86). *M. leprae* is recovered from biopsies taken before and at different intervals during treatment and is inoculated into footpads of mice for assessing their viability. If the proportion of viable organisms in the bacterial population have been carefully titrated by inoculating groups of mice with serial 10-fold diluted inocula prepared from each biopsy, it allows the measurement of the bacteriocidal activity of the treatment up to >99% to >99.9%, depending on the proportion of viable organisms before treatment (44,46,53,70). The sensitivity of measuring the killing cannot be further improved by using immunocompetent (normal) mice because of the limited inoculum size (i.e., 5×10^3 to 1×10^4 organisms per footpad) (10). To improve the sensitivity, one has to inoculate more organisms into footpads of immunocompromised rodents such as thymectomized-irradiated (TR) mice (87), neonatally thymectomized (NT) rat (88), or congenitally athymic (nude) mice (89,90). Up to now, these immunocompromised rodents have been employed only in a limited number of trials, but their superiority in detecting a small proportion of viable organisms has been clearly demonstrated. Because nude mice are extraordinarily susceptible to infection by *M. leprae* (91), as many M. leprae as are available can be inoculated. It is possible to measure the killing rate up to >99.9% to >99.999%, depending on the proportion of viables in the pretreatment samples and the maximal available number of organisms for nude mice inoculation (44,46,53,70). Therefore compared with normal mice, the inoculation of nude mice increased the sensitivity and precision of the measurement of the killing effect by one to three orders of magnitude. However, the use of nude mice enormously increases the costs because of the price of purchase and maintenance. Unless it is crucial to have more precise information with respect to the killing effect of the treatment, it is not always justified to use nude mice for monitoring therapeutic effects in a clinical trial (70).

The disadvantages of using the serial mouse footpad inoculations are evident. It is time-consuming because the results of inoculation will be available only 12 months after inoculation; it requires many animals to monitor a single trial and is therefore expensive. In addition, because no more than 10^6 organisms per milligram can be recovered from biopsies of advanced lepromatous patients and the small size of the mouse footpad, this severely restricts the volume of inoculum, even with nude mice one can at best measure the initial 99.999%, or five orders of magnitude, of killing among lepromatous patients who may have 10^{10} viable organisms before treatment. None of the existing rodent systems are able to monitor precisely the therapeutic effects of any combined regimen containing more than one strong bacteriocidal drug, such as the combination of rifampicin and ofloxacin. Obviously, more rapid, simple, and sensitive systems should be developed for measuring the killing of *M. leprae* by the treatment.

The only available method for monitoring the long-term therapeutic effects, especially for those combined regimens containing more than one strong bacteriocidal drug, is to follow-up the relapse rate after stopping treatment. In multibacillary leprosy, the relapse rate is thought to be proportionally correlated with the number of viable organisms that surviving the treatment, and therefore the relapse rate reflects the bactericidal activity of the treatment. However, because the relapse rate of the current MDT regimen is low and the relapse occurs late, at least 5 ± 2 years on average (92,93), after stopping treatment with any rifampicin-containing combined regimens, in order to prove that the new combined regimen(s) is as good as or even better than the current MDT regimen, the sample size must be sufficiently large (at least 500 patients per group) and the follow-up should be long enough, at least for 7 years after completion of the treatment. Because it is unlikely that such a number of patients may be recruited from any single center within a reasonable period of time (e.g., 24 months), the trial is bound to be multicentric. Apparently, it is very difficult to organize such a large-scale, multicentric, long-term trial, and so far, only a few are being organized by the World Health Organization.

Because the assessment of the therapeutic effects of the regimens depends heavily on the relapse rate, the criteria of relapse must be well defined in advance. In multibacillary leprosy, relapse is diagnosed if two of the following three criteria are met: (i) significant increase of BI (i.e., by at least 2+ over the previous value at any site of skin smears is confirmed); (ii) occurrence of a definite new skin lesion; and (iii) demonstration of viable organisms by mouse footpad inoculation (92,93).

10. New Drugs with Bactericidal Activity Against *M. leprae*

Within the last decade, three different classes of new drugs with promising bactericidal activities against *M. leprae* have been identified: fluoroquinolone derivatives, pefloxacin, ofloxacin and sparfloxacin; a macrolide, clarithromycin; and a tetracycline derivative, minocycline.

10.1. Fluoroquinolone Derivatives

The fluoroquinolones inhibit the bacterial gyrase, a target which has never previously been exploited in leprosy chemotherapy. In view of their strong activities against gram-positive microorganisms and their pharmacokinetic properties, anti-*M. leprae* activities of various fluoroquinolone derivatives have been tested in the mouse footpad (43,45,47,48) as well as in the in vitro system (66). In terms of MIC, ciprofloxacin is more active against most microorganisms than other commercially available fluoroquinolone derivatives, but it is virtually inactive against *M. leprae* in mice even treated with 150 mg/kg daily by the continuous method (43), probably because of its weaker in vitro activity against *M. leprae* (66) and unfavorable pharmacokinetic properties (43). Pefloxacin was the first fluoroquinolone showing promising activity against *M. leprae:* 150 mg/kg daily displayed bactericidal activity in mice (43). Among the commercially available fluoroquinolones, ofloxacin has the strongest in vitro (66) and in vivo (45) activity against *M. leprae;* in the mouse experiment. Ofloxacin 50 mg/kg daily exerted the same degree of bacteriocidal effects as pefloxacin 150 mg/kg daily, and ofloxacin 150 mg/kg displayed profound killing activity (45). These observations, confirmed by other investigators (67–69), represented the first lead to an important new antileprosy drug in many years. It was demonstrated recently that on a weight-to-weight basis, sparfloxacin is more active against *M. leprae* than ofloxacin (47,48). Nevertheless, the greater activity of sparfloxacin is offset by the lower clinical dosage (200 mg daily) recommended by its manufacturer, whereas the clinically tolerated dosage of ofloxacin is 800 mg daily; and in the mouse experiment, if the drugs are administered at dosages equivalent to clinical dosages, the bacteriocidal activity of sparfloxacin is very similar to that of ofloxacin (35). Therefore, the real advantage of sparfloxacin over ofloxacin in the treatment of leprosy remains unclear.

To date, we have conducted three clinical trials of fluoroquinolones among newly diagnosed lepromatous patients (44,46,70). A single dose of 800 mg of pefloxacin or ofloxacin displayed a modest degree of bacteriocidal effect, but, still, a significant degree of killing was observed in three out of eight patients treated with single-dose ofloxacin (46). It was confirmed in the trials that pefloxacin 800 mg or ofloxacin 400 or 800 mg daily all displayed very promising therapeutic effects: after 28 days of treatment, all patients showed remarkable clinical improvement, rapid and very significant decline of the MIs in skin smears, and more than 99.99% of organisms viable on day 0 were killed by 22 doses of treatment, as titrated by serial inoculations into footpads of immunocompetent and nude mice (46,70). Both clinical responses and bacteriocidal activities did not differ significantly between the groups treated with either pefloxacin or ofloxacin (46), and those given monotherapy with ofloxacin 400 or 800 mg daily, or the combination of ofloxacin 400 mg plus dapsone 100 mg, clofazimine 50 mg daily, and clofazimine 300 mg once monthly (70). Therefore, it was concluded that for daily administration, the optimal dosage of ofloxacin appears to be 400 mg, and the combination with dapsone and clofazimine does not enhance its bactericidal activity. The adverse reactions caused by pefloxacin or ofloxacin were rare and mild, even when they were combined with rifampicin, dapsone, and clofazimine, the three components of the current MDT regimen for multibacillary leprosy, and the duration of treatment was extended to 6 months (46). However, we did observe that an older female patient developed psychic disorder after 21 doses of pefloxacin, and a mild to moderate degree of elevation of serum glutamate-pyruvate transaminase (SGPT) in a small proportion of patients after 4 weeks of treatment. All the adverse reactions reverted to normal after stopping treatment. Excellent results were also observed in a clinical trial of sparfloxacin 200 mg daily (49); the impression was that the therapeutic effect of sparfloxacin 200 mg daily was similar to that of ofloxacin 400 mg daily, although both treatments have not yet been compared in the same trial.

The fluoroquinolones are rapidly developing with many new compounds appearing that might be more active against *M. leprae* than pefloxacin/ofloxacin/sparfloxacin. It is important to be on the alert for new compounds with lower MICs against gram-positive microorganisms, or those with favorable pharmacokinetic properties.

10.2. Tetracycline Derivatives

Among the tetracyclines, minocycline is unique in being active against *M. leprae* (50,71), probably because its lipophilic properties allow it to penetrate the cell wall more effectively than other tetracyclines. Bacteriocidal activity of minocycline was consecutively demonstrated by the kinetic method (50,71) and the proportional bacteriocidal method (52). The MIC of minocycline against *M. leprae* was estimated at about 0.2 μg/ml, considerably less than the levels, 2–4 μg/ml, easily achieved in plasma and tissue of patients treated with routine dosage. Recently, it was demonstrated that a single dose of minocycline 25 or 50 mg/kg was active in mice by the kinetic method (36). Since the discovery of the powerful bacteriocidal activity against *M. leprae* in mice and in humans by a single dose of rifampicin, this was probably the first time that significant anti-*M. leprae* activity of a single dose of another antimicrobial has been demonstrated. In an experiment with the proportional bacteriocidal method, a single dose of minocycline 25 mg/kg plus clarithromycin 100 mg/kg killed 96% of viable organisms, which was only slightly inferior to that of rifampicin 10 mg/kg; and the activities of a single dose of combinations consisting of minocycline 25 mg/kg, clarithromycin 100 mg/kg, and ofloxacin 150 mg/kg or sparfloxacin 50 mg/kg were similar to that of rifampicin 10 mg/kg (35).

In two clinical trials, minocycline 100 mg daily displayed powerful bacteriocidal activity in previously untreated lepromatous patients (51,53). All patients showed rapid and remarkable clinical improvement and significant decline of the MIs in skin smears; definite clinical improvement was seen as early as 14 days after treatment (53). More than 99% and >99.9% of the viable *M. leprae* had been killed by 28 and 56 days of treatment, respectively, as measured by serial inoculations of organisms from skin biopsies into the footpads of immunocompetent and nude mice (53). Adverse reactions were rare and mild.

10.3. Macrolides

Among the newer semisynthetic macrolides, clarithromycin is, by far, the most active compound against *M. leprae* in vitro and in vivo (54,72). The bacteriocidal activity of clarithriomycin against *M. leprae* in mice has been well demonstrated by the kinetic (54) and proportional bactericidal method (52). It was more active at a dosage of 25 or 50 mg/kg than it was at 12.5 mg/kg (52). A single dose of clarithromycin 100 or 200 mg/kg also displayed significant anti-*M. leprae* activity (36). Because the activity of a single dose 200 mg/kg in mice, corresponding to 2000 mg in humans, did not differ significantly from that of 100 mg/kg in mice, equivalent to 1000 mg in humans, the optimal dosage for clarithromycin in the clinical trial would be 1000 mg daily.

To date, the results of two clinical trials with clarithromycin 500–1000 mg daily have been published (53,73). The remarkable clinical improvement, significant decline of the MI, and rapid killing of the viable organisms in the skin biopsies, all indicate that clarithromycin displayed powerful bacteriocidal activity against *M. leprae,* and the activity did not differ significantly between clarithromycin 500 mg and 1000 mg daily, nor between clarithromycin 500 mg and minocycline 100 mg daily (53). In addition, one group of patients was treated with the combination of clarithromycin 500 mg plus minocycline 100 mg daily, but its clinical and bacteriological improvements did not differ significantly from patients treated with either clarithromycin 500 mg daily or minocycline 100 mg daily (53), indicating that the additive effect of the combination of clarithromycin plus minocycline, previously demonstrated in mice (52), was not observed in the clinical trial.

10.4. Other Antimicrobial Agents

Clofazimine, a rimino-phenazine derivative, is one of the important components of MDT regimen for multibacillary leprosy, but the problem is the skin coloration it causes. Recently, a series of phenazine derivatives have been synthesized, and several of them are considerably more active in vitro against *M. leprae* than clofazimine (74,75), but none of them is more active in vivo (76). It was thought that the disappointing in vivo results were largely due to the pharmacokinetic properties, in particular to the low lipophilicity of the derivatives, whereas the lipophilicity of phenazines was also responsible for skin coloration. If this assumption is confirmed, it will be difficult to develop a nonpigmenting phenazine with superior in vivo activity against *M. leprae,* which may substitute for clofazimine in the treatment of leprosy.

In the mouse footpad system, on a weight-to-weight basis several rifamycin derivatives, including rifabutin (LM427) (77,78,80), rifapentine (DL473) (79–81), and benzoxazinorifamycin (KRM1648) (82,83), displayed stronger bactericidal activities against *M. leprae* than rifampicin. However, because no difference in bacteriocidal activity could be detected between patients treated with rifampicin 600 mg daily and intermittently, the new rifamycin derivatives could contribute significantly to the treatment of leprosy only if they are active against rifampicin-resistant strains of *M. leprae.* Unfortunately, the earlier claim that rifabutin was active against rifampicin-resistant *M. leprae* (78) is not been confirmed in further experiments (unpublished data). Therefore, it is unlikely that the new rifamycin derivatives may further improve the efficacy of the current MDT regimen.

Fusidic acid has a long half-life, 16 h, in humans, may achieve a very high serum concentration, $>100\,\mu g/ml$, following an oral dosage of 500 mg three times a day, and possesses very little toxicity. It is active against *M. leprae* both in axenic medium and in macrophage culture, as determined in the BACTEC 460 system (84). In a clinical trial, lepromatous patients were treated with fusidic acid at either 500 mg daily for 12 weeks or 750 mg daily for 4 weeks followed by 500 mg daily for 8 weeks. After 8–12 weeks of treatment, all patients showed various degrees of clinical improvement and significant decline of the MIs in skin smears; however, the results of serial mouse footpad inoculations indicated that fusidic acid only displayed weak bactericidal activity against *M. leprae* (85). Therefore, the prospect of fusidic acid as an antileprosy drug remains unclear.

11. Investigational Multidrug Regimens

The available results indicate that the bacteriocidal activities of ofloxacin, clarithromycin, and minocycline against *M. leprae* are very similar. They are less potent than rifampicin but are definitely more active than either dapsone or clofazimine, and their side effects are rare and mild. Therefore, they may be added to the list of effective antileprosy drugs and to be the components of the newer generation of MDT regimens. Because >99.9% or >99.99% of viable organisms were killed by no more than 4 weeks of daily treatment with these drugs, it is assumed that the treatment with MDT regimens containing any of the new drugs may allow rapid elimination of the rifampicin-resistant mutants, these being no more than 10^4 in an untreated multibacillary patient. Combining the new drugs with rifampicin may significantly shorten the duration of treatment.

The long-term therapeutic effect of the following two multidrug regimens are being tested in large-scale multicentric field trials:

1. Daily treatment with rifampicin 600 mg plus ofloxacin 400 mg for 1 month, for both paucibacillary and multibacillary leprosy.
2. Rifampicin 600 mg or, ofloxacin 400 mg plus minocycline 100 mg once monthly for both paucibacillary and multibacillary leprosy; but the duration of treatment for paucibacillary leprosy is 3 or 6 months, respectively, and for multibacillary leprosy is 12 or 24 months, respectively

With respect to rifampicin-resistant leprosy, which is the most serious threat to the success of leprosy control programs, the recommended regimen is the following: treatment begins with ofloxacin 400 mg, clarithromycin 500 mg, minocycline 100 mg, and clofazimine 100 mg daily, supervised, for 4 weeks; followed by clofazimine 50 mg plus minocycline 100 mg daily, self-administered, for minimum 2 years or until skin smear negativity.

12. Management of Leprosy Reactions

Immunologically mediated episodes of acute or subacute inflammation, known as "reaction," may occur in any type of leprosy except indeterminate and can result in permanent deformity. Most reactions belong to one of the two major types, erythema nodosum leprosum (ENL or Jopling's Type 2 reaction) and reversal reaction (Jopling's Type 1 reaction).

12.1. Erythema Nodosum Leprosum

This type of reaction occurs almost exclusively in multibacillary patients. It appears to be less of a problem among patients treated with MDT than those treated with dapsone monotherapy, presumably due to the anti-inflammatory activity of clofazimine in the MDT regimen.

Mild ENL can be treated with analgesics, either aspirin or acetaminophen, or antimonials. The treatment of choice for severe ENL is either corticosteroid or thalidomide. Prednisolone, the least expensive and most widely available corticosteroid, can rapidly control ENL, but many severely ill patients require continuous and often high dosages. The patients are given prednisolone 30–60 mg daily, and ENL is generally controlled within 24–72 h. The dosage of prednisolone can then be reduced gradually. In general, the total duration of the corticosteroid treatment for ENL should preferably not exceed 12 weeks. Thalidomide also acts rapidly in controlling ENL and has fewer side effects than corticosteroids. Patients with severe ENL are given thalidomide 200 mg twice daily, and the reaction is usually controlled within 48–72 h. The dosage can then be reduced gradually. It must, however, be pointed out that because of its well-known teratogenicity, it should be given only to males and postmenopausal females. Women of childbearing age should never be given thalidomide.

If ENL recurs after prednisolone or thalidomide treatment has been stopped, a repeat course will be necessary. In corticosteroid-dependent patients, it may be useful to supplement the prednisolone therapy with higher dosages of clofazimine. Because clofazimine often takes 4–6 weeks to develop its full effect, during that period it may not be used as a sole drug for the treatment of severe ENL. The patients may be put on clofazimine, 100 mg two or three times daily for 8–12 weeks, then gradually reduced to 100 mg daily. In the meantime, prednisolone can be gradually withdrawn. The major problems in the case of higher-dosage clofazimine therapy are the intolerance by some patients due to its gastrointestinal side effects and nonacceptance due to the skin coloration.

12.2. Reversal Reaction

Reversal reaction occurs mostly in borderline (BT, BB, and BL) leprosy and usually soon after the onset of successful chemotherapy, although it may also appear after the completion of MDT. During the reaction, besides the skin lesions, nerve thickening, nerve pain, and tenderness often occur or increase and are accompanied by rapid deterioration of nerve function. For mild reversal reaction, no specific treatment is required except analgesics. However, the patient must be seen at least once every 2 weeks and asked to return at once if the symptoms become more severe. Patients with severe reversal reaction must be hospitalized, and corticosteroids, the most effective treatment for controlling reversal reaction, should be given immediately. The dosage of corticosteroid must be determined according to the severity of the reaction, the body weight of the patient, and the response to treatment, and the initial dosage should be sufficient to relieve both nerve pain and nerve tenderness. Patients usually start with a higher dosage of prednisolone (e.g., 20–30 mg twice daily). As the reaction becomes controlled, the dosage can gradually decrease. A maintenance dosage of prednisolone (e.g., 5–10 mg daily) lasting for several months may be necessary if there is a tendency for recurrence of reaction on complete stoppage of prednisolone.

The antileprosy treatment should continue unchanged during ENL or reversal reaction.

12.3. Quiet Nerve Paralysis

It is well known that neuritis in leprosy, identified by nerve thickening associated with acute or subacute nerve pain and tenderness, is frequently the precursor of irreversible nerve damage. However, thickened nerve trunks quite frequently become paralyzed in leprosy patients "quietly," that is, without pain or tenderness of the concerned nerve. This type of paralysis is referred to as "quiet nerve paralysis" (QNP) or "silent neuritis." Unfortunately, it is usually misdiagnosed in its early stage when chances of recovery are high. Although the frequency of QNP in leprosy is not well documented (16), it appears to be an important cause of nerve damage.

In order to detect QNP as early as possible, it is necessary not only to look for painful, tender nerve trunks but also, more importantly, to look for signs of loss of motor and sensory function. Whenever the onset of QNP is confirmed, corticosteroid therapy should be instituted without delay. The use of high-dosage corticosteroids (e.g., prednisolone 60 mg daily) over a period of 3–6 months is able to prevent the permanent nerve damage in a high proportion of cases, particularly when the condition is identified and treated before the nerve is completely paralyzed (16).

References

1. Levy L (1975) The activity of chaulmoogric acids against *Mycobacterium leprae*. Am Rev Respir Dis 111:703–705.
2. Faget GH, Pogge RC, Johansen FA, Dinan JF, Prejean BM, Eccles CG (1943) The promin treatment of leprosy: a progress report. Publ Hlth Rep 58:1729–1741.

3. Faget GH, Johansen FA, Ross H (1942) Sulfanilamide in the treatment of leprosy. Publ Hlth Rep 57:1982–1999.

4. Cochrane RG, Ramanujam K, Paul H, Russell D (1949) Two-and-a-half years' experimental work on the sulfone group of drugs. Leprosy Rev 20:4–64.

5. Lowe J (1950) Treatment of leprosy with diamino-diphenyl sulfone by mouth. Lancet i:145–150.

6. Browne SG, Hogerzeil LM (1962) 'B663' in the treatment of leprosy: preliminary report of a pilot trial. Leprosy Rev 33:6–10.

7. Levy L, Shepard CC, Fasal P (1976) The bactericidal effect of rifampicin on *M. leprae* in man: a) single doses of 600, 900 and 1200 mg; and b) daily doses of 300 mg. Int J Leprosy 44:183–187.

8. Waters MFR, Rees RJW, Pearson JMH, Laing ABG, Helmy HS, Gelber RH (1978) Rifampicin for lepromatous leprosy: nine years' experience. Br Med J i:133–136.

9. Pattyn SR, Colston MJ (1978) Cross-resistance amongst thiambutosine, thiacetazone, ethionamide and prothionamide with *Mycobacterium leprae*. Leprosy Rev. 49:324–326.

10. Shepard CC (1960) The experimental disease that follows the injection of human leprosy bacilli into footpads of mice. J Exp Med 112:445–454.

11. World Health Organization (1977) WHO Expert Committee on Leprosy. Fifth Report. Geneva: World Health Organization.

12. Ji B (1985) Drug resistance in leprosy—a review. Leprosy Rev 56:265–278.

13. Shepard CC, Rees RJW, Levy L, Pattyn SR, Ji B, Cruz EC (1986) Susceptibility of strains of *Mycobacterium leprae* isolated prior to 1977 from patients with previously untreated lepromatous leprosy. Int J Leprosy 54:11–15.

14. Grosset JH, Guelpa-Lauras CC, Bobin P, Brucker G, Cartel JL, Constant-Desportes M, Flaguel B, Frédéric M, Guillaume JC, Millan J (1989) Study of 39 documented relapses of multibacillary leprosy after treatment with rifampicin. Int J Leprosy 57:607–614.

15. WHO Study Group (1982) Chemotherapy of Leprosy for Control Programmes. Geneva: World Health Organization.

16. World Health Organization (1988) WHO Expert Committee on Leprosy. Sixth Report. Geneva: World Health Organization.

17. Shepard CC (1981) A brief review of experiences with short-term clinical trials monitored by mouse foot pad inoculation. Leprosy Rev 52:299–308.

18. Pieters FAJM, Zuidema J (1987) The absolute oral bioavailability of dapsone in dogs and humans. Int J Clin Pharmacol Therap Toxicol 25:396–400.

19. Aquinas M, Allan WGL, Horsfall PAL, Jenkins PK, Hung-Yan W, Girling D, Tall R, Fox W (1972) Adverse reactions to daily and intermittent rifampicin regimens for pulmonary tuberculosis in Hong Kong. Br Med J 1:765–771.

20. Jamet P, Traore I, Husser JA, Ji B (1992) Short-term trial of clofazimine in previously untreated lepromatous leprosy. Int J Leprosy 60:542–548.

21. Levy L (1974) Pharmacologic studies of clofazimine. Am J Trop Med Hyg 23:1097–1109.

22. Cartel JL, Millan J, Guelpa-Lauras CC, Grosset JH (1983) Hepatitis in leprosy patients treated with a daily combination of dapsone, rifampin and a thioamide. Int J Leprosy 51:461–465.

23. Ji B, Chen J, Wang C, Xia G (1984) Hepatotoxicity of combined therapy with rifampicin and daily prothionamide for leprosy. Leprosy Rev 55:283–289.

24. Pattyn SR, Janssens L, Bourland J, Saylan T, Davies EM, Grillone S, Feracci C (1984) Hepatotoxicity of the combination of rifampin–ethionamide in the treatment of multibacillary leprosy. Int J Leprosy 52:1–6.

25. Cartel JL, Naudillon Y, Artus J, Grosset J (1985) Hepatotoxicity of the daily combination of 5 mg/kg prothionamide + 10 mg/kg rifampin. Int J Leprosy 53:15–18.

26. Ridley DS, Jopling WH (1962) A classification of leprosy for research purposes. Leprosy Rev 33:119–128.

27. WHO Study Group (1994) Chemotherapy of Leprosy. Geneva: World Health Organization.

28. World Health Organization (1994) Progress towards eliminating leprosy as a public health problem. WHO Wkly Epidem Rec vol. 69, (20) 145–151, 153–157.

29. Jesudassan K, Vijayakumaran P, Pannikar VK, Christian M (1988) Impact of MDT on leprosy as measured by selective indicators. Leprosy Rev 59:215–223.

30. World Health Organization Leprosy Unit (1994) Risk of relapse in leprosy. WHO/CTD/LEP/94.1.

31. Collaborative Effort of the U.S. Leprosy Panel (U.S.-Japan Cooperative Medical Science Programme) and the Leonard Wood Memorial (1975) Rifampin therapy of lepromatous leprosy. Am J Trop Med Hyg 24:475–484.

32. Gelber RH, Levy L (1987) Detection of persisting *Mycobacterium leprae* by inoculation of the neonatally thymectomized rat. Int J Leprosy 55:872–878.

33. Husser JA, Traore I, Daumerie D (1994) Activity of two doses of rifampin against *Mycobacterium leprae.* Int J Leprosy 62:359–364.

34. Ellard GA, Pannikar VK, Jesudassan K, Christian M (1988) Clofazimine and dapsone compliance in leprosy. Leprosy Rev 59:205–223.

35. Ji B, Perani EG, Petinon C, Grosset JH (1992) Bactericidal activities of single and multiple doses of various combinations of new antileprosy drugs and/or rifampin against *M. leprae* in mice. Int J Leprosy 60:556–561.

36. Xiong JH, Ji B, Perani EG, Petinon C, Grosset JH (1994) Further study of the effectiveness of single doses of clarithromycin and minocycline against *Mycobacterium leprae* in mice. Int J Leprosy 62:37–42.

37. Kirchheimer WF, Storrs EE (1971) Attempts to establish the armadillo (*Dasypus novemcinctus* Linn.) as a model for the study of leprosy. Int J Leprosy 39:693–702.

38. Report of 5th Meeting of the Scientific Working Group of IMMLEP. Annex 4. TDR/IMMLEP-SWG(5)/80-3.

39. Wheelar PR (1984) Metabolism in *Mycobacterium leprae:* its relation to other research on *M. leprae* and to aspects of metabolism in other mycobacteria and intracellular parasites. Int J Leprosy 52:208–230.

40. Wheelar PR (1986) Metabolism in *Mycobacterium leprae:* possible targets for drug action. Leprosy Rev 57(Suppl 3):171–181.

41. Draper P (1982) The anatomy of *M. leprae.* In: Ratledge C, Stanford JL, eds. The Biology of Mycobacteria, Vol. 1, pp. 9–52. London: Academic Press.

42. Draper P (1984) Wall biosynthesis: a possible site of action for new antimycobacterial drugs. Int J Leprosy 52:527–532.

43. Guelpa-Lauras CC, Perani EG, Giroir AM, Grosset JH (1987) Activities of pefloxacin and ciprofloxacin against *Mycobacterium leprae* in the mouse. Int J Leprosy 55:70–77.

44. N'Deli L, Guelpa-Lauras CC, Perani EG, Grosset JH (1990) Effectiveness of pefloxacin in the treatment of lepromatous leprosy. Int J Leprosy 58:12–18.

45. Grosset JH, Guelpa-Lauras CC, Perani EG, Beoletto C (1988) Activity of ofloxacin against *Mycobacterium leprae* in the mouse. Int J Leprosy 56:259–264.

46. Grosset JH, Ji B, Guelpa-Lauras CC, Perani EG, N'Deli L (1990) Clinical trial of pefloxacin and ofloxacin in the treatment of lepromatous leprosy. Int J Leprosy 58:281–295.

47. Tsutsumi S, Gidoh M (1989) Studies on the development of novel chemotherapeutics using nude mice with special reference to a new quinolone carboxylic acid, AT-4140. Japan Leprosy 58:250–257.

48. Franzblau SG, Parrilla MLR, Chan GP (1993) Sparfloxacin is more bactericidal than ofloxacin against *Mycobacterium leprae* in mice. Int J Leprosy 61:66–69.

49. Chan GP, Garcia-Ignacio BY, Chavez VE, Livelo JB, Jimenez CL, Parrilla MLR, Franzblau SG (1994) Clinical trial of sparfloxacin for lepromatous leprosy. Antimicrob Agents Chemother 38:61–65.

50. Gelber RH (1987) Activity of minocycline in *Mycobacterium leprae*-infected mice. J Infect Dis 186:236–239.

51. Gelber RH, Fukuda K, Byrd S, Murray LP, Siu P, Tsang M, Rea TH (1992) A clinical trial of minocycline in lepromatous leprosy. Br Med J 304:91–92.

52. Ji B, Perani EG, Grosset JH (1991) Effectiveness of clarithromycin and minocycline along or in combination against experimental *Mycobacterium leprae* infection in mice. Antimicrob Agents Chemother 35:579–581.

53. Ji B, Jamet P, Perani EG, Bobin P, Grosset JH (1993) Powerful bactericidal activities of clarithromycin and minocycline against *Mycobacterium leprae* in lepromatous leprosy. J Infect Dis 168:188–190.

54. Franzblau SG, Hastings RC (1988) In vitro and in vivo activities of macrolides against *Mycobacterium leprae.* Antimicrob Agents Chemother 32:1758–1762.

55. Shepard CC, Chang YT (1962) Effect of several anti-leprosy drugs on multiplication of human leprosy bacilli in foot pads of mice. Proc Soc Exp Biol Med 109:636–638.

56. Shepard CC (1967) A kinetic method for the study of activity of drugs against *Mycobacterium leprae* in mice. Int J Leprosy 35:429–435.

57. Shepard CC (1969) Further experience with the kinetic method for the study of activity of drugs against *Mycobacterium leprae* in mice. Activities of DDS, DFD, ethionamide, capreomycin and PAM 1392. Int J Leprosy 37:389–397.

58. Colston MJ, Hilson GRF, Banerjee DK (1978) The 'proportional bactericidal test', a method for assessing bactericidal activity of drugs against *Mycobacterium leprae* in mice. Leprosy Rev 49:7–15.

59. Shepard CC, van Landingham RM, Walker LL (1971) Recent studies of antileprosy drugs. Leprosy Rev 39:340–349.

60. Halvorson HO, Ziegler NR (1933) Application of statistics to problems in bacteriology. I. A means of determining bacterial population by the dilution method. J Bact 25:101–121.

61. Taylor J (1933) The estimation of numbers of bacteria by ten-fold dilution series. J Appl Bact 25:54–68.

62. Shepard CC (1982) Statistical analysis of results obtained by two methods for testing drug activity against *Mycobacterium leprae*. Int J Leprosy 50:96–101.

63. Ramasesh N, Krahenbuhl JL, Hastings RC (1989) In vitro effects of antimicrobial agents on *Mycobacterium leprae* in mouse peritoneal macrophages. Antimicrob Agents Chemother 33:657–662.

64. Ji B, Grosset JH (1990) Recent advances in the chemotherapy of leprosy. Leprosy Rev 61:313–329.

65. Franzblau SG, Biswas AN, Jenner P, Colston MJ (1992) Double-blind evaluation of BACTEC and Buddemeyer-type radiorespirometric assays for in vitro screening of antileprosy agents. Leprosy Rev 63:125–133.

66. Franzblau SG, White KE (1990) Comparative in vitro activities of 20 fluoroquinolones against *Mycobacterium leprae*. Antimicrob Agents Chemother 34:229–231.

67. Saito H, Tomioka H, Nagashima K (1986) In vitro and in vivo activities of ofloxacin against *Mycobacterium leprae* infection induced in mice. Int J Leprosy 54:560–562.

68. Pattyn SR (1987) Activity of ofloxacin and pefloxacin against *Mycobacterium leprae* in mice. Antimicrob Agents Chemother 31:671–672.

69. Banerjee DK, McDermott-Lancaster RD (1992) An experimental study to evaluate the bactericidal activity of ofloxacin against an established *Mycobacterium leprae* infection. Int J Leprosy 60:410–415.

70. Ji B, Perani EG, Petinom C, N'Deli L, Grosset JH (1994) Clinical trial of ofloxacin alone and in combination with dapsone plus clofazimine for treatment of lepromatous leprosy. Antimicrob Agents Chemother 38:662–667.

71. Gelber RH (1986) The use of rodent models in assessing antimicrobial activity against *Mycobacterium leprae*. Leprosy Rev 57(Suppl 3):137–148.

72. Franzblau SG (1988) Oxidation of palmitic acid by *Mycobacterium leprae* in an axenic medium. J Clin Microbiol 26:18–21.

73. Chan GP, Garcia-Ignacio BY, Chavez VE, Livelo JB, Jimenez CL, Parrilla MLR, Franzblau SG (1994) Clinical trial of clarithromycin for lepromatous leprosy. Antimicrob Agents Chemother 38:515–517.

74. Franzblau SG, O'Sullivan JF (1988) Structure–activity relationships of selected phenazines against *Mycobacterium leprae* in vitro. Antimicrob Agents Chemother 32:1583–1585.

75. Franzblau SG, White KE, O'Sullivan JF (1989) Structure–activity relationships of tetramethylpiperidine-substituted phenazines against *Mycobacterium leprae* in vitro. Antimicrob Agents Chemother 33:2004–2005.

76. Van Landingham RM, Walker LL, O'Sullivan JF, Shinnick TM (1993) Activity of phenazine analogs against *Mycobacterium leprae* infections in mice. Int J Leprosy 61:406–414.

77. Hastings RC, Jacobson RR (1983) Activity of ansamycin against *Mycobacterium leprae.* Lancet 2:1079–1080.

78. Hastings RC, Richard VR, Jacobson RR (1984) Ansamycin activity against rifamycin-resistant *Mycobacterium leprae.* Lancet 1:1130.

79. Pattyn SR, Saerens EJ (1977) Activity of three new rifamycin derivatives on the experimental infection by *Mycobacterium leprae.* Ann Soc Belg Med Trop 57:169–173.

80. Pattyn SR (1987) Rifabutin and rifapentine compared with rifampin against *Mycobacterium leprae* in mice. Antimicrob Agents Chemother 31:134.

81. Ji B, Chen J, Lu X, Wang S, Ni G, Hou Y, Zhou D, Tang Q (1986) Antimycobacterial activities of two newer ansamycins: R-76-1 and DL 473. Int J Leprosy 54:563–577.

82. Tomioka H, Saito H (1993) In vivo antileprosy activity of the newly synthesized benzoxazinorifamycin, KRM-1648. Int J Leprosy 61:255–258.

83. Saito H, Tomioka H, Sato K, Dekio S (1994) Therapeutic efficacy of benzoxazinorifamycin, KRM-1648, in combination with other antimicrobials against *Mycobacterium leprae* infection induced in nude mice. Int J Leprosy 62:43–47.

84. Franzblau SG, Biswas AN, Harris EB (1992) Fusidic acid is highly active against extracellular and intracellular *Mycobacterium leprae.* Antimicrob Agents Chemother 36:92–94.

85. Franzblau SG, Chan GP, Garcia-Ignacio BG, Chavez V, Livelo JB, Jimenez CL, Williams DL, Gillis TP (1994) Clinical trial of fusidic acid for lepromatous leprosy. Antimicrob Agents Chemother 38:1651–1654.

86. Levy L (1987) Application of the mouse foot-pad techniques in immunologically normal mice in support of clinical drug trials, and a review of earlier clinical drug trials in lepromatous leprosy. Int J Leprosy 55:823–829.

87. Rees RJW (1966) Enhanced susceptibility of thymectomized and irradiated mice to infection with *Mycobacterium leprae.* Nature 211:657–658.

88. Fieldsteel AH, Levy L (1976) Neonatally thymectomized Lewis rats infected with *Mycobacterium leprae:* response to primary infection, secondary challenge and large inocula. Infect Immun 14:736–741.

89. Colston MJ, Hilson GRF (1976) Growth of *Mycobacterium leprae* and *M. marinum* in congenitally athymic (nude) mice. Nature 262:399–401.

90. Kohsaka K, Mori T, Ito T (1976) Lepromatoid lesion developed in the nude mouse inoculated with *Mycobacterium leprae.* La Lepro 45:177–187.

91. MeCermott-Lancaster RD, Ito T, Kohsaka K, Guelpa-Lauras CC, Grosset JH (1987) Multiplication of *Mycobacterium leprae* in the nude mice, and some applications of nude mice to experimental leprosy. Int J Leprosy 55:889–895.

92. Marchoux Chemotherapy Study Group (1992) Relapses in multibacillary leprosy patients after stopping treatment with rifampin-containing combined regimens. Int J Leprosy 60:525–535.

93. Jamet P, Ji B (1994) Relapse after long-term follow-up of multibacillary patients treated by W.H.O. multidrug regimen. Int J Leprosy 62:662.

94. Waters MFR, Rees RJW (1962) Changes in the morphology of *Mycobacterium leprae* in patients under treatment. Int J Leprosy 30:266–277.

Index